ELEMENTARY TOPOLOGY

ELEMENTARY TOPOLOGY

DICK WICK HALL

Professor of Mathematics

GUILFORD L. SPENCER II

Assistant Professor of Mathematics

University of Maryland

JOHN WILEY & SONS, INC., NEW YORK

CHAPMAN & HALL, LIMITED, LONDON

Library of Congress Catalog Card Number: 55-10237

PRINTED IN THE UNITED STATES OF AMERICA

TO OUR WIVES AND CHILDREN

We gratefully acknowledge our appreciation
for their patience, understanding, and assistance
in the preparation of this text

INTRODUCTION

In recent years it has become increasingly apparent that topology is one of the major branches of modern mathematics. Besides the importance of this subject in its own right, an understanding of it will help the student immeasurably in his courses in advanced calculus, real variables, complex variables, and other courses in analysis. It is therefore essential that students be introduced to this subject at as early a stage as possible in their undergraduate careers. This book is intended as a textbook for such an early introductory course.

A major difficulty in the teaching of advanced undergraduate courses consists in getting the students to follow chains of reasoning containing more than one link. The first two chapters are pointed toward this difficulty. Chapter I starts out rather gently with the elements of set theory, and introduces the concepts of mappings, equivalence, and order relations. It closes with a brief account of finite, infinite, countable, and uncountable sets. Students are usually so intrigued by the infinite that this topic is particularly suitable as a first introduction to a closely argued chain of reasoning. It is also vital to their later work.

Chapter II is devoted to a study of the real number system. In this familiar setting the student is exposed to chains of reasoning typical of many encountered in later chapters. This chapter is fundamental to the remainder of the book. Its purpose is threefold: first, to impart to the student a sound knowledge of real numbers and their fundamental properties; second, to provide a reservoir of examples of topological concepts; third, to furnish a clear motivation for the abstract generalizations of later chapters. In particular, in connection with the third point, we emphasize in Chapter II the fundamental role played by open sets. This motivates clearly the study of general topological spaces in Chapter III.

In Chapter IV we return from these generalizations to a study of metric spaces. This is followed by a long chapter on Arcs and Curves, in which arcs, simple closed curves, and the simple closed surface or two-sphere are characterized. Related material on Peano spaces, cyclic element theory, and property *S* is also included.

Chapter VI is short, and is devoted to the equivalence of partitionability and property *S* in metric spaces. This result is basic to Bing's recent characterization of E^3. Chapter VII contains a detailed treatment of alternative forms of the Axiom of Choice, such as Zorn's lemma, and the Tychonoff and well-ordering theorems. Several applications of these alternative forms to topology are given.

Some additional remarks on our organization of the subject matter, as well as some specific comments on certain theorems, may be helpful to the reader. Some of these remarks will also indicate in more detail the content of the text.

In Chapter IV, the section on Metrizability contains a necessary and sufficient condition for a space to be metrizable. This condition is essentially that found in the study of uniform structures. If desired, this can be omitted, since the next section contains the usual metrization theorem for regular, second countable T_1 spaces. Moreover, this latter result is sufficient for the later chapters. Another alternative would be to present the Smirnov metrization theorem (Yu. M. Smirnov, On Metrization of Topological Spaces, *American Mathematical Society, Translation Number* 91, 1953) which gives a necessary and sufficient condition of a similar nature to the usual theorem for regular, second countable T_1 spaces.

Included as a problem in the last section of Chapter IV is an outline of some of the main theorems on Moore-Smith sequences, the proofs of which are to be worked out as a short term paper by the student. Other problems of the term paper variety may be found in Chapters IV and V. The proofs of theorems in the text are in no way dependent on the solution of these problems.

Depending upon the background of the students in the class, it may or may not be desirable to follow Chapter IV by Chapter VII on the Axiom of Choice, and then to continue with Chapter V. In any event, no use of the results obtained in Chapter VII is made in any of the earlier chapters, although there are some instances where the arguments could have been shortened by such use.

The long Chapter V on Arcs and Curves contains, as previously remarked, characterizations of the arc, the simple closed curve, and the simple closed surface. Algebraic methods of proof have been avoided to keep the text on as elementary a level as possible. The proofs are long, and we have purposely tried to err on the side of giving too many details rather than too few. For this reason, the student should be required to outline the main steps of the arguments so that he will not fail to see the forest for the trees. Indeed, in general, throughout the text it is necessary that the student not merely follow but actually think through the proofs for himself, in order to gain real understanding.

The proof we give of the characterization of the simple closed surface is essentially van Kampen's organization of Zippin's proof. By slight modification of details we obtain at the same time the Jordan-Schoenflies theorem.

It will also be noticed that the concepts of components and local compactness are not treated directly until they occur in Chapter V. This was done to avoid endlessly building new machinery in earlier chapters and having it rust before active use of it was made. Indeed, insofar as possible, this has been our general thought concerning the place to introduce new ideas. On the other hand, it is perfectly possible to give many of our results on local compactness and components at some earlier stage deemed appropriate.

One of our reasons for giving so much material on Peano spaces is that they probably provide the richest single source of illustrations of various mathematical techniques used in topology. This remark also reveals our feeling that the most valuable asset a mathematician can have is a knowledge of a wide variety of techniques and methods of proof.

Our classroom experience indicates that it is reasonable to cover Chapters I, II, III, IV, and the first part of Chapter V, in a two-semester course. The experience cited is with undergraduates who have not had advanced calculus. With more advanced students, Chapters I and II could be covered very rapidly, and a considerably greater portion of the text completed.

We have, of course, drawn heavily and freely upon the existing literature in topology, particularly upon the *American Mathematical Society Colloquium Publications* of G. T. Whyburn and R. L. Wilder. Our primary sources of this nature are listed in the bibliographical comments. We also drew heavily upon the rich experience due to the association with and teachings of our various professors. Professor Hall is particularly indebted to Professors G. T. Whyburn, E. J. McShane, J. R. Kline, and J. D. Tamarkin. Professor Spencer is particularly indebted to Professors R. L. Wilder, Gail S. Young Jr., Hans Samelson, and T. H. Hildebrandt. Our classroom experience with the manuscript has also been most helpful, and we are indebted to our students for suggestions they have made.

We are particularly grateful to Professor R. H. Bing of the University of Wisconsin for valuable suggestions and constructive criticisms of our manuscript.

Dick Wick Hall
Guilford L. Spencer II

October 3, 1955

CONTENTS

CHAPTER ONE

INTRODUCTORY SET THEORY

CHAPTER TWO

THE REAL LINE

CHAPTER THREE

TOPOLOGICAL SPACES

CHAPTER FOUR

METRIC SPACES

CHAPTER ONE

INTRODUCTORY SET THEORY

1. Introduction

In this book we shall be concerned with the study of collections of objects. Each object of a given collection will be called a *point*. A collection of points will be called a *point set* or, more simply, a *set*, and each of the points will be referred to as an element of this set. In any particular discussion the set of all points under consideration will be known as the *universe*. As far as we are concerned in that discussion no other points exist.

To simplify the statements of theorems and problems, we shall always assume, unless otherwise specifically stated, that all the sets mentioned in a single discussion lie in the same universe.

In the following discussion the universe will be denoted by S. This means precisely that, given any point p, p is an element of S. This fact is indicated by the notation $p \in S$. If A is a point set, then, for each point p, exactly one of the following statements is true: (i) p is an element of A (denoted by $p \in A$), or (ii) p is not an element of A (denoted by $p \notin A$). The set of all points p satisfying (i) is called the *subset* A of S. The set of all points p satisfying (ii) is called the *complement* of A in S, or, more simply, the *complement* of A.

Problem 2 shows that the complement of a set depends very strongly on the universe.

Let A and B be subsets of a universe S. We say that "A is contained in B" if and only if every point of A is a point of B. Notationally, "A is contained in B" is written $A \subset B$. The statement "B contains A" (written $B \supset A$) is made if and only if $A \subset B$. Two subsets A and B of a universe S are equal (written $A = B$) if and only if $A \subset B$, and $B \subset A$.

THEOREM 1.1. Let A and B be subsets of a universe S. Then

(*a*) If A is contained in B, then the complement of B is contained in the complement of A.

(*b*) If the complement of B is contained in the complement of A, then A is contained in B.

Proof. (*a*) We must prove that the complement of B is contained in the complement of A. To do this, let p be any point of the complement of B. This means $p \notin B$. But every point of A is a point of B, since $A \subset B$. Thus $p \notin A$; hence p must be a point of the complement of A. We have thus shown that any point of the complement of B is a point of the complement of A, and this is precisely what we mean by the statement "the complement of B is contained in the complement of A."

(*b*) Let $p \in A$. The proof will be complete when we show that $p \in B$. Since $p \in A$, p is not an element of the complement of A; hence p is not an element of the complement of B. Thus $p \in B$.

Theorem 1.1 tells us that either of the conditions "$A \subset B$" and "(complement of B) $\subset$ (complement of A)" implies the other one. For this reason we say that these are equivalent conditions, and that either one of them holds if and only if the other one holds also. We now restate Theorem 1.1 in this "if and only if" language.

THEOREM 1.1. Let A and B be subsets of a universe S. Then $A \subset B$ if and only if (complement of B) $\subset$ (complement of A).

THEOREM 1.2. Let A and B be subsets of a universe S. Then $A = B$ if and only if (complement of A) $=$ (complement of B).

Proof. Since this is an "if and only if" theorem, we must prove two things:

I. If $A = B$, then (complement of A) $=$ (complement of B).

II. If (complement of A) $=$ (complement of B), then $A = B$.

We shall prove I. The proof of II is left as an exercise. Since $A = B$, we know that $A \subset B$; thus (complement of B) $\subset$ (complement of A), by Theorem 1.1. Also, since $A = B$, $B \subset A$; hence (complement of A) $\subset$ (complement of B), again by Theorem 1.1. Thus (complement of A) $=$ (complement of B), and the proof of I is complete.

A subset K of a universe S is defined by dividing the points of S into two classes. The first class consists of all points of S which are not in K, and the set of all these points is called the complement of K. The other class consists of all points of S that are in K. These statements are a repetition of what has been said before. It may not have been noticed, however, that we may pick the first class, namely the complement of K, as the set of all points of S. The set K which is thus uniquely defined contains no points at all. It is called the *null set*, and has as its complement the universe S. We shall use the symbol $\emptyset$ in this text if and only if we mean the null set. The null set is frequently referred to as the empty set.

PROBLEMS

1. Let A be a subset of the universe S, and B the complement of A. Prove that A is the complement of B.

2. (*a*) Let the universe consist of the digits 1, 2, 3, 4, 5, 6, 7, 8, 9, and let A be the set consisting of the digits 1, 3, 5. Find the complement of A.
(*b*) Using the same set A, find the complement of A if the universe is the set consisting of the digits 1, 3, 5, 7, 9.

3. Give the details of the proof of II in Theorem 1.2.

4. If $A \subset B$ and $B \subset C$, show that $A \subset C$.

5. Let A be any set. Then $\emptyset \subset A$.

2. Indexed Sets

Let S be a universe. Suppose that we desire to talk about a collection of subsets of S. To distinguish between the sets of this collection it is convenient to give names to them. For this purpose we shall introduce a collection of names Λ (this is the Greek letter capital lambda). The elements of Λ will be denoted by $\alpha, \beta, \cdots$, and will be called the names of our sets. For example, we might have a collection of subsets of S of which certain members were $A_\alpha, A_\beta, A_\gamma$, etc.

The student has seen this idea before. In calculus we deal with sequences of real numbers $(r_1, r_2, r_3, \cdots, r_n, \cdots)$. This means that we single out certain real numbers and give them the names $1, 2, 3, \cdots, n, \cdots$. Thus our collection of names is in this case the collection of all positive integers.

We see that the process of assigning names from the collection Λ is precisely that of assigning a subscript or index to each set considered. It is for this reason that Λ is known as an *index set*, and the process of assigning names from Λ is called *indexing*, the individual names being known as *indices*. The collection of subsets of S to which names have been assigned is said to have been *indexed by* Λ. We shall indicate this collection by the symbol $\{A_\alpha\}_{\alpha\in\Lambda}$ or, more simply, by $\{A_\alpha\}$ where the index set Λ is understood.

DEFINITION 2.1. Let Λ be an index set, S a universe, and $\{A_\alpha\}$ a collection of subsets of S indexed by Λ. Then

(*a*) the union of the sets $\{A_\alpha\}$ is the set denoted by $\bigcup_{\alpha\in\Lambda} A_\alpha$. A point p of S lies in $\bigcup_{\alpha\in\Lambda} A_\alpha$ if and only if p belongs to A_α for at least one α. If Λ is empty then $\bigcup_{\alpha\in\Lambda} A_\alpha = \emptyset$.

(*b*) The intersection of the sets $\{A_\alpha\}$ is the set denoted by $\bigcap_{\alpha\in\Lambda} A_\alpha$. A point p of S lies in $\bigcap_{\alpha\in\Lambda} A_\alpha$ if and only if p belongs to A_α for every α. If Λ is empty then $\bigcap_{\alpha\in\Lambda} A_\alpha = S$.

When only a few subsets of a universe S are considered, we frequently name them with different letters. Thus we interpret $A \cup B = \bigcup_{i\in\{1,2\}} A_i$ where $A_1 = A$, $A_2 = B$. Similarly, $A \cap B = \bigcap_{i\in\{1,2\}} A_i$ where $A_1 = A$, $A_2 = B$.

As an example of the above concepts, consider a universe whose points are the digits 1, 2, 3, 4, 5, 6, 7, 8, 9. Suppose subsets A, B, C of S are defined as follows: A consists of the points 1, 2, 3; B consists of the points 3, 4, 5; and C consists of the points 5, 6, 7. The set $A \cup B$ consists of the points 1, 2, 3, 4, 5 since each of these points belongs to at least one of the sets A, B and no other point of S belongs to either A or B. In the same way the set $A \cup C$ consists of the points 1, 2, 3, 5, 6, 7. The set that is the intersection of $A \cup B$ and $A \cup C$ is denoted by $(A \cup B) \cap (A \cup C)$. This set consists of all points belonging to both the set $(A \cup B)$ and the set $(A \cup C)$: namely, the points 1, 2, 3, 5. The set $A \cup (B \cap C)$ consists also of the points 1, 2, 3, 5. Thus, in this particular example, $A \cup (B \cap C) = (A \cup B) \cap (A \cup C)$. In the next paragraph we shall find out whether or not this is true in general.

Let A, B, C be subsets of a universe S. We shall try to prove that $A \cup (B \cap C) = (A \cup B) \cap (A \cup C)$. To do this we must show two things: namely, that each of these sets is contained in the other. Let $p \in A \cup (B \cap C)$. Then p belongs to at least one of the sets A and $(B \cap C)$. If $p \in A$, then p belongs to both of the sets $(A \cup B)$ and $(A \cup C)$, and hence to their intersection. If $p \notin A$, then $p \in (B \cap C)$; i.e., p belongs to both the sets B and C. This means again that p is a point of $(A \cup B) \cap (A \cup C)$. We have thus proved that $A \cup (B \cap C) \subset (A \cup B) \cap (A \cup C)$. It remains to be proved that $(A \cup B) \cap (A \cup C) \subset A \cup (B \cap C)$. Let $p \in (A \cup B) \cap (A \cup C)$. If $p \in A$, then $p \in A \cup (B \cap C)$. If $p \notin A$, then $p \in B$ and $p \in C$; thus $p \in B \cap C$. It follows that $p \in A \cup (B \cap C)$, and we have proved that $A \cup (B \cap C) = (A \cup B) \cap (A \cup C)$.

PROBLEMS

6. Let A and B be subsets of a universe S. Prove

   (*a*) $A \cup B = B \cup A$.

   (*b*) $A \cap B = B \cap A$.

   (*c*) $A \subset A \cup B$, $B \subset A \cup B$.

   (*d*) $A \cap B \subset A$, $A \cap B \subset B$.

(*e*) $A \cup B = A$ if and only if $B \subset A$.
(*f*) $A \cap B = A$ if and only if $A \subset B$.
(*g*) $A \cup \emptyset = A$.
(*h*) $A \cap \emptyset = \emptyset$.
(*i*) $A \cup S = S$.
(*j*) $A \cap S = A$.

7. Let Λ be an index set, S a universe, and $\{A_\alpha\}$ a collection of subsets of S indexed by Λ. Prove

(*a*) $\bigcup_{\alpha \in \Lambda} A_\alpha \supset A_\alpha$ for every $\alpha \in \Lambda$, if $\Lambda \neq \emptyset$.

(*b*) $\bigcap_{\alpha \in \Lambda} A_\alpha \subset A_\alpha$ for every $\alpha \in \Lambda$, if $\Lambda \neq \emptyset$.

(*c*) For any subset B of S, $B \cap \bigcap_{\alpha \in \Lambda} A_\alpha = \bigcap_{\alpha \in \Lambda} (B \cap A_\alpha)$, if $\Lambda \neq \emptyset$.

(*d*) For any subset B of S, $B \cup \bigcup_{\alpha \in \Lambda} A_\alpha = \bigcup_{\alpha \in \Lambda} (B \cup A_\alpha)$, if $\Lambda \neq \emptyset$.

8. Prove $A \cap (B \cup C) = (A \cap B) \cup (A \cap C)$.

9. Prove, if $A \cap B = A$ and $A \cup B = A$, then $A = B$.

3. The Theorem of DeMorgan

Let S be a universe and A, B two of its subsets. The set of points that are either in A or in B has been denoted by $A \cup B$. The set of points that are in both A and B has been denoted by $A \cap B$. It is now convenient to introduce a set that will consist of all points of A that are not in B.

DEFINITION 3.1. Let A and B be subsets of a universe S. The set of all points of A that are not in B will be denoted by $A - B$.

We note from this definition that the complement of a subset A of a universe S may be denoted by $S - A$.

THEOREM 3.2 (*DeMorgan's Theorem*). Let Λ be an index set, S a universe, and $\{A_\alpha\}$ a collection of subsets of S indexed by Λ. Then

(*a*) $S - \bigcup_{\alpha \in \Lambda} A_\alpha = \bigcap_{\alpha \in \Lambda} (S - A_\alpha)$.

(*b*) $S - \bigcap_{\alpha \in \Lambda} A_\alpha = \bigcup_{\alpha \in \Lambda} (S - A_\alpha)$.

Proof. If $\Lambda = \emptyset$, both (*a*) and (*b*) are immediate. We thus assume $\Lambda \neq \emptyset$.

(*a*) We must show that each of these sets is contained in the other. This is conveniently done in the following steps.

1. Let $p \in S - \bigcup_{\alpha \in \Lambda} A_\alpha$; then $p \notin \bigcup_{\alpha \in \Lambda} A_\alpha$.
2. Thus p is not in any A_α; hence $p \in S - A_\alpha$ for every α.

3. This means that $p \in \bigcap_{\alpha \in \Lambda} (S - A_\alpha)$; hence $S - \bigcup_{\alpha \in \Lambda} A_\alpha \subset \bigcap_{\alpha \in \Lambda} (S - A_\alpha)$.

4. Let $q \in \bigcap_{\alpha \in \Lambda} (S - A_\alpha)$; then q is not in any A_α.

5. This means $q \in S - \bigcup_{\alpha \in \Lambda} A_\alpha$ so that $\bigcap_{\alpha \in \Lambda} (S - A_\alpha) \subset S - \bigcup_{\alpha \in \Lambda} A_\alpha$, and (*a*) is proved.

(*b*) Define $B_\alpha = S - A_\alpha$. Then, from Problem 1, $S - B_\alpha = S - (S - A_\alpha) = A_\alpha$. Thus, by (*a*), $S - \bigcup_{\alpha \in \Lambda} B_\alpha = \bigcap_{\alpha \in \Lambda} (S - B_\alpha)$, so that $\bigcup_{\alpha \in \Lambda} B_\alpha = S - \bigcap_{\alpha \in \Lambda} (S - B_\alpha)$. But $B_\alpha = S - A_\alpha$; hence $\bigcup_{\alpha \in \Lambda} (S - A_\alpha) = S - \bigcap_{\alpha \in \Lambda} A_\alpha$, and the proof of (*b*) is complete.

We may, of course, apply DeMorgan's theorem to two subsets A and B of a universe S, and obtain the following formulas:

$$S - (A \cup B) = (S - A) \cap (S - B)$$

$$S - (A \cap B) = (S - A) \cup (S - B)$$

The student has noticed the name DeMorgan attached to the previous theorem. The name is given since this theorem is the exact analog of the theorem in logic which is known by the same name. Let us consider briefly the connection between logic and set theory. Let α and β be statements concerning the points of a universe S, such that it can be decided, for each point p of S, whether α is true or false for the point p, and whether β is true or false for the point p. Let A_α denote the set of all points of S for which α is true, and A_β the set of all points of S for which β is true. Consider the statement: "If α is true, then β is true." This is precisely equivalent to the statement: "$A_\alpha \subset A_\beta$." But, by Theorem 1.1, $A_\alpha \subset A_\beta$ if and only if $S - A_\beta \subset S - A_\alpha$; that is, "If β is false, then α is false." This shows that the two statements, "If α is true, then β is true," and "If β is false, then α is false," are equivalent in the sense that, if either has been proved, then the other must also be true. Although they are equivalent in this logical sense, frequently the simplest method of proving the first statement is to show that the second one is true. This is the typical argument by contradiction (reductio ad absurdum).

As another example of the connection between logic and set theory, consider the result of Problem 4, where the sets A, B, C are subsets of S consisting, respectively, of all points for which statements α, β, and γ are true. The statement of Problem 4, namely, "If $A \subset B$ and $B \subset C$, then $A \subset C$," interpreted logically, becomes "If α implies β and β implies γ, then α implies γ."

PROBLEMS

10. Let A_α and A_β be the sets defined above. What is the logical interpretation of $A_\alpha \cup A_\beta$ and $A_\alpha \cap A_\beta$?

11. Interpret logically the result of Problem 8.

12. Interpret logically the result of the application of DeMorgan's theorem to two subsets A and B of a universe S.

4. Cartesian Products. Introduction to Mappings

Given two sets S and T, there will be frequent occasions when we desire to consider a third set W, called the Cartesian product of S and T, written $W = S \times T$. The "points" of W are the pairs (s, t) where s

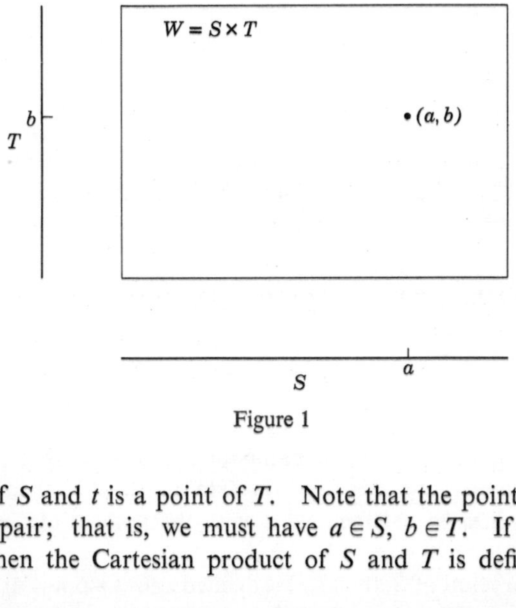

Figure 1

is a point of S and t is a point of T. Note that the point (a, b) of W is an *ordered* pair; that is, we must have $a \in S$, $b \in T$. If either S or T is empty, then the Cartesian product of S and T is defined to be the empty set.

The notion of a Cartesian product first arises in plane analytic geometry. The sets S and T consist of the points on the x and y axes, respectively; thus each of these sets is the set of all real numbers. The Cartesian product $W = S \times T$ is the plane of analytic geometry (sometimes called the Cartesian plane), and each point of this plane is regarded as a pair of real numbers (x, y). The Cartesian product of two arbitrary sets S and T may be represented in a drawing in much the same manner that the plane of analytic geometry is usually shown (see Figure 1).

In analytic geometry the student has become familiar with the problem of graphing a function $y = f(x)$. A point (x, y) of the Cartesian plane is on the graph of this function if and only if its coordinates satisfy the equation $y = f(x)$. It is desirable to generalize these ideas. First of all we wish to introduce the concept of a mapping as the generalization of the idea of a function.

DEFINITION 4.1. Let S and T be sets. A rule f is called a mapping of S into T (written $f: S \to T$) if and only if f associates with each element x of S a unique element y of T. (We may write $y = f(x)$ to indicate this association.) The graph of f is the subset F of the Cartesian product $S \times T$ consisting of all points $(x, f(x))$.

A familiar example of a mapping of the set R of all real numbers into R is $y = f(x)$, where $f(x) = x^2$. Its graph consists of all points (x, x^2) of the Cartesian product $R \times R$: namely, all points on a parabola in the Cartesian plane. One property of the graph of this parabola is that no two distinct points of the graph have the same abscissa and further every real number is the abscissa of some point of the graph. This property is defined in its general form as follows.

omit

DEFINITION 4.2 (*Condition* I). Let F be a subset of the Cartesian product $S \times T$ of a set S and a set T. Then F is said to satisfy condition I if and only if, given points (a, b) and (c, d) of F, $a = c$ implies $b = d$, and, given a point s of S, there exists t of T such that $(s, t) \in F$.

Not only does the graph of the parabola above satisfy this condition, but so also does the graph of any mapping.

omit

THEOREM 4.3. Let f be a mapping of a set S into a set T, and let F be the subset of $S \times T$ which is the graph of f. Then F satisfies condition I.

Proof. Let (a, b) and (c, d) be points of F. Since F is the graph of f, we must have $b = f(a)$ and $d = f(c)$. Thus, if $a = c$, then $f(a) = f(c)$ for f associates a *unique* element of T with the point a of S. Therefore $a = c$ implies $b = d$.

Suppose s is a point of S, then $f(s)$ is defined and is a point of T. Hence, $(s, f(s))$ belongs to F by definition of the graph of f.

Therefore the set F satisfies condition I.

We have seen in Section 3 that, if α and β are statements concerning the points of a universe such that it can be decided for each point of the universe whether α is true or false and whether β is true or false, then the following two statements are equivalent: "If α is true, then β is true," and "If β is false, then α is false." Restating Theorem 4.3 in this equivalent form, we have: "Let S and T be sets and F a subset of the Cartesian product $S \times T$. If F does not satisfy condition I, then there is no mapping f of S into T such that F is the graph of f."

Continuing this line of thought, suppose that F is a subset of $S \times T$ and that F does satisfy condition I. Will there always exist a mapping f of S into T such that F is the graph of f? Will there exist, in some cases, more than one such map? To answer this last question it is necessary that we know precisely what we mean by two mappings being the same.

DEFINITION 4.4. Let f and g be mappings of the set S into the set T. The mappings f and g are said to be the same if and only if $f(s) = g(s)$ for every s in S.

THEOREM 4.5. Let S and T be sets and F a subset of the Cartesian product $S \times T$. If F satisfies condition I, then there is one and only one mapping f of S into T such that F is the graph of f.

Proof. Let F be a subset of $S \times T$ such that F satisfies condition I. We shall first prove that there exists a mapping f of S into T such that F is the graph of f. To produce such a mapping f of S we must associate with each element s of S a unique element t of T. Furthermore, we must set up this association in such a way that the points (s, t) thus defined belong to F and such that every element of F is such a point. That is, we must show that F is the graph of our mapping.

For any point s of S there exists, by condition I, an element t of T such that (s, t) is a point of F. We define our mapping by associating with each element s of S the element t of T thus determined. The graph of our mapping is precisely the set F.

It remains to be shown that, if f and g are any two mappings of S into T such that each of these mappings has F as its graph, then f and g are the same mapping. For any point s of S the points $(s, f(s))$ and $(s, g(s))$ each belong to F. By condition I, $f(s) = g(s)$. Thus f and g are the same mapping by Definition 4.4.

DEFINITION 4.6. Let S and T be sets and $f\colon S \to T$ a mapping. Then, for any subset A of S, we define $f(A) = \bigcup_{x \in A} f(x)$. The subset $f(A)$ of T is called the image of A under f. [If $A = \emptyset$ then, from Definition 2.1, $f(A) = \emptyset$.]

DEFINITION 4.7. Let S and T be sets, $f\colon S \to T$ a mapping, and B a subset of T. We define $f^{-1}(B) = \bigcup_{\substack{x \in S \\ f(x) \in B}} x$. The subset $f^{-1}(B)$ of S is called the source (or inverse) of the set B under the mapping $f\colon S \to T$. [If the set of all $x \in S$ such that $f(x) \in B$ is empty then, from Definition 2.1, $f^{-1}(B) = \emptyset$.]

A mapping $f\colon S \to T$ may be modified in a number of ways. We may change the rule of association f, we may change the set S, we may change the set T, or we may make any suitable combination of these modifications that leaves us with a mapping. The situation frequently arises that we wish to consider f only on a subset A of S; this modification of f is known as the restriction of f to A.

DEFINITION 4.8. Let S and T be sets, $f: S \to T$ a mapping, A a subset of S and $g: A \to T$ a mapping. The mapping g is said to be the restriction of f to A if and only if $g(x) = f(x)$ for every $x \in A$. The restriction of f to A is frequently denoted as f/A.

EXAMPLE 4.9. Let W be the set of all positive integers greater than 1, and V the set of all rational numbers of the form $1/m$, where m is a positive integer. Define a mapping $f: W \to V$ as follows: For each element x of W, $f(x) = 1/p$, where p is the smallest prime factor of x that exceeds 1.

The mapping of Example 4.9 may be changed by changing the set V. Let V_1 be the subset of V consisting of all numbers of the form $1/p$, where p is a prime different from 1. Let $g: W \to V_1$ be the mapping defined by $g(x) = f(x)$ for every $x \in W$. Then the mappings $f: W \to V$ and $g: W \to V_1$ are not the same since $V \neq V_1$ (see Definition 4.4). We note that $f(W) \neq V$, but $g(W) = V_1$. It is for this reason that we say that g is a mapping of W onto V_1.

DEFINITION 4.10. Let S and T be sets and $f: S \to T$ a mapping. Then f is a mapping of S onto T if and only if $f(S) = T$.

It is good mathematical English to state that a mapping f of S into T is onto. Every mapping that is onto is into, but it is false that every mapping that is into is onto.

The mapping of Example 4.9 may be modified by restricting it to the subset W_1 of W consisting of all positive primes. This mapping $f/W_1: W_1 \to V$ has the following properties: (i) with each point x in W_1 the rule f/W_1 associates precisely *one* point of V, (ii) the inverse of each point of $f/W_1(W_1)$ is precisely *one* point of W_1. It is for this reason that this mapping is said to be a one-to-one mapping of W_1 into V. We write our definition as follows:

DEFINITION 4.11. Let S and T be sets and $f: S \to T$ a mapping. Then f is said to be a one-to-one mapping of S into T if and only if, for each point y of $f(S)$, the set $f^{-1}(y)$ is precisely one point of S.

PROBLEMS

13. Let S and T be universes, A a subset of S, and $f: A \to T$ a mapping. Suppose that B and C are subsets of A, and that we are given an index set Λ and a collection of subsets $\{A_\alpha\}$ of A indexed by Λ. Prove

(a) $f(B \cup C) = f(B) \cup f(C)$.

(*b*) $f(B \cap C) \subset f(B) \cap f(C)$, and give an example in which $f(B \cap C) \neq f(B) \cap f(C)$.

(*c*) $f(\bigcup_{\alpha \in \Lambda} A_\alpha) = \bigcup_{\alpha \in \Lambda} f(A_\alpha)$.

(*d*) $f(\bigcap_{\alpha \in \Lambda} A_\alpha) \subset \bigcap_{\alpha \in \Lambda} f(A_\alpha)$.

14. Let $f: A \to T$ be a mapping. Then f is one-to-one if and only if, for every pair of points x, y of A, $f(x) = f(y)$ implies $x = y$.

15. Let $f: S \to T$ be a mapping. Then f is onto if and only if, for every non-empty subset B of T, $f^{-1}(B)$ is non-empty.

16. For the mapping of Example 4.9 give a non-empty subset D of V such that $f^{-1}(D)$ is empty.

17. In Problem 13(*b*) and (*d*), prove that the two sets are equal if the mapping is one-to-one.

18. Modify Example 4.9 to obtain a mapping $h: W \to V$ that is both one-to-one and onto.

19. Let S and T be sets, A a subset of S, and $f: A \to T$ a mapping. Let E be a subset of A, and define $B = f(A)$. Suppose that C and D are subsets of T, and that we are given an index set Λ and a collection of subsets $\{A_\alpha\}$ of T indexed by Λ. Prove

(*a*) $f^{-1}(C \cup D) = f^{-1}(C) \cup f^{-1}(D)$.

(*b*) $f^{-1}(C \cap D) = f^{-1}(C) \cap f^{-1}(D)$.

(*c*) $f^{-1}(\bigcup_{\alpha \in \Lambda} A_\alpha) = \bigcup_{\alpha \in \Lambda} f^{-1}(A_\alpha)$.

(*d*) $f^{-1}(\bigcap_{\alpha \in \Lambda} A_\alpha) = \bigcap_{\alpha \in \Lambda} f^{-1}(A_\alpha)$.

(*e*) $f(f^{-1}(C)) = C \cap B$.

(*f*) $f^{-1}(f(E)) \supset E$. Give an example where $f^{-1}(f(E)) \neq E$.

20. Let S and T be sets, A a subset of S and $f: A \to T$ a mapping. State and prove an "if and only if" condition on f which will insure that $f^{-1}(f(E)) = E$ for every subset E of A.

5. Relations, Mappings, Equivalence, Order

In the previous section we considered the Cartesian product $W = S \times T$ of two sets S and T. Under certain circumstances it was found that we could associate with a subset F of W a mapping $f: S \to T$ such that F was the graph of f. The essential feature of this association was that, on the one hand, the mapping paired with each element of S an element of T, and, on the other, that any subset of $W = S \times T$ consisted of pairs of elements, one from S the other from T. More generally, we think of any pairing of elements of S with elements of T as forming a subset of $S \times T$, and call any such pairing a binary relation.

DEFINITION 5.1. Let S and T be sets and W the Cartesian product $S \times T$. Then a subset F of $W = S \times T$ is said to be a binary relation from S to T. We say, for any two elements a, b satisfying $a \in S$, $b \in T$, that a is in the relation F to b if and only if $(a, b) \in F$. Sometimes we indicate that a is in the relation F to b by writing $a\,F\,b$.

The graph F of any mapping $f: S \to T$ is, of course, an example of a binary relation, and in this case we usually write $b = f(a)$ rather than $a\,F\,b$. It is not true, however, that a mapping $f: S \to T$ can always be associated with a binary relation $F \subset S \times T$. For instance, let F be the subset of the Cartesian plane $R \times R$ consisting of all points above and on the line $y = x$. Notice that a point (x, y) lies in F if and only if $x \leq y$. That is, $x\,F\,y$ in this case has exactly the meaning we ordinarily attribute to $x \leq y$. Furthermore, F does not in this case satisfy condition I; hence this is an example of a binary relation which is not the graph of any mapping.

Throughout mathematics there are three types of binary relations which occur again and again. One is that associated with the graph of a mapping. The other two are the equivalence relation and the order relation.

DEFINITION 5.2. Let S be a set and F a subset of $S \times S$. Then F is said to be an equivalence relation on S if and only if it is reflexive, symmetric, and transitive, i.e., if and only if the following conditions hold:

(i) (*Reflexive*) For any $a \in S$, $a\,F\,a$.
(ii) (*Symmetric*) For any $a, b \in S$, if $a\,F\,b$, then $b\,F\,a$.
(iii) (*Transitive*) For any $a, b, c \in S$; if $a\,F\,b$ and $b\,F\,c$, then $a\,F\,c$.

If F is an equivalence relation, we frequently read $a\,F\,b$ as "a is equivalent to b."

The equal sign used in the ordinary arithmetic of real numbers is a familiar example of an equivalence relation. In this case the set F is the line $y = x$ of the Cartesian plane. In elementary number theory or algebra, different kinds of equal signs or equivalence relations on the set of integers are often considered. For example, to say that two integers m and n are congruent modulo the integer 7 is to say precisely that $m - n$ is divisible by 7. Thus 5 and 12 are equivalent modulo 7. In the arithmetic of modulo 7 we feel free to replace 5 by 12 or 12 by 5 as we choose since they are equivalent, or equal, with respect to 7.

It should be noticed with respect to the congruence of integers, modulo 7, that, if A_i, $i = 0, 1, \cdots, 6$, is the set of integers equivalent to i, then every integer lies in $\bigcup_{i=0}^{6} A_i$ and that $A_i \cap A_j = \emptyset$, for $i \neq j$, $i = 0, 1, \cdots, 6$,

$j = 0, 1, \cdots, 6$. We shall see that similar conditions hold for any equivalence relation.

DEFINITION 5.3. Let S be a set and F an equivalence relation on S. Then a subset A of S is said to be an equivalence class of S (with respect to F) if and only if the following conditions hold:

(i) $x, y \in A$ implies $x\,F\,y$.
(ii) $x \in A$ and $y\,F\,x$ implies $y \in A$.
(iii) $A \neq \emptyset$.

THEOREM 5.4. Let S be a set, F an equivalence relation on S, and M the collection of all distinct equivalence classes of S with respect to F. Then $S = \bigcup_{A \in M} A$, and, for any two elements A, B of M, exactly one of the conditions holds:

(i) $A = B$.
(ii) $A \cap B = \emptyset$.

Proof. It is clear that $S \supset \bigcup_{A \in M} A$ since every element A of M is a subset of S. To see that $S \subset \bigcup_{A \in M} A$, note that, for any $x \in S$, the set of all y of S equivalent to x is an element A of M. Since $x\,F\,x$, $x \in A$. Thus $x \in \bigcup_{A \in M} A$. This proves the equality.

Let $A, B \in M$. If A and B have no points in common, then (ii) holds. Otherwise, there exists a point $p \in A \cap B$. If $y \in A$, then $y\,F\,p$; hence $y \in B$. Similarly, if $y \in B$, then $y \in A$. Thus $A = B$. Thus, if (i) does not hold, then (ii) must hold. Clearly, both (i) and (ii) cannot hold, since no equivalence class can be empty.

The converse of Theorem 5.4, namely, given a collection of subsets of a set S satisfying the conditions of Theorem 5.4, then there corresponds an equivalence relation on S, is also true. This is usually stated in terms of decompositions.

DEFINITION 5.5. Let S be a set and M a collection of subsets A of S. Then M is said to be a decomposition of S if and only if $S = \bigcup_{A \in M} A$, and, for any two elements A, B of M, exactly one of the following conditions holds:

(i) $A = B$.
(ii) $A \cap B = \emptyset$.

THEOREM 5.6. Let S be a set, and M a decomposition of S. Then there exists a unique equivalence relation F on S such that the equivalence classes of S with respect to F are elements of M.

The proof of this theorem is left to the student.

We now consider the third type of binary relation: the order relation.

DEFINITION 5.7. Let S be a set and F a subset of $S \times S$. Then F is said to be a simple order relation on S if and only if the following three conditions hold:

(i) For any two distinct elements a, b of S, either $a\,F\,b$ or $b\,F\,a$.

(ii) For any two elements a, b of S, if $a\,F\,b$, then a and b are distinct.

(iii) For any three elements a, b, c of S, if $a\,F\,b$ and $b\,F\,c$, then $a\,F\,c$.

We frequently read $a\,F\,b$ as "a precedes b" or "b follows a." We often call a relation an order relation when we mean a simple order relation.

The subset F of the Cartesian plane consisting of all points above and not on the line $y = x$ is an example of an order relation. In this example $x\,F\,y$ means that $x < y$.

In studying condition (i) of Definition 5.7, the following question naturally arises: Is it possible that $a\,F\,b$ and $b\,F\,a$ both hold for some order relation F? That the answer to this question is negative follows easily by applying conditions (ii) and (iii).

THEOREM 5.8. Let S be a set and F a simple order relation on S. Then, for any two points a, b of S, it is false that both $a\,F\,b$ and $b\,F\,a$.

The concept of order, as distinct from the concepts of mapping and equivalence, has many modifications, such as the concepts of partial ordering, well-ordering, etc., which occur in topology as well as in other branches of mathematics. The particular kind of order that we have defined, the simple order, is, nonetheless, important, and it is with reference to it that these other types of order are perhaps best understood.

PROBLEMS

21. Let S and T be sets and $f\colon S \to T$ a mapping. Let F be the subset of $S \times S$ consisting of all pairs (a, b) of points of S such that $f(a) = f(b)$. Prove that F is an equivalence relation on S.

22. Let F be the binary relation on the real numbers which is the subset of the Cartesian plane consisting of all pairs (x, y) of real numbers such that x and y are positive integers and $x + y$ is odd.

(*a*) Show that F is not the graph of a mapping of R into R.

(*b*) Show that F is not an equivalence relation.

(*c*) Show that F is not an order relation.

23. Let S be a set and F an order relation on S. Show that the relation G on S defined by $a\,G\,b$ if and only if $b\,F\,a$ is also an order relation on S.

24. Let S be a set and F an equivalence relation on S. Show that F is not an order relation.

25. Let S be the set of all positive integers, and suppose that the relation F on S has the following properties:

(i) If n is even and m is odd, then $n\,F\,m$.

(ii) If n and m are both even or both odd, then $n\,F\,m$ if and only if there exists an integer k such that $n = m + k$.

Decide whether or not F is an order relation on S.

26. Let S be a collection of colored blocks and F the relation on S defined by the property that $x\,F\,y$ if and only if x has the same color as y. Prove that F is an equivalence relation on S. Would separating the blocks into piles, each pile consisting of all blocks of S that have the same color, be a decomposition of S?

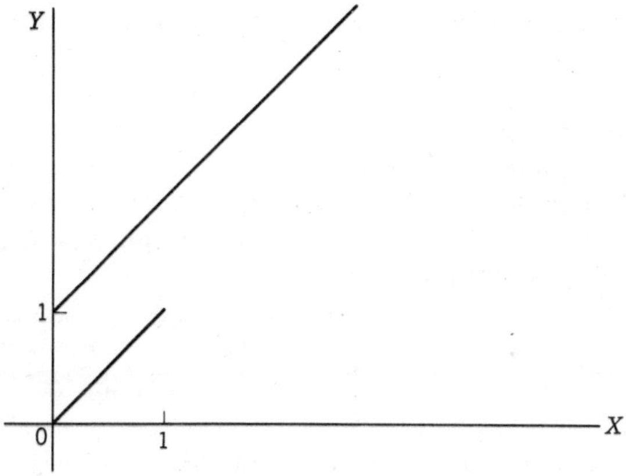

Figure 2

27. Let S be the set of all differentiable functions defined on R, the set of real numbers. Let F be the relation defined by $y\,F\,z$ if and only if $dy/dx = dz/dx$ for all $x \in R$. Is F an equivalence relation?

28. Let G be the subset of the Cartesian plane consisting of all points (x, y) of the form (x, x) for $0 \leq x \leq 1$ or of the form $(x, x + 1)$ for $x \geq 0$. Determine the subset F of the Cartesian plane such that

(1) $G \subset F$.

(2) F is an equivalence relation on R.

(3) For any subset H of the Cartesian plane, if H satisfies 1 and 2, then $F \subset H$.

See Figure 2, and consider the geometric significance of the axioms of equivalence.

29. Let R be the set of real numbers. Let F be the usual order relation on R, namely, all points of the Cartesian plane above and not on the line $y = x$. Let G be the usual equivalence relation on R, namely, all points of the Cartesian plane on the line $y = x$. We then usually write for $x\,H\,y$, where $H = F \cup G$,

$x \leq y$. Discuss the union of two relations generally. Also discuss their intersection.

30. Prove Theorem 5.6.

31. Prove Theorem 5.8.

6. Sequences and Infinite Sets

Let S and T be sets, A a subset of S, f: $A \to T$ a mapping, and $B = f(A)$. Thus, to each point a of A, there corresponds a unique point b of B defined by $b = f(a)$. Suppose we call this point b the point b_a; that is, we give b the name (or index) a. We see that this process assigns at least one name (or index) to each point b of B, and we may write $B = \bigcup_{a \in A} b_a$. Thus we may regard A as an index set and B as a set that has been indexed by A. The rule that assigns indices from A to the points of B is precisely the mapping f. Thus every mapping f of a set A into a set T may be regarded as an indexing of the set $f(A)$ by the set A. It should be noted that, since the mapping is not required to be one-to-one, a point of $f(A)$ may have more than one index. By Problem 14 it follows that points of $f(A)$ that have different indices are necessarily different if and only if the index mapping is one-to-one.

A class of mappings of particular interest arises when we map the set I of all positive integers, 1, 2, 3, $\cdots$ into a set T. Let f: $I \to T$ be such a mapping, and denote for each element n of I the point $f(n)$ of T by p_n. The image set $f(I)$ thus consists of the set of all points p_n of T. Such a collection of points p_n indexed by the set of all positive integers is called a sequence.

Definition 6.1. A sequence is a set A indexed by the set I of all positive integers. The nth element of the sequence is the element a of A which is indexed by the integer n: i.e. a_n. The sequence is symbolized $\{a_n\}$ where $A = \bigcup_{n \in I} a_n$.

It is important to notice that a sequence is not just a set of points, but is a set of points indexed by the set of positive integers. Further the same set of points indexed in two different ways gives rise to two different sequences.

Definition 6.2. Let $\{a_n\}$ and $\{b_n\}$ be two sequences. These sequences are said to be the same if and only if $a_n = b_n$ for every positive integer n.

Let us now consider what we shall mean by saying that a set is infinite. Again we consider the set I of all positive integers. Let f: $I \to I$ be the mapping defined by $f(x) = 2x$ for each $x \in I$. This mapping is one-to-one, and $f(I) = E$ where E denotes the set of all positive even integers. We notice that $E \subset I$ and $E \neq I$; in fact, E is an example of what is called a proper subset of I.

DEFINITION 6.3. Let S be a set and A a subset of S. Then A is said to be a proper subset of S if and only if $A \neq \emptyset$ and $A \neq S$.

In everyday life the statement that two collections of objects have the "same number" of elements is equivalent to the statement that there exists a one-to-one mapping of one collection onto the other. If we apply this notion to the integers, we see that we must say that there are the "same number" of positive integers as positive even integers. The remarkable feature of this example, namely, that the set could be placed in a one-to-one correspondence with a proper subset, is used to define the term infinite.

DEFINITION 6.4. Let S be a set. The set S is said to be infinite if and only if there exists a one-to-one mapping $f: S \to S$ such that $f(S)$ is a proper subset of S. A set is said to be finite if and only if it is not infinite.

We consider some theorems concerning finite and infinite sets.

THEOREM 6.5. The empty set and any set consisting of exactly one point are each finite sets.

Proof. Suppose in either case that a one-to-one mapping of the set into itself exists. The image set is not a proper subset of the given set, since neither the empty set nor a set consisting of one point have any proper subsets. Hence these sets are not infinite; that is, they are each finite sets.

THEOREM 6.6. The set I of all positive integers is infinite.

Proof. The essential details have been given in the discussion above.

THEOREM 6.7. Let A and B be sets, and $A \subset B$. If A is infinite, then B is infinite.

Proof. Assume that A is infinite. To show that B is infinite, we must show that there exists a one-to-one mapping $f: B \to B$ such that $f(B)$ is a proper subset of B.

Since A is infinite, there exists a one-to-one mapping $g: A \to A$ such that $g(A)$ is a proper subset of A. Define $f: B \to B$ as follows:

$$f(x) = x \quad \text{if} \quad x \in B - A$$
$$f(x) = g(x) \quad \text{if} \quad x \in A$$

This mapping $f: B \to B$ is one-to-one. Further, $f(B) \supset f(A) = g(A) \neq \emptyset$; so $f(B) \neq \emptyset$. Also $f(B) = f(B - A) \cup f(A)$, and $f(B - A) = B - A$, but $f(A) = g(A) \neq A$; thus $f(B) \neq B$. Therefore $f(B)$ is a proper subset of B, and B is accordingly infinite.

THEOREM 6.8. Let A and B be sets, and $A \subset B$. If B is finite, then A is finite.

Proof. This theorem is logically equivalent to Theorem 6.7.

THEOREM 6.9. Let A be a set and p be a point of A. If A is infinite, then $A - \{p\}$ is infinite.*

Proof. Assume that A is infinite. Then there exists a one-to-one mapping $f: A \to A$ such that $f(A)$ is a proper subset of A. We distinguish two cases:

(i) $p \in A - f(A)$.
(ii) $p \notin A - f(A)$.

In either case we show how to define a one-to-one mapping $g: A - \{p\} \to A - \{p\}$ such that $g(A - \{p\})$ is a proper subset of $A - \{p\}$.

Case (i). $p \in A - f(A)$. Define a mapping $g: A - \{p\} \to A - \{p\}$ as follows:

$$g(x) = f(x) \quad \text{for all} \quad x \in A - \{p\}$$

This mapping is into $A - \{p\}$ since the condition of $p \in A - f(A)$ assures us that $g(x) = f(x) \neq p$ for any $x \in A - \{p\}$. This mapping is one-to-one since f is one-to-one. The set $g(A - \{p\})$ is non-empty since $A - \{p\}$ is non-empty. (The equality $A - \{p\} = \emptyset$ would imply that $A = \{p\}$ since $p \in A$, and, by Theorem 6.5, A would then be finite, contradicting our assumption.) Further, $g(A - \{p\}) \neq A - \{p\}$ since $f(p) \in A - \{p\}$ and $f(p) \notin g(A - \{p\})$. Thus $g(A - \{p\})$ is a proper subset of $A - \{p\}$; hence $A - \{p\}$ is infinite in case (i).

Case (ii). $p \notin A - f(A)$. In this case $p \in f(A)$, and, since f is one-to-one, there is a unique point $a \in A$ such that $f(a) = p$. Let b be a point of the non-empty set $A - f(A)$. Define $h: A \to A$ as follows:

$$h(x) = f(x) \quad \text{for} \quad x \in A \quad \text{if} \quad x \neq a$$
$$h(x) = b \quad \text{for} \quad x \in A \quad \text{if} \quad x = a$$

This mapping h is one-to-one from A into A and $p \in A - h(A)$. The argument given for case (i) applied to this mapping h yields the result that $A - \{p\}$ is infinite.

We now apply these theorems and methods to obtain information about sets of integers.

THEOREM 6.10. Let A be a subset of the set of all positive integers I. Then A is finite if and only if there exists a positive integer m such that $x \in A$ implies $x \leq m$ (i.e., A is a bounded set of positive integers).

* The symbol $\{p\}$ is frequently employed to designate the set consisting of the single point p.

Proof. We first prove that, if A is finite, then A is bounded. The proof is by contradiction and makes use of the fact that every non-empty subset of I has a least element. Suppose A is not bounded; then, given any positive integer n, there exists $x \in A$ such that $x > n$. We define a one-to-one mapping f of I onto A as follows:

$$f(1) = \text{the least integer in } A$$

$$f(k+1) = \text{the least integer in } A - \bigcup_{j=1}^{k} f(j)$$

This mapping is well defined since $A - \bigcup_{j=1}^{k} f(j) \neq \emptyset$ for any k. For, if not, there exists an integer k_0 such that $A - \bigcup_{j=1}^{k_0} f(j) = \emptyset$, and then $f(k_0)$ is such that $x \in A$ implies $x \leq f(k_0)$, which is impossible. Let $g\colon I \to I$ be the mapping defined by $g(x) = 2x$. Define the mapping $h\colon A \to A$ as follows:

$$h(a) = f[g(f^{-1}(a))] \quad \text{for every} \quad a \in A$$

This mapping is one-to-one. The set $h(A)$ is non-empty since $f(2) \in h(A)$. The set $h(A) \neq A$ since the least element of A does not belong to $h(A)$. Thus A is infinite, a contradiction. Hence, if A is finite it is bounded; that is, there exists a positive integer m such that $x \in A$ implies $x \leq m$.

We now prove that, if A is bounded, then A is finite. Suppose m is an integer such that $x \in A$ implies $x \leq m$. We shall show that the set I_m consisting of all positive integers less than or equal to m is finite. Once this is done, the conclusion that A is finite follows from Theorem 6.8. Let I_k denote the set of all positive integers less than or equal to the integer k. The set I_1 is finite by Theorem 6.5. The set I_2 is finite, since, if it were infinite, then, by Theorem 6.9, the set $I_1 = I_2 - \{2\}$ would be infinite, a contradiction. Similarly, if I_j is finite then I_{j+1} is finite. Hence I_k is finite for every positive integer k. In particular, then, I_m is finite, but $A \subset I_m$; hence A is finite by Theorem 6.8.

Our next theorem shows that every infinite set contains a one-to-one image of the set of all positive integers I. The proof is based on the fundamental postulate that, given any non-empty subset A of a set S, a point of A can be chosen or picked out. The choice of a point of A associates with A a *unique* point of A. Thus this fundamental postulate may be stated formally as an axiom as follows.

Axiom of Choice. Let S be a set and $\mathscr{A}$ a non-empty collection of non-empty subsets of S. Then there exists a mapping $f\colon \mathscr{A} \to S$ such that $f(A) \in A$ for every subset A of S of the collection $\mathscr{A}$.

In almost every instance that we choose elements of sets we rely on this axiom for the justification of our procedure.

THEOREM 6.11. Let S be a set and I be the set of all positive integers. Then, if S is infinite, there exists a one-to-one mapping $f: I \to S$.

Proof. Assume that S is infinite. Let $\mathscr{A}$ be the collection of all non-empty subsets of S; then, by the axiom of choice, there exists a mapping $g: \mathscr{A} \to S$ such that $g(A) \in A$ for every set A of the collection $\mathscr{A}$. Define the mapping $f: I \to S$ as follows:

$$f(1) = g(S)$$
$$f(k+1) = g\left(S - \bigcup_{j=1}^{k} f(j)\right)$$

This mapping is well defined since $S \neq \emptyset$ and $S - \bigcup_{j=1}^{k} f(j) \neq \emptyset$ for any integer k. For otherwise S would be finite. The mapping f is one-to-one since, if $f(m) = f(n)$, then each of the assumptions $m > n$ and $m < n$ leads to a contradiction.

PROBLEMS

32. Consider the proof of Theorem 6.7.
    (*a*) Why is $f: B \to B$ one-to-one?
    (*b*) Why does $f(B) = f(B - A) \cup f(A)$?
    (*c*) Why is $g(A) \neq \emptyset$?
    (*d*) Why is $f(B) \neq B$?

33. Consider the proof of Theorem 6.10.
    (*a*) Show that the mapping $f: I \to A$ is *onto*.
    (*b*) Show that the mapping $h: A \to A$ is one-to-one.

34. Consider the proof of Theorem 6.11.
    (*a*) Suppose $S - \bigcup_{j=1}^{k} f(j) = \emptyset$; why is S finite?
    (*b*) Give the details of the contradictions at the end of the proof.

35. Show that the set of all even integers is infinite.

36. Let A and B be sets, $f: A \to B$ a one-to-one mapping. Prove that, if A is infinite, then B is infinite.

37. Let A and B be sets, $f: A \to B$ be a mapping onto. Prove that, if A is finite, then B is finite.

38. Prove that a set K is infinite if and only if there exists a sequence $\{p_n\}$ of distinct points of K.

39. Let K be a non-empty set. Prove that K is finite if and only if there exists a finite set of integers L and a mapping $f: L \to K$ which is onto.

40. Let K be a non-empty set, and I_n be the set of all positive integers less than or equal to n. Prove that K is finite if and only if there exists an integer n_0 and a one-to-one mapping of I_{n_0} onto K.

7. Countable and Uncountable Sets

Let I be the set of positive integers, E the set of positive even integers, and R the set of all real numbers. We have seen that there exists a one-to-one mapping $f: I \to E$ such that $f(I) = E$. We shall prove that there exists no mapping of I onto R.

THEOREM 7.1. There exists no mapping $f: I \to R$ such that $f(I) = R$.

Proof. Suppose such a mapping exists. We shall obtain a contradiction by constructing a real number r such that $r \neq f(n)$ for any $n \in I$. To do this, we may assume that each element of R has been assigned a unique decimal representation. The number r is then defined to be $5.r_1 r_2 r_3 \cdots r_n \cdots$, where r_n is defined to be 6 if the nth decimal place in the expansion of $f(n)$ is 5, and $r_n = 5$ if this is not the case. The number r then differs from $f(n)$ in its nth decimal place. Now r cannot be a number that has two decimal representations since it contains no 9's and no 0's. It is thus clear not only that r differs in at least one digit from $f(n)$ for every n, but also that r is a real number different from $f(n)$ for every n. This completes the proof.

Since the integers are used in counting, it is natural to say that a set K is countable if and only if it is empty or there exists a mapping of I onto K. Intuitively this means that we would not run out of integers if we attempted to assign a different integer to each element of K; such an assignment would be necessary in any attempt to "count" K. Theorem 7.1 tells us that any attempt to make such an assignment of the positive integers to R must fail. It is for this reason that the set of all real numbers is said to be not countable or uncountable.

DEFINITION 7.2. A set K is said to be countable if and only if it is empty or there exists a mapping of I onto K, where I is the set of all positive integers. A set K is said to be uncountable if and only if it is not countable.

Let X be the set of all positive rational numbers. We know that $I \subset X \subset R$. These inclusions do not enable us to determine whether X is countable or uncountable. The fact that X is actually a countable set will be a corollary to the next theorem. Before stating this theorem, we remark that a positive rational number is a real number which can be written in the form p/q, where p and q are elements of I. Thus we may regard the set X of all positive rational numbers as the set of all ordered pairs (p, q), where p and q are relatively prime positive integers; that is, the set X may be regarded as a subset of the Cartesian product $I \times I$. We shall prove that the set $I \times I$ is countable. It will then follow that X is countable, since X is contained in $I \times I$.

THEOREM 7.3. Let I be the set of all positive integers. Then the set $I \times I$ is countable.

Proof. Consider the set K of all positive integers x which may be written in the form $x = 2^p 3^q$, where p and q are integers. Then K is countable, since K is a subset of I (see Problem 43*c*). Define a mapping f of K into $I \times I$ as follows: for $x = 2^p 3^q$, a point of K, $f(x) = (p, q)$, a point of $I \times I$. This mapping is clearly onto. Thus $I \times I$ is countable by Problem 44.

COROLLARY 7.4. The set of all positive rational numbers is countable.

This corollary is immediate from the theorem and Problem 45.

The theorem above also provides us with the information that the union of a countable collection of countable sets is a countable set.

THEOREM 7.5. The union of a countable collection of countable sets is countable.

Proof. If the union is the empty set, it is countable by definition. If the union is not empty, then we may index the collection with the set of positive integers I. Let the collection thus indexed be $\{A_i\}$. The subcollection of $\{A_i\}$ consisting of all non-empty sets A_i is a countable collection by Problem 45; hence we can denote this subcollection by $\{B_i\}$. Clearly, $\bigcup_{i \in I} A_i = \bigcup_{i \in I} B_i$.

Now each B_i is a non-empty countable set, and thus its points may be indexed by the set I. We may therefore denote the jth point of B_i by $p_{i,j}$, where $j = 1, 2, 3, \cdots$. We now define a mapping f of $I \times I$ into $\bigcup_{i \in I} B_i$, by $f((i, j)) = p_{i,j}$. This mapping is clearly onto. Thus $\bigcup_{i \in I} B_i$ is countable, since the image of every countable set is countable (see Problem 44).

PROBLEMS

41. Exhibit two infinite decimals that differ in at least one digit but represent the same real number.

42. Show how the number r in the proof of Theorem 7.1 might depend on the choice of the decimal expansion of $f(21)$: i.e., produce a real number x having two different decimal expansions such that, if $f(21) = x$, the numbers r obtained from these two expansions are different.

43. Prove that each of the following sets is countable:
(*a*) Every finite set.
(*b*) The set I.
(*c*) Every subset of I.

44. Let A and B be sets and $f: A \to B$ a mapping onto B. Prove that, if A is countable, then B is countable. Can this same theorem be stated in the form "If B is uncountable, then A is uncountable"?

45. Every subset of a countable set is countable.

46. Prove that the collection of all rational numbers (positive, negative and zero) is countable.

47. Prove that the collection of all pairs of rational numbers is countable.

48. Prove that the collection of all circles in the Cartesian plane that have rational centers and rational radii is countable.

49. Let K be any countable set and B any set indexed by K. Then B is countable.

50. Sharpen Corollary 7.4 by exhibiting a one-to-one correspondence between the set of all positive rational numbers and the set of all positive integers.

Bibliographical Comment

§§ 5, 6, 7. More extensive treatment of the subjects considered in these sections may be found in

R. L. Wilder, *The Foundations of Mathematics*, John Wiley & Sons, New York, 1952.

E. Kamke, *Theory of Sets*, Dover Publications, New York, 1950.

CHAPTER TWO

THE REAL LINE

1. Real Numbers and Sequences

Throughout this chapter the universe will be the set of all real numbers, and this universe will always be denoted by *R*. The most familiar representation of a real number is as a decimal. Among the class of all decimals, two types are commonly distinguished:

(*a*) Repeating decimals.

(*b*) Non-repeating decimals.

The integers, both positive and negative, are examples of repeating decimals. Other examples are —6.345*0*, 2.981*0*, 2.345*345*, —7.897676*76*, where the italic denotes the group of digits that is to be repeated. It will be recalled that the decimal expansion of every rational number is a repeating decimal. Conversely, every repeating decimal can be expressed as a rational number: i.e., as the quotient of two integers. The class of non-repeating decimals comprises the class of irrational numbers.

It is a familiar fact that the two repeating decimals 1.99*9* and 2.00*0* represent the same real number. Indeed it is true in general that, if two different decimals represent the same number, then each of these decimals is repeating, one ending in all zeros, the other in all nines. Thus the number represented by this decimal is rational. If we agree never to write a decimal ending in all nines, but always to use its equivalent form ending in all zeros, then our assumption is that there exists a one-to-one correspondence between the real numbers and the set of all infinite decimals. We also assume the familiar rules for addition, multiplication, division, at least for decimals terminating in all zeros. Our remaining assumption is that the real numbers are simply ordered.

In numerical work with decimals we usually round off our result to a certain number of places and say that it approximates the answer. The number of places to which we round off is controlled by the accuracy that the particular problem requires. For example, consider the problem of rounding off the repeating decimal 0.33*3*. This problem consists exactly of choosing a term of the sequence $(0.3, 0.33, 0.333, \cdots)$ which approximates the answer to the required degree of accuracy. We know that 0.333 is the decimal expansion of the real number $\frac{1}{3}$. Suppose we desire to round off the decimal to an accuracy of $\frac{1}{1000}$. This means that we must find a term x in the sequence $(0.3, 0.33, 0.333, \cdots)$ such that $\left|x - \frac{1}{3}\right| < \frac{1}{1000}$. It is easily seen that we may choose for x the term 0.333 or *any term* that follows it in the sequence.

DEFINITION 1.1. A sequence $\{a_n\} = (a_1, a_2, a_3, \cdots)$ of real numbers converges to a real number A if and only if, given any real number $\varepsilon > 0$, there exists an integer N such that $n > N$ implies $|A - a_n| < \varepsilon$. The number A is called the limit of the sequence, and we write $\lim a_n = A$.

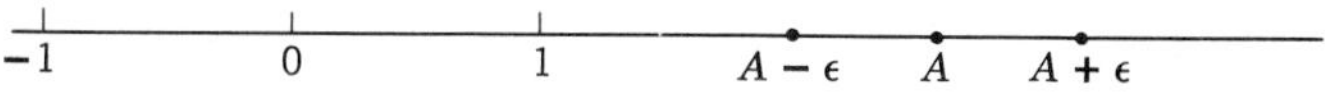

Figure 3

On the real line where A, $A - \varepsilon$, $A + \varepsilon$ are plotted, as in Figure 3, Definition 1.1 states that, for *all* $n > N$, a_n will lie between $A - \varepsilon$ and $A + \varepsilon$. In what follows, the student may be considerably aided in his understanding by figures, such as Figure 3, which he draws himself.

DEFINITION 1.2. A pair of real numbers (a, b) is said to confine a sequence $\{a_n\}$ if and only if there exists an integer N such that $a < a_n < b$ for every $n > N$.

THEOREM 1.3. The sequence $\{a_n\}$ converges to the real number A if and only if the sequence $\{a_n\}$ is confined by every pair (a, b) such that $a < A < b$.

Proof. Suppose $\{a_n\}$ converges to A. Let (a, b) be any pair of real numbers such that $a < A < b$. Let ε be the minimum of the positive numbers $b - A$ and $A - a$. By Definition 1.1 there exists N such that $n > N$ implies $A - \varepsilon < a_n < A + \varepsilon$. But $a \leq A - \varepsilon$ and $A + \varepsilon \leq b$. Thus $n > N$ implies $a < a_n < b$; hence (a, b) confines $\{a_n\}$.

Suppose on the other hand that every pair of real numbers (a, b), $a < A < b$, confines $\{a_n\}$. Thus in particular, for $\varepsilon > 0$, the pair $(A - \varepsilon, A + \varepsilon)$ confines $\{a_n\}$. In other words, there exists N such that $n > N$ implies $A - \varepsilon < a_n < A + \varepsilon$; that is, $|A - a_n| < \varepsilon$. Thus $\lim a_n = A$.

In Figure 4 we have the graph of the index mapping $f\colon I \to R$, where I is the set of all positive integers $1, 2, 3, \cdots$ and f is defined by $f(n) = a_n$ for each integer n. Thus the points of the graph are the pairs (n, a_n).

The statement that $\{a_n\}$ is confined by the pair (a, b) says precisely that there exists an integer N such that all points (n, a_n) of the graph lie between the lines $y = a$ and $y = b$ for $n > N$.

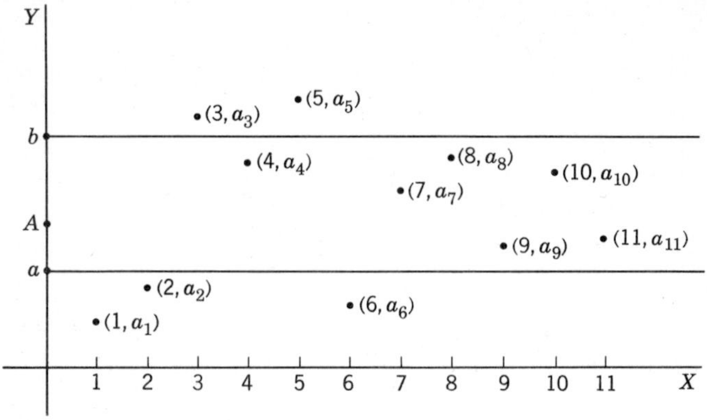

Figure 4

THEOREM 1.4. (*a*) $\text{Lim } a_n = A$, $\lim a_n = B$ implies $A = B$.
(*b*) If $a_n = A$ for every n, then $\lim a_n = A$.

Proof. (*a*) Suppose $A \neq B$. Let $d = \frac{1}{3}|A - B|$. Then each of the pairs $(A - d, A + d)$, $(B - d, B + d)$ must confine the sequence $\{a_n\}$. This is impossible since there exists no real number x satisfying both the inequalities $A - d < x < A + d$ and $B - d < x < B + d$.

The proof of (*b*) is an immediate application of Theorem 1.3.

PROBLEMS

1. Prove that the sequence $(0.3, 0.33, 0.333, \cdots)$ converges to $\frac{1}{3}$.

2. Give the details of the contradiction in Theorem 1.4*a*.

3. Given a sequence $\{a_n\}$ and a real number A such that $\lim a_n = A$. Prove that $\{a_n\}$ is *bounded*; i.e., there exist real numbers B and C such that $B \leq a_n \leq C$ for every n.

4. Find the limit of each of the following sequences $\{a_n\}$ showing all details of the proof that $\{a_n\}$ converges to the number you found:
(*a*) $a_n = 1/n$ for all n.
(*b*) $a_n = \frac{1}{2}$ for all n.

5. Given $\lim a_n = A$, $\lim b_n = B$, and any real number k, prove
(*a*) $\lim ka_n = kA$.
(*b*) $\lim (a_n + b_n) = A + B$.
(*c*) $\lim (a_n - b_n) = A - B$.
(*d*) $\lim a_n b_n = AB$.

6. Given a sequence $\{a_n\}$ and a real number A such that $\lim a_n = A$. Let (b, c) be any pair of real numbers that confines $\{a_n\}$. Prove that $b \leq A \leq c$. Give an example to show you could not prove $b < A < c$.

7. Given a sequence $\{a_n\}$ which converges to A, and a real number $\varepsilon > 0$. Prove that there exists a pair (a, b) which confines $\{a_n\}$ such that $|b - a| < \varepsilon$.

8. Given a sequence $\{a_n\}$ which converges to A. Prove that, given any real number $\varepsilon > 0$, there exists an integer N such that, for any pair of integers h and k, each greater than N, it is true that $|a_h - a_k| < \varepsilon$.

9. State clearly what the statement $\lim a_n \neq A$ means.

2. Monotone Sequences

The following problem frequently arises: Given a sequence $\{a_n\}$, does there exist a real number A such that $\lim a_n = A$. For a certain class of sequences this question is answered by the next theorem.

THEOREM 2.1. Let $\{a_n\}$ be a sequence with the following properties:

(i) $a_n \geq a_{n+1}$ for every n.

(ii) there exists a number B such that $a_n \geq B$ for every n.

Then there exists a number A such that $\lim a_n = A$.

Proof. Let C_0 be the greatest integer satisfying $C_0 \leq a_n$ for every n. Since C_0 is the greatest such integer, there must exist an integer h_0 such that $a_{h_0} < C_0 + 1$.

Let c_1 be the greatest integer such that $C_0 + \frac{c_1}{10} \leq a_n$ for every n. Clearly $0 \leq c_1 \leq 9$. Furthermore, there exists an h_1 such that $a_{h_1} < C_0 + \frac{c_1}{10} + \frac{1}{10}$.

Let c_2 be the greatest integer such that $C_0 + \frac{c_1}{10} + \frac{c_2}{10^2} \leq a_n$ for every n. Clearly, $0 \leq c_2 \leq 9$. Furthermore, there exists an h_2 such that $a_{h_2} < C_0 + \frac{c_1}{10} + \frac{c_2}{10^2} + \frac{1}{10^2}$.

Suppose that $C_0, c_1, c_2, \cdots, c_k$ have been defined such that $C_0 + \frac{c_1}{10} + \frac{c_2}{10^2} + \cdots + \frac{c_k}{10^k} \leq a_n$ for every n, and that there exists an integer h_k such that $a_{h_k} < C_0 + \frac{c_1}{10} + \frac{c_2}{10^2} + \cdots + \frac{c_k}{10^k} + \frac{1}{10^k}$. We then define c_{k+1} as the greatest integer such that

$$C_0 + \frac{c_1}{10} + \frac{c_2}{10^2} + \cdots + \frac{c_k}{10^k} + \frac{c_{k+1}}{10^{k+1}} \leq a_n$$

for every n. We note that $0 \leq c_{k+1} \leq 9$, and that there exists an integer h_{k+1} such that

$$a_{h_{k+1}} < C_0 + \frac{c_1}{10} + \frac{c_2}{10^2} + \cdots + \frac{c_k}{10^k} + \frac{c_{k+1}}{10^{k+1}} + \frac{1}{10^{k+1}}.$$

Thus by induction the integer c_n has been defined for every n. Let A be the real number $C_0 + 0.c_1c_2c_3 \cdots$ thus defined. The sequence $\{a_n\}$ converges to A. We can see this as follows. We know that $A \leq a_n$ for every n. Thus, using (i), the convergence is established if we show that, for any given $\varepsilon > 0$, there exists an integer h such that $a_h < A + \varepsilon$. Take n sufficiently large so that $10^{-n} < \varepsilon$. Then, by our definition of h_n, we have $a_{h_n} < C_0 + 0.c_1c_2c_3 \cdots c_n + \dfrac{1}{10^n} \leq A + \dfrac{1}{10^n} < A + \varepsilon$. Thus $\{a_n\}$ converges to A.

The concepts involved in Theorem 2.1 are of frequent occurrence. The definitions that follow give these concepts names.

Definition 2.2. The sequence $\{a_n\}$ is monotone if and only if it satisfies one of the following conditions:

(i) $a_n \geq a_{n+1}$ for every n.

(ii) $a_n \leq a_{n+1}$ for every n.

These cases are distinguished by saying that the sequence is monotone non-increasing if and only if (i) holds, and monotone non-decreasing if and only if (ii) holds. The sequence is said to be monotone if and only if it is either monotone non-increasing or monotone non-decreasing.

Definition 2.3. Let H be a set of real numbers. Then b is a lower bound of H if and only if $b \leq x$ for every x in H; and b is an upper bound of H if and only if $x \leq b$ for every x in H. The set H is bounded below if and only if H has a lower bound; and H is bounded above if and only if H has an upper bound. The set H is bounded if and only if H is bounded above and bounded below.

Definition 2.4. Let H be a set of real numbers. Then b is a greatest lower bound of H if and only if (i) b is a lower bound of H, and (ii) if $c > b$; then c is not a lower bound of H. Also b is a least upper bound of H if and only if (i) b is an upper bound of H, and (ii) if $c < b$; then c is not an upper bound of H.

Several important results on upper and lower bounds are immediate consequences of Definition 2.3 and 2.4. These are embodied in the theorem that follows.

THEOREM 2.5. Let H be a set of real numbers.

(a) A point p is not a lower bound of H if and only if there exists a point q of H such that $q < p$.

(b) A point p is not an upper bound of H if and only if there exists a point q of H such that $q > p$.

(c) If p is a greatest lower bound of H, then, for every point c such that $p < c$, there exists a point q of H such that $p \leq q < c$.

(d) If p is a least upper bound of H, then, for every point c such that $p > c$, there exists a point q of H such that $p \geq q > c$.

Note that we may phrase Theorem 2.1 in terms of bounds as follows: "Let $\{a_n\}$ be a monotone non-increasing sequence. If the set $\bigcup_{n \in I} a_n$ has a lower bound, then it has a greatest lower bound. If A is the greatest lower bound of $\bigcup_{n \in I} a_n$, then $\lim a_n = A$."
This rephrasing suggests the following theorem. The proof, which follows closely that of Theorem 2.1, is left to the student.

THEOREM 2.6. Let K be a non-empty set of real numbers. If K has a lower bound, then K has a greatest lower bound.

PROBLEMS

10. Show in the proof of Theorem 2.1 why there must exist an integer C_0 satisfying the given condition.

11. In the proof of Theorem 2.1 consider the number A. Prove that, given any integer k, it is false that $c_n = 9$ for every $n > k$.

12. Let $\{a_n\}$ be a sequence having the following properties:
(i) $a_n \leq a_{n+1}$ for every n.
(ii) There exists a number B such that $a_n \leq B$ for every n.
Prove that there exists a number A such that $\lim a_n = A$. (*Hint.* Consider the sequence $\{-a_n\}$.)

13. Consider again the proof of Theorem 2.1. Let D be any number such that $D \leq a_n$ for every n. Prove $D \leq A$. State and prove the analogous statement for the theorem of Problem 12.

14. Give the details of the proof of Theorem 2.6.

15. Prove that, if K is a non-empty set of real numbers which has an upper bound, then K has a least upper bound.

16. Prove that a sequence $\{a_n\}$ which is both monotone non-increasing and monotone non-decreasing has the property that $a_h = a_k$, for every pair of integers h, k.

17. Let H be the empty set and x any real number. Prove that x is both an upper bound and a lower bound of H. Prove that H has no least upper bound and no greatest lower bound.

18. Let H be a non-empty set having b as its greatest lower bound and a as its least upper bound. If $a = b$, prove that H consists of exactly one point, and that this point is a.

19. Prove that every monotone non-decreasing sequence that is bounded above is bounded.

20. Prove that every monotone non-increasing sequence that is bounded below is bounded.

21. Given a sequence $\{a_n\}$ which is not bounded above. Prove that $\{a_n\}$ is not monotone non-increasing.

22. Given a monotone sequence $\{a_n\}$. Prove that there exists a number A to which $\{a_n\}$ converges if and only if $\{a_n\}$ is bounded.

23. (*a*) Let c and d each be greatest lower bounds of a set H. Prove that $c = d$.

(*b*) Let c and d each be least upper bounds of a set H. Prove that $c = d$ (*Remark*. This problem gives the result that greatest lower bounds and least upper bounds of a set are unique when they exist. We may thus replace the expression "a greatest lower bound of H" by "the greatest lower bound of H," and similarly for least upper bounds.)

24. Given a set H. If b is a lower bound of H, then b is the greatest lower bound of H if and only if, for every lower bound c of H, $c \leq b$. Prove this theorem, and then state and prove the corresponding theorem for least upper bounds.

25. Every bounded monotone sequence has a least upper bound and a greatest lower bound and converges to one of these.

3. Cauchy Sequences

We have been studying the question: "Given a sequence $\{a_n\}$, under what conditions will there exist a number A to which this sequence converges?" This question has been answered completely for monotone sequences by Problem 22. The task remains of finding if and only if conditions for the existence of a number A to which an arbitrary sequence converges. The discussion of this question will be based on our knowledge of monotone sequences. We shall state conditions under which a sequence $\{a_n\}$ is called a Cauchy sequence, and shall prove that, given a sequence $\{a_n\}$, there will exist a number A to which $\{a_n\}$ converges if and only if $\{a_n\}$ is a Cauchy sequence.

DEFINITION 3.1. A sequence $\{a_n\}$ is a Cauchy sequence if and only if, for every $\varepsilon > 0$, there exists an integer N such that $h > N$ and $k > N$ implies $| a_h - a_k | < \varepsilon$.

We have seen that, if $\lim a_n = A$, then $\{a_n\}$ is a Cauchy sequence (see Problem 8). Thus the problem of this section is to show that every

Cauchy sequence converges. We shall show this by making use of the idea of a subsequence. The general concept of a subsequence is based on that of a subsequence of the sequence of all positive integers in their natural order namely 1, 2, 3, 4, $\cdots$.

DEFINITION 3.2. The sequence $\{n_i\}$ is a subsequence of *the* sequence of positive integers if and only if the following conditions hold:

(i) Each n_i is a positive integer.

(ii) For every positive integer i, $n_i < n_{i+1}$.

From condition (ii) it follows at once that $n_i \geq i$ for every $i \in I$.

DEFINITION 3.3. The sequence $\{b_n\}$ is a subsequence of $\{a_n\}$ if and only if there exists a subsequence $\{n_i\}$ of the sequence of positive integers such that $b_i = a_{n_i}$ for every i.

The student may already see how we intend to prove that, given a Cauchy sequence $\{a_n\}$, there exists a number A such that $\lim a_n = A$. In broad outline our method will be the following. We first show that every sequence has a monotone subsequence. In particular, the given Cauchy sequence $\{a_n\}$ has a monotone subsequence $\{b_n\}$. Since every Cauchy sequence is bounded, it follows that $\{b_n\}$ is bounded (see Problems 26 and 28). Consequently, by Problem 22, there exists a number A to which the sequence $\{b_n\}$ converges. The proof is then completed by showing that the sequence $\{a_n\}$ must itself converge to A.

THEOREM 3.4. Every sequence $\{a_n\}$ has a monotone subsequence.

Proof. Exactly one of the following statements is true:

(i) There exists an integer N such that no element of the sequence $(a_{N+1}, a_{N+2}, a_{N+3}, \cdots)$ is a lower bound of this sequence.

(ii) For every integer N, there exists an element of the sequence $(a_{N+1}, a_{N+2}, a_{N+3}, \cdots)$, which is a lower bound of this sequence.

Suppose statement (i) is true. We shall define a monotone non-increasing subsequence $\{b_n\}$ of $\{a_n\}$. Let $b_1 = a_{N+1}$. There exists an integer k_2 such that $a_{N+k_2} < b_1$, and we define $b_2 = a_{N+k_2}$. There must exist an integer $k_3 > k_2$ such that $a_{N+k_3} < b_2$, and we define $b_3 = a_{N+k_3}$. In general, if b_i has been defined as a_{N+k_i}, there must exist an integer $k_{i+1} > k_i$ such that $a_{N+k_{i+1}} < b_i$, and we define $b_{i+1} = a_{N+k_{i+1}}$. The sequence $\{b_n\}$ thus defined is a monotone non-increasing subsequence of $\{a_n\}$.

Suppose, on the other hand, that statement (ii) is true. Let a_{k_1} be an element of the sequence $\{a_n\}$ which is a lower bound of this sequence. By (ii) there is an element a_{k_2} of the sequence $(a_{k_1+1}, a_{k_1+2}, \cdots)$ which is a lower bound of this sequence. If a_{k_j} has been defined then by (ii), there exists an element $a_{k_{j+1}}$ of the sequence $(a_{k_j+1}, \cdots)$, which is a lower bound of this sequence. Thus we have defined a subsequence $\{a_{k_j}\}$ of $\{a_n\}$. This sequence is monotone non-decreasing.

THEOREM 3.5. Given a Cauchy sequence $\{a_n\}$ there exists a number A such that $\{a_n\}$ converges to A.

Proof. The sequence $\{a_n\}$ is bounded (by Problem 26). This means, using Problem 27, that there exists a number A and a monotone subsequence $\{b_n\}$ of $\{a_n\}$ such that $\lim b_n = A$.

It remains to be shown that $\lim a_n = A$. Precisely stated, we must show that, for $\varepsilon > 0$, there exists an integer N such that $n > N$ implies $|a_n - A| < \varepsilon$. We suppose such a number ε has been given, and show how to find the integer N. We note the following facts:

I. There exists an integer N_1 such that $h > N_1$ and $k > N_1$ implies $|a_h - a_k| < \varepsilon/3$.

II. There exists an integer N_2 such that $n > N_2$ implies $|b_n - A| < \varepsilon/3$.

(Statement I is true since the sequence $\{a_n\}$ is a Cauchy sequence, and II holds since $\lim b_n = A$.)

Let $N = N_1 + N_2$, and $n > N$. Then, since $\{b_n\}$ is a subsequence of $\{a_n\}$, $b_N = a_m$ where $m \geq N$. Since $n > N_1$, $m > N_1$, and $N > N_2$, we must have

$$|a_n - A| \leq |a_n - a_m| + |a_m - A| = |a_n - a_m| + |b_N - A| < \frac{\varepsilon}{3} + \frac{\varepsilon}{3} < \varepsilon$$

Consequently, $\lim a_n = A$.

PROBLEMS

26. Prove that every Cauchy sequence is bounded.

27. Given a bounded sequence $\{a_n\}$. Prove that there exists a number A and a monotone subsequence $\{b_n\}$ of $\{a_n\}$ such that $\lim b_n = A$.

28. Let $\{a_n\}$ be a sequence and $\{b_n\}$ a subsequence of $\{a_n\}$. Prove that, if $\{a_n\}$ has any one of the properties listed below, then $\{b_n\}$ has the same property:

(i) Bounded.

(ii) Monotone.

(iii) The sequence is a Cauchy sequence.

Give examples in each case of sequences $\{a_n\}$ where the subsequence $\{b_n\}$ has the given property, but where the sequence $\{a_n\}$ does not.

29. In the proof of Theorem 3.4 show that statement (i) is true if and only if statement (ii) is false.

30. In the second part of the proof of Theorem 3.4

(*a*) Show that the sequence $\{b_n\}$ where $b_j = a_{k_j}$ for each j is a subsequence of $\{a_n\}$.

(*b*) Show that this sequence $\{b_n\}$ is monotone non-decreasing.

31. Consider the statements (i) and (ii) of the proof of Theorem 3.4. Replace the word "lower" by the word "upper" in each of these statements. Prove that exactly one of the new statements is true, and use this result to obtain another proof of the theorem.

32. If $\lim a_n = A$ and $\{b_n\}$ is a subsequence of $\{a_n\}$, prove $\lim b_n = A$.

33. In the last part of the proof of Theorem 3.5, suppose that the definition of N is changed to $N =$ the greater of N_1 and N_2. Will the remaining statements of the proof still be true?

34. Let $\{a_n\}$ be a sequence and $\{b_n\}$ a subsequence of $\{a_n\}$. Suppose $\lim b_n = B$; then $\lim a_n = B$ if and only if $\{a_n\}$ is a Cauchy sequence.

35. (*a*) Given a sequence $\{a_n\}$ and a number A such that $\{a_n\}$ does not converge to A. Prove that there exists a number $d_0 > 0$ such that, for any integer N, there is an $n > N$ for which $|a_n - A| > d_0$.

(*b*) Using the result of (*a*), prove that there exists a subsequence $\{b_n\}$ of $\{a_n\}$ satisfying one of the following conditions: (i) $b_n \leq A - d_0$, for every n, or (ii) $b_n \geq A + d_0$, for every n.

36. Given a sequence $\{a_n\}$ which is bounded and for which there exists no number A such that $\{a_n\}$ converges to A. We know that there exists a number B and a subsequence $\{b_n\}$ of $\{a_n\}$ such that $\{b_n\}$ converges to B. Making use of Problem 35, show that there exists a number $C \neq B$ and a subsequence $\{c_n\}$ of $\{a_n\}$ such that $\{c_n\}$ converges to C.

37. Give a proof of Theorem 3.5 by contradiction, using the ideas developed in Problems 27, 35, and 36.

38. Let H be a non-empty set having a least upper bound B such that $B \notin H$. Prove that there exists a monotone non-decreasing sequence of distinct points of H which converges to B.

4. Open Sets

The student should be convinced by this time that the arbitrary positive number ε and the integer N, depending on ε, are not easy to manipulate without considerable practice. They are nonetheless the standard tools of the analyst and are essential in certain studies. We intend to avoid the explicit use of ε by replacing it with an equivalent geometrical idea.

If the student has made drawings similar to Figures 3 and 4 to aid his interpretation of ε and N, he has already realized that the geometry of the situation is simple. Consider this arbitrary positive number ε. When we first used it, we were thinking geometrically of a point A and a set of points near A, more precisely those points x such that $A - \varepsilon < x < A + \varepsilon$. (It will be understood that "point" means real number.) By use of sets of the form $a < x < b$, of which $A - \varepsilon < x < A + \varepsilon$ is an example, the explicit use of ε can be avoided.

DEFINITION 4.1. Given two points a and b, $a < b$. The set of all points x such that $a < x < b$ is called an open-interval. It will be denoted as the interval (a, b). The points a and b are called the end points of the interval.

DEFINITION 4.2. A set K is called an open-left-ray if and only if there exists a real number A such that K consists of all $x < A$. A set K is called an open-right-ray if and only if there exists a real number A such that K consists of all $x > A$. In either case A is called the end point of the ray. A set K is called an open-ray if and only if K is an open-left-ray or an open-right-ray.

THEOREM 4.3. The sequence $\{c_n\}$ converges to C if and only if, for every open-interval (a, b) containing C, there exists an integer N, such that c_n is in (a, b) for every $n > N$.

Proof. This is a restatement of Theorem 1.3.

Theorem 4.3 enables us to talk about the convergence of a sequence $\{c_n\}$ to a point C without explicit use of the arbitrary positive number ε. Note that the use of an arbitrary ε has been replaced by the use of an arbitrary open-interval (a, b) which contains C.

The above theorem is usually stated in terms of open sets rather than open-intervals.

DEFINITION 4.4. A set G of real numbers is open if and only if, for each point p of G there exists an open-interval (a, b) such that $p \in (a, b) \subset G$. We use the expression "G is a neighborhood of p" if and onlv if "G is an open set containing p."

THEOREM 4.5. The sequence $\{c_n\}$ converges to C if and only if, for every neighborhood G of C, there exists an integer N such that $c_n \in G$ for every $n > N$.

Proof. Suppose the condition holds. We must show that $\{c_n\}$ converges to C. To do this let (a, b) be any open-interval containing C. By Definition 4.4, the set (a, b) is a neighborhood of C. Thus, by our hypothesis, there exists an integer N such that $c_n \in (a, b)$ for every $n > N$. It follows from Theorem 4.3 that $\{c_n\}$ converges to C.

To complete the proof, suppose that $\{c_n\}$ converges to C, and that G is any neighborhood of C. Then there exists an open-interval (a, b) such that $C \in (a, b) \subset G$. By Theorem 4.3 there is an integer N such that $c_n \in (a, b) \subset G$ for every $n > N$.

Several important results on open sets are immediate consequences of Definition 4.4. These are embodied in the next theorem.

THEOREM 4.6. (*a*) Give an open set G and a point p of G, there exist points a and b of G such that $a < p < b$. In particular, no point p of G is an upper bound of G or a lower bound of G.

(*b*) A non-empty set G is open if and only if G is the union of a collection of open-intervals. In particular, every open-interval is an open set, as is also every open-ray.

(*c*) A set G is not open if and only if there exists a point p in G such that every neighborhood of p contains at least one point q not in G. In particular, the empty set is open and R is open.

We have seen in Theorem 4.6*b* that every open-interval is an open set as is also every open-ray. For this reason we shall henceforth drop the hyphen and replace the single word open-interval by the two words open interval, and the single word open-ray by the two words open ray. It is, of course, not true that every open set is an open interval or an open ray.

Before stating the next theorem, we introduce the convenient terminology of disjoint sets.

DEFINITION 4.7. Two sets H and K are said to be disjoint if and only if $H \cap K = \emptyset$. A collection of sets is said to be disjoint if and only if no two sets of the collection have a point in common.

THEOREM 4.8. There do not exist two non-empty disjoint open sets G_1 and G_2 such that $G_1 \cup G_2 = R$.

Proof. We shall assume the existence of two such sets and obtain a contradiction. The sets being non-empty, there must exist two points a and c such that $a < c$, and a and c are not in the same set. We may choose our notation in such a way that G_1 is the set containing a, and G_2 is the set containing c.

Let A be the set of all points x such that $x < c$, and define $H = A \cap G_1$. Since A and G_1 are open sets, the set H is open (see Problem 45). The point c is an upper bound of H, and H is non-empty since $a \in H$. Consequently H has a least upper bound which we denote by b. We note by Theorem 4.6*a* that $b \notin H$. Since $b \leq c$ and $b \notin H = A \cap G_1$, we see that $b \notin G_1$. Hence $b \in G_2$. Thus G_2 is a neighborhood of b. By Theorem 2.5*d*, $G_2 \cap H \neq \emptyset$. Therefore $G_2 \cap G_1 \neq \emptyset$, a contradiction.

Theorem 4.8 may be used to prove things that at first sight seem far removed from it. Suppose, for example, that we desire to show that the equation $x^2 = 2$ has a real root. We assume that it does not and express R as the union of two sets G_1 and G_2 which we now define. The set G_1 consists of all numbers $x \leq 0$ and all positive numbers x such that $x^2 < 2$. The set $G_2 = R - G_1$. It is easily seen that the set G_1 is open, and the set G_2 can be shown to be open if we use our assumption that there exists no x such that $x^2 = 2$. We will thus have a contradiction to Theorem 4.8 if we show that the sets G_1 and G_2 are non-empty. This is immediate, since G_1 contains the point $x = 1$ and G_2 contains the point $x = 2$.

PROBLEMS

39. Prove that the intersection of two open-intervals is an open-interval or the empty set.

40. A set H is open if and only if, for each point p of H, there exists a neighborhood G of p such that $G \subset H$.

41. A set V is called a *vicinity* of a point p if and only if there exists a neighborhood G of p such that $G \subset V$. Replace the word "neighborhood" in the statement of Theorem 4.5 by the word "vicinity," and prove the theorem which results.

42. Prove that every open interval contains infinitely many distinct points. Thus no non-empty finite set is open.

43. Prove that every open interval contains an infinite sequence of distinct rational points.

44. Prove that the union of any collection of open sets is an open set.

45. Prove that the intersection of a finite number of open sets is an open set.

46. Consider the collection of open intervals I_n, defined by $I_n = \left(-\frac{1}{n}, \frac{1}{n}\right)$ for each n. Prove that the intersection of this collection of open sets is not open. Thus it is not true that the intersection of every collection of open sets is an open set.

47. Let G be any open set and K any finite set. Prove that $G - K$ is an open set.

48. If $R = H_1 \cup H_2$, where H_1 and H_2 are disjoint non-empty sets, then there exists a sequence of points of one of these sets which converges to a point of the other.

49. If b is the least upper bound of a set K, and $b \notin K$, then there exists a sequence $\{p_n\}$ of distinct points of K such that $\lim p_n = b$.

5. Connected Sets

An open interval has the property that, if it contains two distinct points, then it must contain every point between them. It is for this reason that we say that an open interval is a connected set.

DEFINITION 5.1. A subset K of R is said to be connected if and only if, given any two points a and b of K, $(a < b)$, then K contains every point of the open interval (a, b).

There are only a few forms that a connected subset of R can take. These are listed in the next theorem.

THEOREM 5.2. A subset K of R is connected if and only if it has one of the following forms:

(*a*) K is empty.
(*b*) K is a single point.
(*c*) K is an open interval.
(*d*) K is the union of an open interval and one or both of its end points.
(*e*) K is an open ray.
(*f*) K is the union of an open ray and its end point.
(*g*) K is the set R.

Proof. It is immediate from Definition 5.1 that each of the sets listed is connected. We shall prove that, given a connected set K, the set K must have one of the forms listed above. This follows at once if K is empty, a single point, or the set R. We may thus suppose that K contains at least two points, and that there exists a point c of R not in K.

Clearly, since K is connected, the point c is an upper bound of K or a lower bound of K. Thus there are two cases to be considered. We shall discuss only the case in which c is an upper bound of K, since the discussion of the other case is so similar. The set K, being non-empty and having the upper bound c, has a least upper bound d. If K is not bounded below, it is of the form (*e*) or (*f*). If K is bounded below, it is of the form (*c*) or (*d*).

We are now in a position to obtain more precise information about the structure of open sets.

THEOREM 5.3. Let $\{G_\alpha\}$ be a collection of disjoint non-empty open sets. Then $\{G_\alpha\}$ is countable.

Proof. Let K be the set of all rational numbers contained in $\cup G_\alpha$. Then K is countable. The set K is non-empty by Problem 43. We define a mapping f of K into the set whose elements are the members of the collection $\{G_\alpha\}$. The sets $\{G_\alpha\}$ are disjoint. Thus, for each element x of K, there exists a unique member of the collection $\{G_\alpha\}$ which contains x. We define $f(x)$ to be this member of the collection. We see further that the mapping f is onto, since every non-empty open set contains a rational point (see Problem 43). It follows that the set whose elements are the members of the collection $\{G_\alpha\}$ is countable since it is the image of the countable set K (see Problem I, 44).

THEOREM 5.4. A non-empty subset G of R is open if and only if G is the union of a countable collection of disjoint open connected sets.

Proof. If G is the union of a countable collection of disjoint open connected sets, then G is open. It remains to be shown that, given a non-empty open set G, we may express G as the union of a countable collection of disjoint open connected sets. To do this we first define, for each point p of G, the set C_p to be the union of all open intervals containing p and contained in G. The sets C_p have the following properties:

(i) Each C_p is an open connected set containing p and contained in G. This is a consequence of Problem 50 and the definition of C_p.

(ii) If C_p and C_q have a point in common, then $C_p = C_q$.

Let x and y be arbitrary points of C_p and C_q respectively, and r a point of $C_p \cap C_q$. Then there exist open intervals (a, b), (c, d), (e, f), (g, h) contained in G such that $x \cup p \subset (a, b)$, $p \cup r \subset (c, d)$, $r \cup q \subset (e, f)$, $q \cup y \subset (g, h)$. The union of these four open intervals is an open interval contained in G and containing the points p, q, r, x, and y. Thus $x \in C_q$, hence $C_p \subset C_q$, and $y \in C_p$, hence $C_q \subset C_p$. Consequently, $C_p = C_q$.

(iii) If p and q are points of G, then C_p and C_q are equal or disjoint.

Using Theorem 5.3, we see that the collection of distinct C_p is countable. It is evident that G is the union of the sets of this collection.

PROBLEMS

50. The union of any collection of open intervals, each containing the point p, is an open connected set containing p.

51. The union of any collection of connected sets, each containing the point p, is a connected set containing p.

6. Limit Points and Closed Sets

Geometrically it is natural to proceed from our notion of the limit of a sequence (see Definition 1.1) to the idea of a limit point of a set.

DEFINITION 6.1. A point p is a limit point of a set H if and only if there exists a sequence $\{p_n\}$ of distinct points of H such that $\{p_n\}$ converges to p.

EXAMPLE 6.2. Let the set H be the union of the open interval $(0, 1)$ and the set consisting of the single point 3. We see that, among the limit points of H are the points 0, $\frac{1}{2}$, and 1. The points 0 and 1 do not belong to H, whereas the point $\frac{1}{2}$ does. We note, however, that the point 3, which belongs to H, is not a limit point of H. This shows clearly the statement "p is a limit point of H" gives us no information whatever as to whether the point p does or does not belong to H.

The next theorem enables us to talk about limit points of a set without the idea of a sequence.

THEOREM 6.3. A point p is a limit point of a set H if and only if every neighborhood of p contains a point q of H distinct from p.

Proof. Suppose first that p is a limit point of H, and let $\{p_n\}$ be a sequence of distinct points of H such that $\{p_n\}$ converges to p. Let U be an arbitrary neighborhood of p. Using Theorem 4.5, there exists an integer N such that $p_n \in U$ for $n > N$. Thus the two points p_{N+1} and p_{N+2} are in U. Since these points are different, at least one of them must be different from p, and we take this point as our point q.

On the other hand, suppose that every neighborhood of p contains a point of H distinct from p. Let U_n be the neighborhood of p that is the open interval $(p - 1/n, p + 1/n)$. Let p_1 be a point of $U_1 \cap H$ distinct from p. Define $U_2' = U_2 - p_1$; then U_2' is a neighbourhood of p by Problem 47. Let p_2 be a point of $U_2' \cap H$ distinct from p. In general, suppose that the points $p_1, p_2, \cdots, p_k$ have been defined. Let F_k be the finite set consisting of these k points. Define $U_{k+1}' = U_{k+1} - F_k$. The set U_{k+1}' is a neighborhood of p by Problem 47, and thus contains a point p_{k+1} of H distinct from p. This defines a sequence $\{p_n\}$ of distinct points of H which converges to p. Consequently, p is a limit point of H.

DEFINITION 6.4. A set H is said to be closed if and only if every limit point of H is contained in H.

THEOREM 6.5. A set H is closed if and only if its complement is open.

The proof is left as an exercise for the reader.

In ordinary English usage the words "open" and "closed" are antonyms. This is not true in mathematics. We shall now give examples of sets satisfying each of the following conditions: (i) A set that is both open and closed, (ii) a set that is neither open nor closed, (iii) a set that is open but not closed, (iv) a set that is closed but not open.

EXAMPLE 6.6. Let H be the null set and R the set of all real numbers. By Theorem 4.6, each of these sets is open. But H is the complement of R, and R is the complement of H. Thus, by Theorem 6.5, each of the sets H and R is closed. We have thus found two sets, namely, the null set H and the set R, each of which is both open and closed.

EXAMPLE 6.7. Let H be the set defined in Example 6.2. This set is not closed since the point 0 is a limit point of H which does not belong to H. It is not open, since it contains a point, 3, which is a limit point of its complement.

EXAMPLE 6.8. Let K be any non-empty finite set, and define $H = R - K$. Then H is open by Problem 47. But H is not closed, since each point of K is a limit point of H.

EXAMPLE 6.9. Let K be any non-empty finite set. We have seen in Example 6.8 that the complement of K is open but not closed. It follows at once from Theorem 6.5 that K is closed but not open.

Since closed sets are the complements of open sets, many theorems concerning open sets can be translated into theorems concerning closed sets by use of DeMorgan's theorem.

THEOREM 6.10. (*a*) The intersection of any collection of closed sets is a closed set.

(*b*) The union of any finite collection of closed sets is a closed set.

Proof. (*a*) Let $\{H_\alpha\}$ be the collection of closed sets, and set $K = \bigcap_\alpha H_\alpha$. We must show that every limit point of K is contained in K. To do this, let p be any limit point of K. By Problem 56, p is a limit point of the closed set H_α for every α. Thus $p \in H_\alpha$ for every α; i.e. $p \in K$.

(*b*) Let $H_1, H_2, \cdots, H_n$ be the finite collection of closed sets, and $K = \bigcup_{i=1}^{n} H_i$. To show that K is closed is equivalent to showing that $R - K$ is open (Theorem 6.5). Using DeMorgan's theorem I, 3.2, we have at once

$$R - K = R - \bigcup_{i=1}^{n} H_i = \bigcap_{i=1}^{n} (R - H_i)$$

which is an open set by Problem 45 and Theorem 6.5.

In the proof of Theorem 6.10 we used different techniques in parts (*a*) and (*b*). Either of these techniques could have been used to prove both parts.

We have at our disposal a large class of sets which are known to be open: namely, the sets of Theorem 4.6. The next theorem gives us a similar class of closed sets.

THEOREM 6.11. The following sets are closed:

(i) The set R of all real numbers.
(ii) The null set.
(iii) For any fixed B, the set of all x such that $x \leq B$.
(iv) For any fixed A, the set of all x such that $x \geq A$.
(v) For any fixed A and B, the set of all x such that $A \leq x \leq B$.
(vi) Every finite set.

Proof. The sets in (i), (ii), (iii), (iv) are closed by Theorem 6.5 and Theorem 4.6. The set in (v) is closed by Theorem 6.10, since it is the intersection of the sets in (iii) and (iv). That every finite set is closed follows at once as in Example 6.9. Another method of proving this same fact is to take $A = B$ in (v). This gives the result that every point is a closed set. This fact, together with Theorem 6.10*b*, tells us at once that every finite set is closed.

EXAMPLE 6.12. We give here an example of sets G and K having the property that both G and $G - K$ are open, but K is not closed. To do this let G be the open interval (0, 1) while K is the union of the open interval $(\frac{1}{2}, 2)$ and the point $\frac{1}{2}$.

PROBLEMS

52. Prove that no finite set has a limit point.

53. In the second half of the proof of Theorem 6.3 give the details of the proof that the sequence $\{p_n\}$ converges to p.

54. Given a point p and a set H. Then p is a limit point of H if and only if every neighborhood of p contains infinitely many distinct points of H.

55. If p is a limit point of A and $A \subset B$, then p is a limit point of B.

56. If p is a limit point of $\bigcap_{\alpha} A_\alpha$, then p is a limit point of A_α for every α.

57. Let p be a limit point of the union of two sets A and B. Prove, without using sequences, that p is a limit point of A or a limit point of B.

58. Give an alternative proof of Problem 57, using sequences.

59. Let H be a set and K the complement of H. Then K is open if and only if no sequence of points of H converges to a point of K.

60. Prove Theorem 6.10*a*, using DeMorgan's theorem.

61. Prove Theorem 6.10*b* without using DeMorgan's theorem.

62. We have seen in Problem 46 an example of an infinite collection of open sets whose intersection is not open. Use this example and DeMorgan's theorem to find an example of an infinite collection of closed sets whose union is not closed.

63. Let $a \in R$. Then no point $b > a$ is a limit point of a set H if every point x of H satisfies the condition $x \leq a$.

64. Let $a \in R$. Then no point $b < a$ is a limit point of a set H if every point x of H satisfies the condition $x \geq a$.

65. Let $\{a_n\}$ be a bounded monotone sequence. Define $H = \bigcup_{n \in I} a_n$, and let A and B be the least upper bound and greatest lower bound of H, respectively. Prove that the set K which is the union of the set H and the points A and B is a closed set.

66. Let G be an open set and K a closed set. Prove that $G - K$ is open and $K - G$ is closed. (It should be noted that Problem 47 is a special case of Problem 66 in view of Theorem 6.11vi.) Compare this result with Example 6.12.

67. Let A be a connected set. Let B be any set containing A and having the property that B is contained in every closed set that contains A. Prove that B is connected.

68. Prove that R is not the union of two disjoint non-empty closed sets.

69. Let A be a set. Let A' be the set of all limit points of A. Prove that each of the sets A' and $A \cup A'$ is closed. Prove also that $A \cup A'$ is contained in every closed set containing A.

70. Let $\{p_n\}$ be a sequence of points that converges to a point p. Prove that, if $q \neq p$, then q is not a limit point of H where $H = \bigcup_{n \in I} p_n$.

7. Closed and Bounded Sets

It follows from Definition 6.1 that every set H that has a limit point contains infinitely many distinct points. It is not true that every set containing infinitely many distinct points has a limit point, as is easily seen by considering the set of all positive integers. If, however, we require that the set be bounded, we can prove the following fundamental theorem of analysis.

THEOREM 7.1. Every infinite bounded set H has at least one limit point.

Proof. Since H is bounded and infinite, it must contain a bounded monotone sequence $\{a_n\}$ of distinct points. Clearly this sequence converges to a point A; thus A is a limit point of H.

THEOREM 7.2. Every unbounded set H contains an unbounded monotone sequence of distinct points $\{a_n\}$. The infinite set $K = \bigcup_{n \in I} a_n$ has no limit point.

The proof is left as an exercise for the reader.

The Bolzano-Weierstrass theorem gives us an important way of characterizing closed and bounded sets of real numbers.

THEOREM 7.3 (*Bolzano-Weierstrass Theorem*). A set H is closed and bounded if and only if every infinite subset of H has at least one limit point in H.

Proof. Suppose that H is closed and bounded, and let K be any infinite subset of H. Then, by Theorem 7.1, K has at least one limit point. Let p be any limit point of K. Then p is a limit point of H, by Problem 55. Thus $p \in H$, since H is closed.

To complete the proof suppose that every infinite subset of H has at least one limit point in H. We must show that H is closed and bounded. We show first that H is closed. Let p be any limit point of H. Then there exists a sequence $\{p_n\}$ of distinct points of H such that $\lim p_n = p$.

Let $K = \bigcup_{n \in I} p_n$. Then K is an infinite subset of H; hence K has at least one limit point in H. By Problem 70 this limit point must be p. Thus $p \in H$; hence H is closed. The conclusion that H is bounded is immediate from Theorem 7.2.

We have seen that, for a given set A, the property of being closed and bounded is equivalent to the property that every infinite subset of A has a limit point in A. It is frequently useful to be able to phrase the property of being closed and bounded in terms of "coverings."

DEFINITION 7.4. A collection of sets $\{D_\alpha\}$ is said to be a covering of a set A, or to cover A, if and only if $A \subset \bigcup_\alpha D_\alpha$. If, in addition, each of the sets D_α is required to be an open set, then the collection $\{D_\alpha\}$ is known as an open covering, or as a covering by open sets.

DEFINITION 7.5. A set A is said to be compact if and only if for every open covering $\{U_\alpha\}$ of A, there exists a finite collection of open sets $U_{\alpha_1}, U_{\alpha_2}, \cdots, U_{\alpha_n}$, each belonging to the covering, such that $A \subset \bigcup_{i=1}^{n} U_{\alpha i}$.

THEOREM 7.6. Every compact set K is closed and bounded.

Proof. We first show that K is bounded. Let G_n be the open interval $(-n, n)$. Then K is covered by the collection of open sets $\{G_n\}$, for $n = 1, 2, 3, \cdots$. Since K is compact, a finite subcollection of the $\{G_n\}$ must also cover K. Let N be the largest subscript of any G_n in this finite subcollection. Clearly K is contained in the open interval $(-N, +N)$; hence K is bounded.

To prove that K is closed, we shall show that $R - K$ is open. Let p be an arbitrary point of $R - K$. Let G_n be the complement of the closed interval $[p - 1/n, p + 1/n]$. Then K is covered by the collection of open sets $\{G_n\}$, for $n = 1, 2, 3, \cdots$. It follows, as in the previous paragraph, that $K \subset G_N$ for some integer N. Thus the open interval $(p - 1/N, p + 1/N)$ contains p and is contained in $R - K$. Consequently, $R - K$ is open.

The following lemma will be useful in the proof of our next theorem.

LEMMA 7.7. Let K be a closed and bounded set and $\{G_\alpha\}$ an open covering of K. Then there exists a fixed $\delta > 0$ such that, for any point x of K, we can find an index α such that the open interval $(x - \delta, x + \delta)$ is contained in G_α.

Proof. Suppose that no such positive number δ exists. Let n be a positive integer. Then $\delta = 1/n$ will not satisfy the conditions of the theorem. This means that there exists a point x_n of K such that the open interval $(x_n - 1/n, x_n + 1/n)$ is not contained in G_α for any index α. Allowing n to take on the successive values $1, 2, 3, \cdots$, we obtain a sequence $\{x_n\}$ of points of K. Since K is closed and bounded, it follows that there exists a subsequence $\{x_{n_i}\}$ of $\{x_n\}$ which converges to a point p of K. But $\{G_\alpha\}$ is a covering of K by open sets; hence there exists an index α_p such that $p \in G_{\alpha_p}$. From the definition of an open set, we see that there exists a positive number ε such that the open interval $(p - \varepsilon, p + \varepsilon)$ is contained in G_{α_p}.

Pick an integer N sufficiently large so that $\frac{1}{N} < \frac{1}{3}\varepsilon$. We note that the sequence $\{x_{n_i}\}$ converges to p. Consequently, there exists an integer $n_i > N$ such that x_{n_i} lies in the open interval $(p - \frac{1}{3}\varepsilon, p + \frac{1}{3}\varepsilon)$. It is easily seen that the open interval $(x_{n_i} - \frac{1}{3}\varepsilon, x_{n_i} + \frac{1}{3}\varepsilon)$ is contained in $(p - \varepsilon, p + \varepsilon)$ and thus in G_{α_p}. This means that the open interval $(x_{n_i} - 1/n_i, x_{n_i} + 1/n_i)$ is contained in G_{α_p}, which contradicts the definition of the point x_{n_i}. This contradiction completes the proof of the lemma.

Theorem 7.8 (*Heine-Borel Theorem*). A set K is closed and bounded if and only if it is compact.

Proof. Suppose that K is closed and bounded. Let $\{G_\alpha\}$ be a collection of open sets covering K, and let δ be the positive number given by Lemma 7.7. Since K is closed and bounded, there exist a finite number of points $x_1, x_2, \cdots, x_n$ of K such that the finite collection of open intervals $(x_1 - \delta, x_1 + \delta), \cdots, (x_n - \delta, x_n + \delta)$ covers K. But, by the lemma, each of these open intervals is contained in a G_α. Thus we may find a collection of n of the sets G_α which is an open covering of K; i.e. K is a compact set. The fact that if K is compact then it is closed and bounded has been proved as Theorem 7.6.

PROBLEMS

71. Given a point p and a closed set F not containing p, prove that there exist two disjoint open sets U and V such that U contains p and V contains F.

72. Let K be a compact set and H a closed set contained in K. Prove that H is compact.

73. Let A be the set of all x such that $0 < x \leq 1$, and define a sequence of sets A_n as follows:

$$A_1 = \left(\frac{1}{3}, 2\right), A_2 = \left(\frac{1}{4}, \frac{1}{2}\right), \cdots, A_n = \left(\frac{1}{n+2}, \frac{1}{n}\right) \text{ for } n \geq 2$$

(*a*) Prove that $\{A_n\}$ is an open covering of A.

(*b*) Prove that no finite subcollection of the $\{A_n\}$ covers A. Thus conclude that A is not compact.

(*c*) Define B as the union of A and the single point $x = 0$, and, for each positive number b, set $B_b = (-1, b)$. Prove that the collection of sets consisting of all the A_n and B_b, for any fixed b, forms an open covering of B.

(*d*) Prove that, from the covering described in (*c*), we can find a finite number of sets which form an open covering of B.

(*e*) State why the result of (*d*) does not prove that B is a compact set.

74. Let $\{A_n\}$ be a collection of non-empty compact sets such that $A_n \supset A_{n+1}$ for every n. Prove $\bigcap_{n \in I} A_n \neq \emptyset$. (*Hint.* Assume $\bigcap_{n \in I} A_n = \emptyset$, and show that $\{R - A_n\}$ is an open covering of A_1.)

8. Continuous Functions

In courses in the calculus the student has made frequent use of continuous functions such as $f(x) = x$, $f(x) = 3x$, and $f(x) = 17x^3 - 25x + 9$. The three functions just mentioned are special cases of continuous mappings of the set R of all real numbers into itself.

DEFINITION 8.1. Let $f: R \to R$ be a mapping, and a a point of R. Then f is continuous at the point a if and only if, given any sequence $\{a_n\}$ of points of R converging to a, the sequence $\{f(a_n)\}$ converges to $f(a)$. The mapping f is continuous if and only if f is continuous at the point a for every a in R.

From Definition 8.1 the following theorem is immediate.

THEOREM 8.2. Each of the following mappings $f: R \to R$ is continuous:

(*a*) The mapping defined by $f(x) = x$, for every x in R.

(*b*) The mapping defined by $f(x) = k$, for every x in R, where k is any constant.

THEOREM 8.3. Let $f: R \to R$ and $g: R \to R$ be continuous mappings. Then each of the following mappings is continuous:

(*a*) The mapping $h: R \to R$ defined by $h(x) = f(x) + g(x)$, for every x in R.

(*b*) The mapping $h: R \to R$ defined by $h(x) = f(x) \cdot g(x)$ for every x in R.

Proof. The proof is immediate from Problem 5.

The importance of Theorem 8.3 lies in the fact that it allows us to build up new continuous functions from functions that we know are continuous. For example, Theorem 8.2 tells us that each of the following functions is continuous: $f(x) = x$, $f(x) = 17$, $f(x) = -25$, $f(x) = 9$. Using Theorem 8.3, we see at once that each of the following functions is continuous also: $f(x) = x^2$, $f(x) = x^3$, $f(x) = 17x^3$, $f(x) = -25x$, $f(x) = 17x^3 - 25x + 9$. This same method gives us at once the corollary:

COROLLARY 8.4. Every polynomial $f(x)$ is a continuous mapping $f: R \to R$.

It is frequently desirable to consider mappings $f: M \to R$, where M is a subset of the real numbers. We now define the concept of continuity for such mappings.

DEFINITION 8.5. Let $f: M \to R$ be a mapping, and a a point of M. Then f is continuous at the point a of M if and only if, for any sequence $\{a_n\}$ of points of M that converges to the point a, the sequence $\{f(a_n)\}$ converges to the point $f(a)$. The mapping f is continuous on M if and only if f is continuous at a for every a in M.

THEOREM 8.6. Let $f: M \to R$ be continuous and H a subset of M. Let $g: H \to R$ be the restriction of f to H (i.e. the mapping $g: H \to R$ defined by $g(x) = f(x)$ for every $x \in H$). Then $g: H \to R$ is continuous.

THEOREM 8.7. Let A be a subset of R and $f: A \to R$ a continuous mapping such that $f(A) = B$. Suppose that $g: B \to R$ is continuous. Then the mapping $h: A \to R$ defined by $h(x) = g(f(x))$, for each x in A, is continuous.

Proof. Let p be any point of A and $\{p_n\}$ any sequence of points of A converging to p. Since f is continuous on A, the sequence $\{f(p_n)\}$ converges to $f(p)$. Since g is continuous on $B = f(A)$, the sequence $\{g(f(p_n))\}$ converges to $g(f(p))$. Hence the sequence $\{h(p_n)\}$ converges to $h(p)$ and h is continuous on A.

PROBLEMS

75. Supply the few details needed for a proof of Theorem 8.6.

76. Let M be the subset of R consisting of all points x of R except the point $x = 0$. Define the mapping $f: M \to R$ by $f(x) = 1/x$ for each x in M. Prove that f is continuous.

77. Let M and N be subsets of R and $H = M \cap N$. Suppose that $f: M \to R$ and $g: N \to R$ are continuous. Define $h(x) = f(x) + g(x)$, $r(x) = f(x) \cdot g(x)$, for every x in H. Prove that the mappings $h: H \to R$, $r: H \to R$ are continuous.

78. Let I be the subset of R consisting of all positive and negative integers, and suppose that $f: I \to R$ is a mapping. Prove that f is continuous.

79. Let M and N be closed subsets of R and $H = M \cap N$. Suppose that $f: M \to R$ and $g: N \to R$ are continuous and that $f(x) = g(x)$ for all $x \in H$. Define $h: M \cup N \to R$ by $h(x) = f(x)$ for $x \in M$ and $h(x) = g(x)$ for $x \in N$. Prove that h is continuous.

80. Let M be a connected set and $f: M \to R$ a continuous mapping. Let a and b be points of M, $a < b$, and let H be the closed interval $a \leq x \leq b$. Define a mapping $g: R \to R$ as follows:

$$g(x) = f(x) \quad \text{for} \quad x \in H$$
$$g(x) = f(a) \quad \text{for} \quad x < a$$
$$g(x) = f(b) \quad \text{for} \quad x > b$$

Prove that g is continuous mapping.

9. Properties of Continuous Functions

Let $f\colon M \to R$ be continuous. We have seen that f preserves the property of being a convergent sequence of points; i.e. if $\{a_n\}$ is any sequence of points of M that converges to a point a of M, then the sequence $\{f(a_n)\}$ converges to the point $f(a)$. There are other properties which are preserved by continuous mappings. To obtain these we shall make use of the following theorem.

THEOREM 9.1. Let $f\colon M \to R$ be a mapping and x_0 a point of M. Then f is continuous at x_0 if and only if, given any open set G of R such that $f(x_0) \in G$, there exists an open set V containing x_0 such that $f(V \cap M) \subset G$.

Proof. Suppose the condition holds; we must show that f is continuous at x_0. To do this, let $\{x_n\}$ be any sequence of points of M that converges to x_0. Let G be an arbitrary open set containing $f(x_0)$, and let V be the open set given by the condition. Since $\{x_n\}$ converges to x_0, there exists an integer N such that $n > N$ implies $x_n \in V$, and thus $x_n \in V \cap M$. It follows that, for any $n > N$, $f(x_n) \in G$. Consequently $\{f(x_n)\}$ converges to $f(x_0)$, and f is continuous at x_0.

To complete the proof we shall assume that the condition does not hold and prove that f is not continuous at x_0. The assumption that the condition does not hold tells us that there exists an open set G_0 such that $f(x_0) \in G_0$ and that, for any open set V containing x_0, there exists a point $x \in V \cap M$ such that $f(x) \notin G_0$. Consequently, for every positive integer n, there exists a point x_n of M such that $x_0 - \frac{1}{n} < x_n < x_0 + \frac{1}{n}$ and $f(x_n) \notin G_0$. Clearly $\{x_n\}$ converges to x_0, but $\{f(x_n)\}$ does not converge to $f(x_0)$. Thus f is not continuous at x_0.

COROLLARY 9.2. Let $y = f(x)$ be a function defined on a subset M of R, and x_0 a point of M. Then f is continuous at x_0 if and only if, given any $\varepsilon > 0$, there exists a $\delta > 0$ such that, if $x \in M$ and $|x - x_0| < \delta$, then $|f(x) - f(x_0)| < \varepsilon$.

The condition of this corollary is used as the definition of continuity at a point by most writers of texts on the calculus.

THEOREM 9.3. Let $f\colon M \to R$ be a mapping. Then f is continuous on M if and only if, given any open set G of R, there exists an open set V such that $f^{-1}(G) = V \cap M$.

Proof. Suppose that f is continuous on M: that is, that f is continuous at every point $x \in M$. Let G be any open subset of R, and let x be any point of $f^{-1}(G)$. By Theorem 9.1, there exists a neighborhood V_x of x such that $f(V_x \cap M) \subset G$. Clearly

$$f^{-1}(G) = \bigcup_{x \in f^{-1}(G)} (V_x \cap M) = (\bigcup_{x \in f^{-1}(G)} V_x) \cap M.$$

The set $\bigcup_{x \in f^{-1}(G)} V_x$ is the required open set V.

On the other hand, if the condition holds, then f is continuous at each point of M by Theorem 9.1. Thus f is continuous on M.

Corollary 9.4. Let $f: R \to R$ be a mapping. Then f is continuous if and only if for every open set G of R, the set $f^{-1}(G)$ is open.

Theorem 9.3 will be important to us in the sequel. Several of its many applications will be given in this section.

Theorem 9.5. Let $f: M \to R$ be continuous, and K a closed and bounded subset of R such that $K \subset M$. Then $f(K)$ is a closed and bounded subset of R.

Proof. By Theorem 7.8, the proof will be complete if we show that $f(K)$ is compact. To do this let $\{G_\alpha\}$ be any open covering of $f(K)$. By Theorem 9.3, there exists, for each index α, an open set V_α of R satisfying $f^{-1}(G_\alpha) = V_\alpha \cap M$. The collection of these sets $\{V_\alpha\}$ is an open covering of K, so that there exists a finite subcollection $V_{\alpha_1}, \cdots, V_{\alpha_n}$ which covers K. Now $f(V_{\alpha_i} \cap K) \subset G_{\alpha_i}$ for $i = 1, \cdots, n$. Hence $f(\bigcup_{i=1}^{n} (V_{\alpha_i} \cap K)) \subset \bigcup_{i=1}^{n} G_{\alpha_i}$. Since $K \subset \bigcup_{i=1}^{n} (V_{\alpha_i} \cap K)$, it follows that $f(K) \subset \bigcup_{i=1}^{n} G_{\alpha_i}$. Thus $f(K)$ is compact.

Corollary 9.6. Let $y = f(x)$ be a continuous function defined on a closed bounded non-empty set K. Then there exists x_1 and x_2 in K such that $f(x_1) \leq f(x) \leq f(x_2)$ for every x in K.

Theorem 9.7. A subset M of R is connected if and only if there exists no continuous mapping $f: M \to R$ such that $f(M) = a \cup b$ where a and b are distinct points.

Proof. We show first that, if such a mapping $f: M \to R$ exists, the set M is not connected. We know there must exist points p and q in M such that $f(p) = a$ and $f(q) = b$, and we may choose the notation in such a way that $p < q$. Assume that M is a connected set. Then the closed interval $[p, q]$ is contained in M. Define a mapping $g: R \to R$ as follows:

$$g(x) = f(p) \quad \text{if} \quad x < p$$

$$g(x) = f(x) \quad \text{if} \quad p \leq x \leq q$$

$$g(x) = f(q) \quad \text{if} \quad x > q$$

Then g is a continuous mapping by Problem 80. Let G_1 be the open set consisting of all points $x < (a + b)/2$ and G_2 the open set consisting of all points $x > (a + b)/2$. It follows from the definition of g and Corollary 9.4 that R is the union of the two disjoint non-empty open sets $g^{-1}(G_1)$ and $g^{-1}(G_2)$. This contradiction to Theorem 4.8 proves that M is not a connected set.

To complete the proof we show that, if M is not connected, then such a mapping can be defined. Since M is not connected, there must be three points p, q, and r such that $p < q < r$ and $p \cup r \subset M$, $q \notin M$. Let a and b be two distinct points of R, and define a mapping $f: M \to R$ as follows:

$$f(x) = a \quad \text{if} \quad x < q$$

$$f(x) = b \quad \text{if} \quad x > q$$

It is evident that $f(M) = a \cup b$, and that f is a continuous mapping.

THEOREM 9.8. Let $f: M \to R$ be a continuous mapping. If M is connected then $f(M)$ is connected.

Proof. Let $K = f(M)$. We show that, if K is not connected, then M is not connected. Since K is not connected, there exists a continuous mapping $g: K \to R$ such that $g(K) = a \cup b$, $a \neq b$. Let $h: M \to R$ be the mapping defined by $h(x) = g(f(x))$ for each x in M. The mapping $h: M \to R$ is continuous by Theorem 8.7. Moreover $h(M) = a \cup b$. Consequently, by Theorem 9.7, M is not connected.

One of the most frequently used theorems of analysis is a corollary to Theorem 9.8.

COROLLARY 9.9 (*The Intermediate-Value Theorem*). Let $y = f(x)$ be defined and continuous on a connected subset M of R. Let x_1 and x_2, $x_1 < x_2$, be two points of M such that $f(x_1) \neq f(x_2)$, and let C be any number between $f(x_1)$ and $f(x_2)$. Then there exists x_3 such that $x_1 < x_3 < x_2$ and $f(x_3) = C$.

The student first employed the intermediate-value theorem in college algebra when he made use of the fact that, if $f(x)$ is a polynomial such that $f(a)$ and $f(b)$ have opposite signs, then $f(x) = 0$ has a real root between a and b.

10. Summary

Our detailed consideration of the real number system is now complete. The outstanding facts developed in this chapter are the following:

1. Every sequence of real numbers has a monotone subsequence.

2. A sequence of real numbers $\{a_n\}$ converges to a real number A if and only if $\{a_n\}$ is a Cauchy sequence.

3. The Bolzano-Weierstrass characterization of closed and bounded sets.

4. The Heine-Borel theorem.

5. The classification of the connected subsets of R.

6. Every non-empty set that has an upper bound has a least upper bound. Every non-empty set that has a lower bound has a greatest lower bound.

7. R is not the union of two disjoint non-empty open sets.

8. The continuous image of a connected set is connected.

9. The continuous image of a compact set is compact.

PROBLEMS

81. Prove the theorem: Let $f: M \to R$ be a mapping. Then f is continuous if and only if, for every closed set K of R, there exists a closed set F of R such that $f^{-1}(K) = F \cap M$.

82. Give a proof of Theorem 9.5, using the Bolzano-Weierstrass theorem.

83. Let H be a closed set and $f(H) = A \cup B$ a continuous mapping, where A and B are distinct points. Let H_1 be all points x of H for which $f(x) = A$, and H_2 all points x of H for which $f(x) = B$. Prove

(i) $H = H_1 \cup H_2$.
(ii) $H_1 \neq \emptyset$, $H_2 \neq \emptyset$.
(iii) $H_1 \cap H_2 = \emptyset$.
(iv) Each of the sets H_1 and H_2 is closed.

84. Prove that a closed set H is connected if and only if H cannot be written as the union of two non-empty closed sets having no points in common.

85. Let L_1 and L_2 be non-empty subsets of R having no point in common, such that $R = L_1 \cup L_2$. Prove that L_1 contains a limit point of L_2 or L_2 contains a limit point of L_1.

86. Using the notation of Problem 85, suppose that L_1 and L_2 satisfy the following additional condition: If x is in L_1 and y is in L_2, then $x < y$. Prove that there exists a point z which is the last point of L_1 or the first point of L_2. (The result of this problem is known as the Dedekind cut theorem.)

87. Let H be a non-empty set. Define K as the set of all points x that are upper bounds of H, and M as the set of all points that are lower bounds of H. Prove

(*a*) Each of the sets K and M is a closed set.
(*b*) The set R is the union of K and M if and only if H is a single point.

88. Give as many examples as you can of continuous mappings $f(H) = K$, where H is a connected set and K is a connected set of a different type; e.g. H might be an open interval and K a half-open interval, that is the union of an open interval and one of its end points; H might be the set R and K a closed interval; etc.

89. Let H and K be disjoint non-empty closed sets. Prove that there exists a continuous mapping f of $H \cup K$ onto a set consisting of two distinct points.

90. Prove that there exists no continuous mapping of R onto the set of all rational numbers.

91. Prove that every equation $f(x) = 0$, where $f(x)$ is a polynomial of odd degree, has at least one real root.

92. Let J be the closed interval $0 \leq x \leq 1$, and f a continuous mapping of J into J. Prove that there exists a point x in J such that $f(x) = x$. (*Hint.* Consider the mapping $g: J \to R$ defined by $g(x) = f(x) - x$.)

93. Given two distinct points a and b of R, there exist open sets U, V such that $a \in U$, $b \in V$, $U \cap V = \emptyset$.

94. Given a point a of R and any neighborhood U of a. Prove that there exists a neighborhood V of a, such that $V \subset U$, and every limit point of V is a point of U.

95. Consider the subset J of R where J is the closed unit interval. Define K_1 to be the subset of J consisting of all points of J that do not lie in the open interval $(\frac{1}{3}, \frac{2}{3})$. Thus K_1 is obtained by deleting the open middle third of the interval J. Define K_2 to be the subset of K_1 obtained by deleting the open middle third of each of the two intervals that form K_1. Continuing in this manner, define K_n for each $n \in I$.

(*a*) Prove that the set $K = \bigcap_{n \in I} K_n$ is

1. Closed.
2. Uncountable.
3. *Perfect*; i.e., each point of K is a limit point of K.
4. *Non-dense*; i.e., between every two distinct points of K, there is an open interval of $J - K$.

The set K just defined is known as the Cantor ternary set.

(*b*) Does the same result hold for a set obtained in the same manner except that the *closed* middle third of each interval is deleted at each stage?

Bibliographical Comment

For a detailed discussion of the nature of our assumptions concerning real numbers as infinite decimals, the reader is referred to

R. L. Wilder, *The Foundations of Mathematics*, John Wiley & Sons, New York, 1952, particularly to Chapters VI and VII.

CHAPTER THREE

TOPOLOGICAL SPACES

1. Definitions and Examples

In the study of plane geometry, "point" and "line" are taken as undefined objects about which nothing is known except the relations given in a system of axioms. One axiom states that every line is a collection of points. Note that this axiom does not state that every collection of points is a line. Indeed, it gives us no criterion by which we can determine whether a given collection of points is a line or is not a line. This axiom does not assert the existence of any lines or any points. The existence of such objects is guaranteed by another axiom which states: "There are at least two different points, and, given any two different points, there is exactly one collection of points that contains these points and is a line."

In the study of topology, the undefined objects are usually called "point" and "open set." The relationship between these objects is given by the following axioms:

Axiom 1. Every open set is a set of points.

Axiom 2. The empty set is an open set.

Axiom 3. For each point p, there exists at least one open set containing p.

Axiom 4. The union of any collection of open sets is an open set.

Axiom 5. The intersection of any finite collection of open sets is an open set.

It follows immediately from Definition I, 2.1 that Axiom 4 implies that the empty set is open. Similarly, Axiom 5 implies that S is open. Thus, from Axioms 1, 4, and 5, Axioms 2 and 3 may be derived as theorems. We have included Axioms 2 and 3 for emphasis.

We recall that, in Definition II, 4.4, we stated: "A set G of real numbers is open if and only if, for each point p of G, there exists an open-interval (a, b) such that $p \in (a, b) \subset G$." We were able to prove that, with this definition of an open set, Axioms 1 through 5 are satisfied, for the set R of real numbers. It is for this reason that we say that R, with this collection of open sets, is a topological space.

DEFINITION 1.1. A set S, together with a collection of subsets called open sets, is called a topological space if and only if the collection of open sets satisfies Axioms 1 through 5. The collection of open sets is called the topology of the topological space.

We frequently denote a topological space by the same symbol used to specify its set of points. Thus the symbol R will be used to denote the topological space having R (i.e., the set of all real numbers) as its set of points, and having as its topology the collection of open sets defined above. In this book, the word space will always be considered as an abbreviation for the two words topological space.

DEFINITION 1.2. Two topological spaces are said to be the same if and only if they consist of the same set of points with the same collection of subsets as open sets.

Following Definition 1.1, we see that, in order to determine a topological space, it is necessary to specify not only its set of points but also the collection of all open subsets of this set.

EXAMPLE 1.3. Let S be a set consisting of three distinct points a, b, c. Let the open subsets of S be the following: $\emptyset$, a, $a \cup b$, $a \cup c$, S. Clearly these open sets satisfy Axioms 1 through 5. Thus S, with the collection of open sets just listed, is a topological space.

In Example 1.3, we listed all the open subsets of S. We can frequently avoid listing all these subsets by considering them as being "generated" by a certain one of their subcollections, known as a basis.

DEFINITION 1.4. Let S be a set and σ a collection of subsets of S. Then σ is said to generate the collection τ of subsets of S defined as follows: A subset K of S is an element of τ if and only if K is the union of a collection of elements of σ. The collection σ is said to be a basis for the collection τ which it generates. [Note that, for *any* collection σ, the collection τ generated by σ will always contain the null set as one of its elements.]

Given a set S and two collections of subsets σ_1 and σ_2 of S, we frequently wish to know whether the collections τ_1 and τ_2 generated by σ_1 and σ_2, respectively, are the same. Clearly, τ_1 and τ_2 will be the same if and only if every element of σ_1 is the union of a collection of elements of σ_2 and every element of σ_2 is the union of a collection of elements of σ_1. The form in which we shall use this criterion is given in the next theorem.

THEOREM 1.5. Let S be a set. Then two collections σ_1 and σ_2 of subsets of S generate the same collection τ if and only if the following conditions hold:

(i) Given $U \in \sigma_1$ and $p \in U$, there exists an element V of σ_2 such that $p \in V \subset U$.

(ii) Given $V \in \sigma_2$ and $p \in V$, there exists an element U of σ_1 such that $p \in U \subset V$.

To illustrate these ideas, let R be the space of real numbers and σ_1 the collection consisting of all open intervals of R. Clearly σ_1 generates the collection of all open subsets of R (see Theorem II, 4.6). Let σ_2 be the collection consisting of all open connected subsets of R. This collection σ_2 also generates the collection of all open subsets of R (see Theorem II, 5.4).

Let S be a set and τ a collection of subsets of S called the open sets of S. In order to determine whether or not S, together with this collection, is a topological space, we must determine whether or not the collection τ satisfies Axioms 1 through 5. We now show that this question can be decided by consideration of properties of the sets of a basis for τ without direct consideration of all the sets of τ.

THEOREM 1.6. Let S be a set and τ a collection of subsets of S called open sets. Let σ be a basis for the collection τ. Then S is a topological space with the topology τ if and only if the following conditions hold:

(i) Given $p \in S$, there exists $U \in \sigma$ such that $p \in U$.

(ii) Given U, $V \in \sigma$, and any point $p \in U \cap V$, there exists an element W of σ such that $p \in W \subset U \cap V$.

Proof. Suppose first that S is a topological space with σ as a basis for its topology τ. We must show that conditions (i), (ii) are satisfied. These two conditions follow at once from Axiom 5, and the fact that every element of τ is the union of a collection of elements of σ.

To complete the proof, we suppose that σ satisfies conditions (i), (ii) and show that τ satisfies Axioms 1 through 5. The fact that Axioms 1 through 4 are satisfied is evident. We shall show that Axiom 5 is satisfied. To do this we note that, if G and H are elements of τ, then G is the union of a collection $\{G_\alpha\}_{\alpha\in\Lambda}$ of sets of σ, and H is the union of a collection $\{H_\beta\}_{\beta\in\Sigma}$ of sets of σ. Now $\left(\bigcup_{\alpha\in\Lambda} G_\alpha\right)\cap\left(\bigcup_{\beta\in\Sigma} H_\beta\right) = \bigcup_{\substack{\alpha\in\Lambda\\ \beta\in\Sigma}} (G_\alpha \cap H_\beta)$. Also, from (ii), for each α and β, the set $G_\alpha \cap H_\beta$ is the union of a collection of elements of σ. Hence, $G \cap H$ is the union of a collection of sets, each of which is the union of sets of σ. Thus $G \cap H$ is the union of a collection of elements of σ and, consequently, is itself an element of τ. It follows easily by induction that the intersection of any finite number of elements of τ is again an element of this collection. Therefore, Axiom 5 is satisfied.

The reader should observe that the basis for the space R consisting of the open intervals of R has exactly the property (ii) of Theorem 1.6.

Indeed, as in the proof above, this was important in checking that Axiom 5 was satisfied.

In our discussion thus far, we have given only two specific examples of topological spaces: namely, the space R of real numbers and the space S of Example 1.3. If we recall the concept of the Cartesian product of two sets (Section I, 4), and make use of the idea of a basis, which we have just introduced, we may describe other topological spaces. The discussion of these spaces will be facilitated by the introduction of some convenient notation.

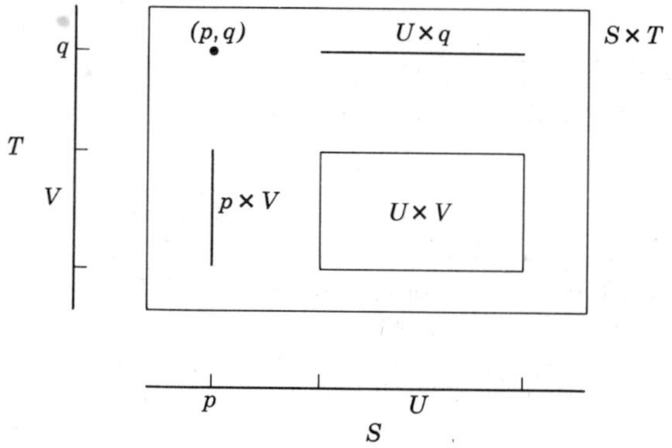

Figure 5

DEFINITION 1.7. Let S be a set, σ a collection of subsets of S, and τ the collection of subsets of S generated by σ. The symbol (S, σ) denotes the set S, together with the collection τ of its subsets. Whenever we write the symbol (S, σ), we shall consider σ as a basis for the collection of all open subsets of S. Thus, by definition, a subset G of S is open if and only if G is an element of τ.

Using this notation, we see that two topological spaces each having the same set of points S may be distinguished by being written in the form (S, σ_1) and (S, σ_2). Of course, if the bases σ_1 and σ_2 for the topologies of (S, σ_1) and (S, σ_2), respectively, are equivalent in the sense of Theorem 1.5, then these two topological spaces are the same.

DEFINITION 1.8. Let σ_1 and σ_2 be collections of sets. The Cartesian product $\sigma_1 \times \sigma_2$ is a collection of sets. A set F is an element of $\sigma_1 \times \sigma_2$ if and only if there exist sets H_1 and H_2, such that $H_1 \in \sigma_1$, $H_2 \in \sigma_2$, and $F = H_1 \times H_2$.

Suppose that σ_1 consists of the subsets p and U of a set S, while σ_2 consists of the subsets q and V of a set T. Then $\sigma_1 \times \sigma_2$ consists of the sets (p, q), $p \times V$, $U \times V$, $U \times q$ (see Figure 5).

We now extend the idea of the Cartesian product of two sets to that of the topological product of two topological spaces.

DEFINITION 1.9. Let (S_1, σ_1) and (S_2, σ_2) be topological spaces. The topological product of (S_1, σ_1) by (S_2, σ_2) is defined to be $(S_1 \times S_2, \sigma_1 \times \sigma_2)$.

Under the supposition that (S_1, σ_1) and (S_2, σ_2) are topological spaces, the following natural questions arise:

1. Is $(S_1 \times S_2, \sigma_1 \times \sigma_2)$ a topological space?
2. If σ_1' and σ_2' are different bases for (S_1, σ_1) and (S_2, σ_2), respectively, do $(S_1 \times S_2, \sigma_1 \times \sigma_2)$ and $(S_1 \times S_2, \sigma_1' \times \sigma_2')$ have the same collection of open sets (i.e., if they are spaces, are they the same space?)?

THEOREM 1.10. If (S_1, σ_1) and (S_2, σ_2) are topological spaces, then $(S_1 \times S_2, \sigma_1 \times \sigma_2)$ is a topological space.

Proof. To prove this theorem, we need only show that conditions (i), (ii) of Theorem 1.6 hold for $(S_1 \times S_2, \sigma_1 \times \sigma_2)$. We know that these conditions hold for σ_1 and σ_2 individually, since (S_1, σ_1) and (S_2, σ_2) are topological spaces.

Proof that (i) holds: Let $(p, q) \in S_1 \times S_2$. We know that there exist elements U_1 and U_2 of σ_1 and σ_2, respectively, such that $p \in U_1$, $q \in U_2$. The set $U_1 \times U_2$ is an element of $\sigma_1 \times \sigma_2$ containing (p, q).

Proof that (ii) holds: Let U, V be elements of $\sigma_1 \times \sigma_2$. Then there exist elements U_1, V_1 of σ_1 and elements U_2, V_2 of σ_2, such that $U = U_1 \times U_2$, $V = V_1 \times V_2$. It is easily seen that $U \cap V = (U_1 \cap V_1) \times (U_2 \cap V_2)$, and that $U \cap V \subset S_1 \times S_2$. Let p be an arbitrary point of $U \cap V$; thus $p = (p_1, p_2)$, where $p_1 \in S_1$, $p_2 \in S_2$. Since (S_1, σ_1) and (S_2, σ_2) are topological spaces, condition (ii) must hold for σ_1 and σ_2. Thus there must exist elements W_1 and W_2 of σ_1 and σ_2, respectively, such that $p_1 \in W_1 \subset U_1 \cap V_1$ and $p_2 \in W_2 \subset U_2 \cap V_2$. Consequently, $p = (p_1, p_2) \in (W_1 \times W_2) \subset (U_1 \cap V_1) \times (U_2 \cap V_2) = U \cap V$. Since $(W_1 \times W_2) \in \sigma_1 \times \sigma_2$, (ii) must hold for $\sigma_1 \times \sigma_2$. Consequently, by Theorem 1.6, $(S_1 \times S_2, \sigma_1 \times \sigma_2)$ is a topological space.

In connection with the theorem just proved, notice that, if $U_1, V_1 \in \sigma_1$, $U_2, V_2 \in \sigma_2$, then $W = (U_1 \times U_2) \cup (V_1 \times V_2)$ is an open subset of $(S_1 \times S_2, \sigma_1 \times \sigma_2)$, but that W *cannot in general* be written as the product of two sets, let alone the product of sets of σ_1 and σ_2. For an example of such a situation see Figure 6.

We now know, for example, that, if σ is a basis for the space R of real numbers, then $(R \times R, \sigma \times \sigma)$ is a topological space. We do not as yet know whether we obtain the same product space if we use different bases for the space R. This is precisely in line with our question 2 above.

THEOREM 1.11. Let (S_1, σ_1), (S_2, σ_2) be topological spaces and σ_1', σ_2', respectively, different bases for these spaces. Then $(S_1 \times S_2, \sigma_1 \times \sigma_2)$ and $(S_1 \times S_2, \sigma_1' \times \sigma_2')$ are the same topological space.

Proof. The proof follows easily by showing that conditions (i) and (ii) of Theorem 1.5 are satisfied. We know that σ_1 and σ_1' satisfy these conditions as do also σ_2 and σ_2'. If $U \in \sigma_1 \times \sigma_2$, then there exist $U_1 \in \sigma_1$, $U_2 \in \sigma_2$, such that $U = U_1 \times U_2$. Let $p \in U$; then $p = (p_1, p_2)$ where $p_1 \in U_1$, $p_2 \in U_2$. Since σ_1' and σ_2' are bases for the topological spaces

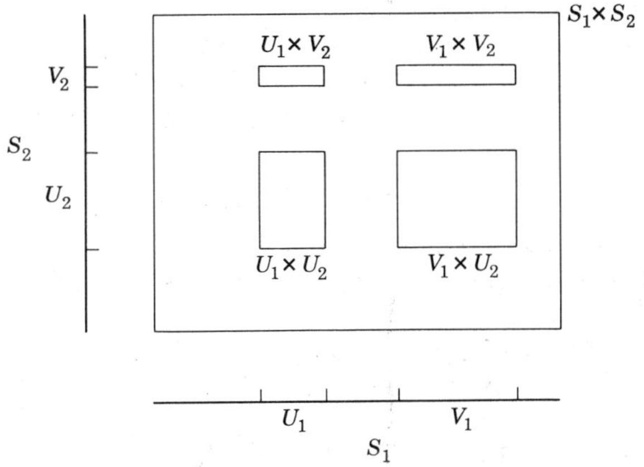

Figure 6

(S_1, σ_1) and (S_2, σ_2), respectively, there must exist $U_1' \in \sigma_1'$, $U_2' \in \sigma_2'$, such that $p_1 \in U_1' \subset U_1$ and $p_2 \in U_2' \subset U_2$. Consequently, $U_1' \times U_2' \in \sigma_1' \times \sigma_2'$ and $p \in U_1' \times U_2' \subset U_1 \times U_2$. We have thus proved that condition (i) of Theorem 1.5 holds. The proof that (ii) holds is similar and will be left to the student.

We have already remarked that the plane of analytic geometry is the Cartesian product of the point set R of all real numbers with itself. In view of the discussion just given, it is a simple matter to make this plane into a topological space. We need only consider it as the topological product of the space of all real numbers R with itself. The space thus obtained will be denoted in this book by the standard notation E^2. The collection of all open intervals forms a basis for the space R; hence, all sets, each of which is a Cartesian product of two open intervals, form a basis for the space E^2. The Cartesian product of two open intervals is what we shall call an open rectangle. The information given in this paragraph will be needed repeatedly. Consequently, we formalize it for easy reference.

DEFINITION 1.12. The symbol E^2 denotes the topological space which is the topological product of the space R of all real numbers with itself.

DEFINITION 1.13. Let a, b, c, d be four real numbers, $a < b$, $c < d$. An open rectangle (a, b, c, d) in E^2 consists of all points (x, y) of E^2 such that $a < x < b$, $c < y < d$ and only these points. [Clearly the open rectangle (a, b, c, d) is the Cartesian product of the open interval (a, b) with the open interval (c, d).]

THEOREM 1.14. The topological space E^2 has a basis consisting of all open rectangles of E^2.

In the next section we shall consider not only how the distance formula of analytic geometry relates to the topology of E^2, but also how the distance between two points of a set relates to a topology for the set.

PROBLEMS

1. Let S denote the topological space of Example 1.3. List the open sets of the topological product $S \times S$, and verify that these open sets satisfy Axioms 1 through 5.

2. Let Λ and Σ be two index sets, $\{G_\alpha\}_{\alpha \in \Lambda}$ and $\{H_\beta\}_{\beta \in \Sigma}$ two collections of subsets of a set S. Prove

$$(\bigcup_{\alpha \in \Lambda} G_\alpha) \cap (\bigcup_{\beta \in \Sigma} H_\beta) = \bigcup_{\substack{\alpha \in \Lambda \\ \beta \in \Sigma}} (G_\alpha \cap H_\beta)$$

3. Let S and T be sets, $W = S \times T$. If F is a subset of W, do there exist sets $F_1 \subset S$, $F_2 \subset T$, such that $F = F_1 \times F_2$? Give examples.

4. Let S and T be sets, $\emptyset_s$ the empty set of S, $\emptyset$ the empty set of $S \times T$, and A an arbitrary subset of T. Does $\emptyset_s \times A = \emptyset$?

5. Let S and T be sets, $W = S \times T$. Let A, B be subsets of S; C, D subsets of T. Consider the statements

(a) $(A \times C) \cup (B \times D) = (A \cup B) \times (C \cup D)$.

(b) $(A \times C) \cap (B \times D) = (A \cap B) \times (C \cap D)$.

(c) If $A \subset B$, $C \subset D$, then $(A \times C) \subset (B \times D)$.

In each case, either prove that the statement always holds or give an example to illustrate a case in which it does not.

6. Show that the intersection of two open rectangles is an open rectangle or is empty.

√7. Let (a, b, c, d) be an open rectangle in E^2 and p a point in (a, b, c, d). Prove that there exists a positive number r such that the circle with center at p and radius r is contained in the open rectangle (a, b, c, d).

8. Let p be a point of E^2, and r a positive number. Let C be the circle with center at p and radius r. Prove that, given any point q of E^2 which is inside C (i.e., the distance from p to q is less than r), then there exists an open rectangle containing q, every point of which is inside C.

9. Let S be the set of all real numbers, and τ the collection of all subsets of S which consists of the null set and every set K, such that $S - K$ is a finite set. Prove that (S, τ) is a topological space. Also prove that (S, τ) is not the same as the space R of all real numbers.

2. Metric Spaces

In the previous section we introduced the formal definition of a topological space. Our examples of such spaces have been few. To indicate the wealth of examples, we make one or two remarks on one aspect of the development of topology. Analysts, in studying the real line, the Cartesian plane, the set of all complex numbers, the convergence of sequences of functions, four-dimensional Euclidean space, etc., became interested not only in finding what theorems were true but also in discovering exactly on what key facts the truth of these theorems rested. Time and time again they were led, in this investigation, to the properties of the function defining the distance between two points. Time and time again they found that, if they replaced phrases peculiar to a certain subject by the phrase "open set," then the proofs from several different fields of specialization read alike. It turned out that these open sets satisfied Axioms 1 through 5, and were closely related to the function defining distance. It is the connection between open sets and the distance function that we wish to study in this section.

The student is familiar with the distance formula of plane analytic geometry: namely,

$$\rho(p, q) = \sqrt{(x_1 - x_2)^2 + (y_1 - y_2)^2}$$

where the coordinates of the points p, q are (x_1, y_1) and (x_2, y_2), respectively. This distance function $\rho(p, q)$ is a mapping $\rho\colon E^2 \times E^2 \to R$. It is clear that, for any $(p, q) \in E^2 \times E^2$, $\rho(p, q)$ is *non-negative*, and that $\rho(p, q) = 0$ if and only if $p = q$. Furthermore, $\rho(p, q) = \rho(q, p)$ for any $(p, q) \in E^2 \times E^2$; this property is known as the *symmetric* property of ρ. It is also a fact that, for any three points p, q, r of E^2, $\rho(p, q) + \rho(q, r) \geq \rho(p, r)$.

This is the familiar result known as the *triangle inequality*, that the sum of the lengths of two sides of a triangle is at least as great as the length of the third side (the details of the proof of this statement will be an exercise). The properties of the distance function ρ of E^2, which we have just listed, are characteristic of those of any distance function or metric.

DEFINITION 2.1. A set S is said to be metric if and only if there is associated with S a mapping $\rho: S \times S \to R$ (where R is the space of all real numbers) having the following properties for every x, y, z in S:

(i) (Non-negative): $\rho(x, y) \geq 0$.
(ii) $\rho(x, y) = 0$ if and only if $x = y$.
(iii) (Symmetric): $\rho(x, y) = \rho(y, x)$.
(iv) (Triangle property): $\rho(x, z) \leq \rho(x, y) + \rho(y, z)$.

The mapping ρ is called the metric for the set S.

DEFINITION 2.2. Two metric sets are said to be the same if and only if the sets are the same, and the mappings which are the respective metrics are the same.

Intimately associated with a metric set S are the subsets of S known as spherical neighborhoods.

DEFINITION 2.3. Let K be a metric set. Then, with each point p of K and each real number $r > 0$, we associate a subset $S_r(p)$ called a spherical neighborhood of p (or sometimes the r neighborhood of p). A point q of K is in $S_r(p)$ if and only if $\rho(p, q) < r$.

THEOREM 2.4. Let S be a metric set with ρ as metric. Denote by τ the collection of subsets of S consisting of all spherical neighborhoods of S. Then (S, τ) is a topological space.

Proof. The conclusion will follow if we show that the collection satisfies conditions (i) and (ii) of Theorem 1.6. Condition (i) is immediate. To establish (ii), let $S_r(p)$ and $S_t(q)$ be spherical neighborhoods in the set S, and let $w \in S_r(p) \cap S_t(q)$. Define δ as the smaller of the two positive numbers $t - \rho(w, q)$ and $r - \rho(w, p)$. Then $S_\delta(w) \subset S_r(p) \cap S_t(q)$. To prove this, let $x \in S_\delta(w)$. We have at once $\rho(x, p) \leq \rho(x, w) + \rho(w, p) < \delta + \rho(w, p) \leq r$. Consequently, $\rho(x, p) < r$; hence, $x \in S_r(p)$. Similarly, $x \in S_t(q)$. The proof that condition (ii) holds is thus complete.

We know that the Cartesian plane E^2 is a topological space, and that the topology for this space can be defined by taking as a basis the collection of subsets of E^2 consisting of all open rectangles of E^2. But E^2 is also a metric set, with the familiar metric of analytic geometry. Thus we may also make E^2 into a topological space by taking as a basis the collection of subsets of E^2 consisting of all spherical neighborhoods in E^2, and making use of the result of Theorem 2.4. It is an interesting and important fact that these two topological spaces are the same. This follows at once from Theorem 1.5 if we use the results established in Problems 7 and 8. Since the topology which we obtained for the set E^2 using spherical neighborhoods was precisely the same as the original topology for the topological space E^2, we say that the Cartesian plane E^2 is a metric space.

DEFINITION 2.5. A metric set S is said to be a metric space if and only if the topology of S is that which is generated by the collection of subsets of S consisting of all spherical neighborhoods in S. This topology of S is said to be induced by the metric ρ of S.

When we say that a set S is a metric space, we imply, by the word metric, that S has a specific metric ρ. We imply by the word space that S is a topological space. The statement "S is a metric space" refers specifically to the topology induced by the particular metric ρ.

Let us re-examine some of our earlier results in terms of what we have just done. In the first place, we expect the student to realize that the set R of real numbers has a metric, namely, $\rho(x, y) = |x - y|$, and that the topology of R induced by this metric is precisely that which is generated by the collection of all open intervals. In the second place, the topological product E^2 of R with itself leads to exactly the same topology of E^2 as that of plane analytic geometry.

Metric spaces have a number of properties that are not possessed by topological spaces in general. We introduce only one of these properties in this section; namely, that of being a Hausdorff space.

DEFINITION 2.6. A topological space S is said to be a Hausdorff space if and only if, given any two distinct points p, q of S, there exist disjoint open sets U and V of S such that $p \in U$, $q \in V$.

THEOREM 2.7. Every metric space is a Hausdorff space.

Proof. Let S be a metric space with metric ρ, and x, y two distinct points of S. Thus $\rho(x, y) > 0$. Let r be any real number, such that $0 < 2r < \rho(x, y)$. It is easily seen that the spherical neighborhoods $S_r(x)$ and $S_r(y)$ are disjoint open sets of S containing x and y, respectively. Thus S is a Hausdorff space.

Let S be the topological space described in Example 1.3. We have seen that S consists of exactly three points a, b, c and that a non-empty subset of S is open if and only if it contains the point a. Thus there do not exist two disjoint open sets of S containing the points a and b, respectively. Consequently, S is not a Hausdorff space. Therefore, there exists no metric ρ for S which will induce the topology of S.

The example just cited shows us that, given a topological space S, it is not true that we can always define a metric for S which will induce the given topology of S. Making S into a metric set is a simple matter; for example, we can define a metric for S by

$$\rho(x, y) = 1 \quad \text{if} \quad x \neq y$$

$$\rho(x, y) = 0 \quad \text{if} \quad x = y$$

Obtaining a metric for S that will induce the original topology of S is a far deeper problem which we shall study later. Such a metric may be obtained only if the given space S is what is known as a metrizable space.

DEFINITION 2.8. Let S be a topological space with topology τ. Then S is said to be metrizable if and only if it is possible to define a metric for S which will induce the topology τ. In particular, a metric space is a metrizable space for which a metric has been specified.

imp (cf chap IV)

PROBLEMS

10. Prove that $\rho(x, y) = |x - y|$ is a metric for R.

11. Let ρ be a metric for a set S. Prove that the function $\rho_1(x, y)$ defined by $\rho_1(x, y) = \rho(x, y)[1 + \rho(x, y)]^{-1}$ is also a metric for S, and that these two metric spaces are the same topological space.

12. Give the details of the proof that

$$\rho[(x_1, y_1), (x_2, y_2)] = \sqrt{(x_1 - x_2)^2 + (y_1 - y_2)^2}$$

is a metric for E^2. This metric is called the standard metric for E^2.

13. Let a metric ρ_1 for the set E^2 be defined in terms of the standard metric ρ by

$$\rho_1[(x_1, y_1), (x_2, y_2)] = \begin{cases} \rho[(0, 0), (x_1, y_1)] + \rho[(0, 0), (x_2, y_2)] & \text{if } x_2y_1 \neq x_1y_2 \\ \rho[(x_1, y_1), (x_2, y_2)] & \text{if } x_2y_1 = x_1y_2 \end{cases}$$

Show that ρ_1 is a metric for the set E^2. Show also that the topology of the space E^2 is not the same as the topology induced by ρ_1.

14. Fill in the missing details in the proof of Theorem 2.7.

15. Let S be a topological space, K a subset of S, and σ the collection of all subsets of K, each of which can be written as the intersection of K with an open set of S. Prove that (K, σ) is a topological space.

3. Elementary Properties of Topological and Metric Spaces

Thus far we have been concerned with the definitions of topological and metric spaces. The space R of real numbers satisfies both these definitions. We shall make use of our knowledge of R not only to suggest the formulation of definitions but also to guide us in the proofs of theorems. In the following definitions we note in parentheses the related result of Chapter 2.

DEFINITION 3.1 (II, 4.4). Let S be a space and p a point of S. A subset K of S is said to be a neighborhood of p if and only if K is an open set of S containing p.

DEFINITION 3.2 (II, 6.5). A subset H of a space S is said to be closed if and only if $S - H$ is open.

DEFINITION 3.3 (II, 6.3). A point p of a topological space S is said to be a limit point of a subset H of S if and only if every neighborhood of p contains a point of H distinct from p. The set of all limit points of H is denoted by H' and is frequently referred to as the derived set of H.

The following characterization of open sets is immediate from Axiom 4. It is frequently useful.

THEOREM 3.4. A subset G of a space S is open if and only if, for each point p of G, there exists a neighborhood U of p contained in G.

The next two theorems follow at once from Axioms 2, 4, and 5, Definition 3.2, and the theorem of DeMorgan.

THEOREM 3.5. Let S be a space. Then each of the subsets $\emptyset$ and S of S is both open and closed.

THEOREM 3.6. Let S be a space. Then
(i) The intersection of any collection of closed subsets of S is closed.
(ii) The union of any finite collection of closed subsets of S is closed.

In Definition II, 6.4, we used the statement of the next theorem as our definition of a closed set of real numbers.

THEOREM 3.7. A subset H of a space S is closed if and only if every limit point of H belongs to H; i.e., if and only if $H' \subset H$.

Proof. We first suppose that H is a closed set, and must prove that every limit point of H belongs to H; i.e., that no point of $S - H$ is a limit point of H. This is immediate from Definition 3.2 and 3.3.

To complete the proof, suppose that every limit point of H belongs to H. Thus, for each point p of $S - H$, we can find a neighborhood U_p of p such that $U_p \subset S - H$. Consequently, $S - H = \bigcup_{p \in S-H} U_p$, which is an open set by Theorem 3.4.

Let A be a subset of a topological space S. We know, by Theorem 3.5, that there exists at least one closed subset of S containing A; namely, the set S itself. We shall now define a set, called the closure of A, which is, in one sense of the word, the smallest closed subset of S containing A.

DEFINITION 3.8. Let A be a subset of a space S, and $\{A_\alpha\}$ the collection of all closed subsets of S, each of which contains A. Then the closure of the set A is denoted by $\bar{A}$, and defined by the equation $\bar{A} = \cap A_\alpha$.

THEOREM 3.9. Let A be a subset of a space S; then
(i) $\bar{A}$ is a closed set containing A.
(ii) $\bar{A}$ is the smallest closed set containing A in the sense that, given any closed set F of S which contains A, we must have $\bar{A} \subset F$.
(iii) $\bar{A} = A \cup A'$.

Proof. (i) follows at once from Definition 3.8 and Theorem 3.6, whereas (ii) is immediate from Definition 3.8 if we observe that $F = A_{\alpha_0}$ for some α_0. To prove (iii), we note that $\bar{A} \subset A \cup A'$ by (ii) and Problem 16*c*. But $\bar{A} \supset A \cup A'$ since $\bar{A}$ is a closed set containing A; hence $\bar{A}$ contains every limit point of A (see Problem II, 55).

A combination of the results of Theorems 3.7 and 3.9 gives us the following useful fact.

THEOREM 3.10. A subset A of a space S is closed if and only if $A = \bar{A}$.

We have seen that the space S described in Example 1.3 is not a Hausdorff space. It is easily seen that, for this space S, consisting of the three distinct points a, b, c, the space S is the closure of the point a. Thus it need not be true in a space S that every subset of S consisting of a single point is a closed set. This is true for Hausdorff spaces, as we shall see from the next theorem, the proof of which is immediate.

THEOREM 3.11. Given a point p of a space S. Then $p = \bar{p}$ if and only if, for each point q of $S - p$, there exists a neighborhood U of q not containing p.

COROLLARY 3.12. Each point p of a Hausdorff space S is a closed subset of S.

In Definition 2.6 we introduced the concept of a Hausdorff space, which is one of several properties enjoyed by metric spaces but not by topological spaces in general. The next such property that we wish to discuss will now be defined.

DEFINITION 3.13. A space S is regular if and only if, given any closed subset F of S and any point p of S not in F, then there exist disjoint open subsets U and V of S containing F and p, respectively.

It will be seen in Problem 20 that every metric space is regular. It is not true that every topological space is regular. As a matter of fact, a topological space that is a Hausdorff space need not be regular, and a topological space that is regular need not be a Hausdorff space. We give several examples to illustrate these points.

EXAMPLE 3.14. This is an example of a space S that is regular but is not a Hausdorff space. The space S consists of three distinct points a, b, c and a topology consisting of the following open sets: $\emptyset$, a, $b \cup c$, S. We note that a subset of S is closed if and only if it is open. The space S is not a Hausdorff space since it contains no pair of disjoint open sets containing b and c, respectively. This space is easily seen to be regular.

Example 3.15. This is an example of a space S which is Hausdorff but not regular. We shall define the space S by making use of the Cartesian plane consisting of all pairs of real numbers (x, y), where this plane is regarded as a metric set with the metric of plane analytic geometry. Let S consist of all (x, y) such that $y \geq 0$, and L be the subset of S consisting of all (x, y) such that $y = 0$. For each point p of S and each real number $r > 0$, we define a set $U_r(p)$ as follows:

$$U_r(p) = S_r(p) \cap S \qquad \text{if } p \in S - L$$
$$U_r(p) = [S_r(p) \cap (S - L)] \cup p \quad \text{if } p \in L$$

The topology of S is generated by the collection of subsets of S consisting of all sets of the form $U_r(p)$, where $p \in S$, and $r > 0$. It is evident that S, with this topology, is a topological space. Moreover, if p and q are distinct points of S, we may define $2r = \rho(p, q)$. Then $r > 0$, and $U_r(p)$, $U_r(q)$ are disjoint open subsets of S containing p and q, respectively. Consequently, S is a Hausdorff space. The fact that S is not regular follows easily if we consider the point $p = (0, 0)$, and note that the subset $L - p$ is closed. Clearly, there do not exist two disjoint open subsets of S containing p and $L - p$, respectively. (The space S is not metric, since, by Problem 20, every metric space is regular.)

A criterion for the regularity of a topological space, which is somewhat easier to use than Definition 3.13, will now be given.

Theorem 3.16. A space S is regular if and only if, given any point p of S, and any neighborhood U of p, there exists a neighborhood V of p such that $p \in \bar{V} \subset U$.

Proof. Suppose first that S is regular, and that the point p and the neighborhood U of p are given. Define $F = S - U$. Then F is a closed set not containing p. Since S is regular, there exist disjoint open subsets V and G of S containing p and F, respectively. Clearly, $p \in \bar{V} \subset U$.

To complete the proof, suppose that the condition holds. We must prove that S is regular. Let p be any point of S, and F any closed subset of S not containing p. Define $U = S - F$; then U is a neighborhood of p. Thus there exists a neighborhood V of p such that $p \in \bar{V} \subset U$. Clearly the sets V and $S - \bar{V}$ are disjoint open subsets of S containing p and F, respectively. Thus S is regular.

We introduce, precisely as in Definitions II, 7.4 and 7.5, the definition of a covering of a set, and the definition of compact sets in terms of coverings.

Definition 3.17. A collection of sets $\{D_\alpha\}$ is said to be a covering of a set A, or to cover A, if and only if $A \subset \bigcup_\alpha D_\alpha$. If, in addition, each of the sets D_α is required to be an open set, then the collection $\{D_\alpha\}$ is known as an open covering of A or as a covering of A by open sets. Any subcollection of a collection $\{D_\alpha\}$ covering A which also covers A is said to be a subcovering of A.

DEFINITION 3.18. Let A be a subset of a space S. Then A is said to be compact if and only if every open covering of A contains a finite subcovering of A.

If we recall the space S of Example 1.3, we note that the topology of this space consists of only a finite number of open sets. Thus every subset of S is a compact set. In particular, every point of S is a compact subset of S. But S contains a point a having as its closure the entire space S. It follows that a is a compact set which is not closed. Consequently, it is not true that every compact set is closed. We have, however, the following result.

THEOREM 3.19. Every compact subset A of a Hausdorff space S is a closed subset of S.

Proof. Let q be any point of $S - A$. The proof will be complete, by Definition 3.2 and Theorem 3.4, if we can find a neighborhood V of q containing no point of A. This may be done as follows. For each point p of A we find, since S is a Hausdorff space, two disjoint neighborhoods U_p of p and V_q of q. Then $A \subset \bigcup_{p \in A} U_p$; thus $\{U_p\}$ is an open covering of A. Since A is compact, this open covering of A has a finite subcovering, which we denote by $U_1, U_2, \cdots, U_n$. Let $V_1, V_2, \cdots, V_n$ be the neighborhoods of q corresponding to $U_1, U_2, \cdots, U_n$, and define $V = \bigcap_{j=1}^{n} V_j$. The set V is a neighborhood of the point q, since it is the intersection of a finite collection of open sets, each of which contains the point q. It is easily seen that V contains no point of the set A. Thus $S - A$ is open; hence A is closed.

An important property possessed by all compact sets is known as countable compactness. The reason for the term "countable compactness" will be apparent later.

DEFINITION 3.20. A subset A of a space S is said to be countably compact if and only if every infinite subset of A has at least one limit point in A.

THEOREM 3.21. Every compact subset A of a space S is countably compact.

Proof. Assume that A is compact, and let K be any infinite subset of A. We shall show that the assumption that K has no limit point in A leads to a contradiction. We know that there exists a sequence $\{x_n\}$ of distinct points of K, and we define $H = \bigcup_{n \in I} x_n$. By our assumption, the set H has no limit point in A; thus, in particular, for each $n \in I$, the point x_n is not a limit point of H. Hence there exists, for each $n \in I$, a neighborhood V_n of x_n, such that $V_n \cap H = x_n$. For each point p of $A - H$, there exists a neighborhood V_p of p, such that $V_p \cap H = \emptyset$. Thus $U = \bigcup_{p \in A - H} V_p$ is an open set. The collection of open sets $\{V_n\}$, together with the set U is an open covering of A. Manifestly, no finite subcollection of this covering covers A. This contradiction to the assumption that A was compact establishes the conclusion that A is countably compact.

Actually, we have proved the following corollary.

Corollary 3.22. Every space S having the property that every countable open covering of S contains a finite subcovering of S is countably compact.

In the space R of all real numbers, the concepts of compactness and countable compactness have been seen to be identical (Section II, 7). The next example shows rather vividly how different these concepts may be in a topological space.

Example 3.23. We will show

(*a*) There exists a countably compact space that is not compact.

(*b*) There exists a countably compact space and a countable open covering of this space that contains no finite subcovering of the space.

Let S be the set of all positive integers: $1, 2, 3, \cdots$. For each positive integer n, define V_n as the subset of S consisting of the two points $2n - 1$ and $2n$. Take as a basis for S the collection of all the sets V_n. It is easily seen that, with the topology defined by this basis, the set S becomes a topological space. This space S is covered by a countable collection of open sets: namely, the collection of all the V_n. Since no finite subcollection of $\{V_n\}$ covers S, it follows that S is not compact. But S is certainly countably compact, since every non-empty subset of S has a limit point in S. To see this, note that, for every n, the point $2n$ is a limit point of the set consisting of the point $2n - 1$, and the point $2n - 1$ is a limit point of the set consisting of the point $2n$. Thus S is countably compact. We have thus established (*a*) and (*b*).

In our discussion of DeMorgan's theorem in Chapter 1, we have seen that there is a close relationship between the union of a collection of sets and the intersection of the complements of these sets. Because of this relationship, it is not surprising that the idea of compactness, which we have thus far expressed in terms of open coverings, i.e., unions of sets, can also be expressed in terms of the intersections of sets.

DEFINITION 3.24. Let $\{A_\alpha\}$ be a collection of subsets of a set S. This collection is said to have the finite intersection property if and only if every finite non-empty subcollection of $\{A_\alpha\}$ has a non-empty intersection.

THEOREM 3.25. A space S is compact if and only if, for any family $\{A_\alpha\}$ of closed subsets of S having the finite intersection property, there exists a point $p \in S$, such that $p \in A_\alpha$ for every α.

Proof. Suppose first that S is compact, but that there exists a family $\{A_\alpha\}$ of closed subsets of S having the finite intersection property and such that $\cap A_\alpha = \emptyset$. Using DeMorgan's theorem, we see that the collection of all open subsets of S of the form $S - A_\alpha$ is an open covering of S. By compactness, a finite number of these sets, namely, $S - A_{\alpha_1}, S - A_{\alpha_2}, \cdots, S - A_{\alpha_k}$, is also an open covering of S. Thus $S = (S - A_{\alpha_1}) \cup (S - A_{\alpha_2}) \cup \cdots \cup (S - A_{\alpha_k}) = S - \bigcap_{i=1}^{k} A_{\alpha_i}$, i.e., $\bigcap_{i=1}^{k} A_{\alpha_i} = \emptyset$. But, this is impossible since the collection of sets $\{A_\alpha\}$ has the finite intersection property.

Conversely, let $\{G_\alpha\}$ be any open covering of S, and suppose that the condition holds. If no finite subcollection of $\{G_\alpha\}$ covers S, then the collection of closed sets $\{S - G_\alpha\}$ has the finite intersection property; hence $\cap (S - G_\alpha) \neq \emptyset$. Therefore, $S \neq S - \cap (S - G_\alpha) = \cup G_\alpha$ which is impossible. Thus some finite subcollection of $\{G_\alpha\}$ covers S, and S is compact.

For a space in which each point is a closed set, we can relate the idea of countable compactness with that of countable open coverings. It is this relationship that leads to the name "countable compactness."

DEFINITION 3.26. A space S is said to be T_1 (or a T_1 space) if and only if every point of S is a closed subset of S.

THEOREM 3.27. A T_1 space S is countably compact if and only if every countable collection of open sets that covers S has a finite subcollection that also covers S.

Proof. If the condition holds, then S must be countably compact, by Corollary 3.22. We thus suppose that S is countably compact, and that $\{G_n\}$ is a countable collection of open sets covering S. We shall show that some finite subcollection of $\{G_n\}$ also covers S. Suppose that this is not true. Then, for every integer n, the set $K_n = S - \bigcup_{i=1}^{n} G_i$ is closed and non-empty. We denote by p_n an arbitrary point of this set and define $H = \bigcup_{n \in I} p_n$. Suppose first that the set H is infinite. Then, since S is countably compact, there is a point p in S that is a limit point of H. Since S is a T_1 space, it follows, by Problem 26, that every neighborhood of p contains infinitely many points of H. Consequently, p is a limit point of the set $H_n = \bigcup_{\substack{i \in I \\ i > n}} p_i$. But H_n is a subset of the closed set K_n. It follows that $p \in K_n$ for every n; hence $p \notin G_n$ for every $n \in I$, and this contradicts the hypothesis that $\{G_n\}$ is a covering of S. We conclude that the set H is finite.

It follows, from the fact that H is finite, that there exists $p \in H$, such that, given any integer N, there exists $n > N$, such that $p_n = p$. Thus $p \in K_n$ for every n, and this contradicts the hypothesis that $\{G_n\}$ is a covering of S. Thus some finite subcollection of $\{G_n\}$ covers S, and the proof of the theorem is complete.

By using Theorem 3.27 together with the method of proof of Theorem 3.25, the following theorem is easily obtained.

THEOREM 3.28. A T_1 space S is countably compact if and only if every countable collection of closed subsets of S that has the finite intersection property has a non-empty intersection.

A frequently used form of this theorem is stated in terms of montone sequences of sets and is called the Cantor product theorem.

DEFINITION 3.29. Let $\{A_n\}$ be a sequence of sets. Then

(*a*) The sequence is said to be monotone decreasing if and only if $A_n \supset A_{n+1}$ for every n.

(*b*) The sequence is said to be monotone increasing if and only if $A_n \subset A_{n+1}$ for every n.

THEOREM 3.30 (*The Cantor Product Theorem*). Let $\{A_n\}$ be a monotone decreasing sequence of non-empty closed subsets of a T_1 space S, and suppose that A_1 is countably compact. Then there exists a point p of S such that p lies in A_n for every n.

Proof. This theorem is a corollary to Theorem 3.28.

PROBLEMS

16. Let A and B be subsets of a space S. Prove

(*a*) If $A \subset B$, then $A' \subset B'$.

(*b*) $(A \cup B)' = A' \cup B'$.

(*c*) $A \cup A'$ is a closed set.

17. Let S be a space, A, B subsets of S, and $\{A_\alpha\}$ any collection of subsets of S. Prove

(*a*) $\bar{\emptyset} = \emptyset$.

(*b*) $\bar{S} = S$.

(*c*) $A \subset \bar{A}$.

(*d*) If $A \subset B$, then $\bar{A} \subset \bar{B}$.

(*e*) $\overline{A \cup B} = \bar{A} \cup \bar{B}$.

(*f*) $\bar{\bar{A}} = \bar{A}$.

(*g*) $\overline{A \cap B} \subset \bar{A} \cap \bar{B}$.

(*h*) $\overline{\cap A_\alpha} \subset \cap \bar{A}_\alpha$.

(*i*) $\cup \bar{A}_\alpha \subset \overline{\cup A_\alpha}$.

(*j*) $\overline{A - B} \supset \bar{A} - \bar{B}$.

18. Given a metric space S and a subset H of S. Prove that a point p of S is a limit point of H if and only if, given any $\varepsilon > 0$, there exists a point $q \in H$ such that $0 < \rho(p, q) < \varepsilon$.

19. Given a metric space S, a point p of S, and real numbers r, s such that $0 < s < r$. Prove $p \in S_s(p) \subset \overline{S_s(p)} \subset S_r(p)$. Show also that, if $q \in \overline{S_r(p)}$, then $\rho(p, q) \leq r$, but the converse is not always true.

20. Prove that every metric space is regular.

21. Given a metric space S and a point p of S. Prove that there exists a monotone decreasing sequence $\{U_n\}$ of neighborhoods of p such that $U_n \supset \bar{U}_{n+1}$ for every n, and $p = \cap \bar{U}_n = \cap U_n$.

22. Given a subset K of a metric space S, and a real number $r > 0$. Define $S_r(K)$ to consist of all points x of S such that $\rho(x, y) < r$ for at least one point y of K.

(*a*) Prove that the set $S_r(K)$ is open.

(*b*) If K_1 and K_2 are disjoint closed subsets of S and K_1 is countably compact, prove that there exists a real number $r > 0$, such that the sets $S_r(K_1)$ and $S_r(K_2)$ are disjoint.

23. Prove that, in a space S, every closed subset of a countably compact set is countably compact.

24. Prove that, in a space S, every closed subset of a compact set is compact.

25. Let K be a subset of a T_1 space S. If K is given the topology described in Problem 15, prove that K is a T_1 space.

26. Prove that, in a T_1 space, a point p is a limit point of a set H if and only if every neighborhood of p contains infinitely many distinct points of H.

27. Prove that, in a T_1 space, no finite set has a limit point.

28. Set up a definition of what you would mean by the statement: The sequence of points $\{p_n\}$ of the metric space S converges to the point p of S. Prove, by using your definition, that a point p is a limit point of a subset H of a metric space S if and only if there exists a sequence $\{p_n\}$ of distinct points of H which converges to the point p.

4. Mappings and Subspaces

The last section was concerned with elementary properties of topological and metric spaces. Here we shall consider the relationship between topological spaces. We are therefore interested in mappings of one topological space into another.

DEFINITION 4.1. Let S and T be spaces and $f: S \to T$ a mapping. Then f is said to be continuous at the point s of S if and only if, given any open subset G of T such that $s \in f^{-1}(G)$, there exists an open subset V of S such that $s \in V \subset f^{-1}(G)$.

DEFINITION 4.2. Let S and T be spaces and $f: S \to T$ a mapping. Then f is said to be continuous on S if and only if f is continuous at every point s of S.

THEOREM 4.3. Let S and T be spaces and $f: S \to T$ a mapping. Then f is continuous on S if and only if, for every open subset G of T, $f^{-1}(G)$ is an open subset of S.

Proof. Suppose first that f is continuous on S, and that G is any open subset of T. Then, for each point s of $f^{-1}(G)$, there exists a neighborhood V_s of s contained in $f^{-1}(G)$. The set $f^{-1}(G)$ is the union of all such neighborhoods, for s in $f^{-1}(G)$; hence $f^{-1}(G)$ is an open subset of S, by Axiom 4.

Suppose, on the other hand, that the condition holds. Let s be any point of S, and G any open subset of T containing $f(s)$. Define $V = f^{-1}(G)$. Then V is an open subset of S such that $s \in V \subset f^{-1}(G)$. Thus f is continuous at s, by Definition 4.1. Since s was an arbitrary point of S, f is continuous on S, by Definition 4.2.

We proved in Chapter 2 that, if $f: R \to R$ is a continuous mapping, and K is a subset of R, then the restriction of f to K is a continuous mapping. In Definition 4.2, we have defined continuity of a mapping only when the mapping is from one topological space into another. Thus, if we wish to extend our theorem to topological spaces, it is natural to use the open sets of the space S in such a way that a subset K of S becomes a topological space, and the restriction of any continuous mapping on S to the set K is continuous.

DEFINITION 4.4. Let S be a set, K a subset of S, and σ a collection of subsets of S. Then $\sigma \cap K$ denotes the collection of sets, each of whose elements is of the form $H \cap K$, where H is an element of σ.

THEOREM 4.5. Let S be a space and σ_1, σ_2 bases for S. Then, for any subset K of S, each of the collections $\sigma_1 \cap K$ and $\sigma_2 \cap K$ is a basis for a topology for K. These topologies are the same.

Proof. See Problem 15.

DEFINITION 4.6. Let (S, σ) be a topological space and K a subset of S. A topological space having K as its set of points is said to be a subspace of the space S if and only if the topology of K is the topology of

$(K, \sigma \cap K)$. A subset M of K is said to be open in K relative to S if and only if M is open in $(K, \sigma \cap K)$.

We customarily say "M is open in K" when we mean "M is open in K relative to S." We say "M is open" when we mean "M is an open subset of S." Also, when we use a subset K of a space S in a context that requires a topology for K, then K is to be taken as a subspace of S.

THEOREM 4.7. Let S and T be spaces, K a subset of S, and $f: S \to T$ a continuous mapping. Then the restriction of f to K is a continuous mapping.

Proof. By definition, the restriction of f to K is the mapping $g: K \to T$ defined by $g(x) = f(x)$ for every x in K. It is clear that, for every subset G of T, $g^{-1}(G) = f^{-1}(G) \cap K$. Hence, for every open subset G of T, $g^{-1}(G)$ is open in K. Thus g is a continuous mapping.

THEOREM 4.8. Let S, T, W be spaces; $f: S \to T$, $g: T \to W$ continuous mappings. Then the composition mapping $h: S \to W$, defined by $h(x) = g(f(x))$ for each x of S, is continuous.

Proof. Let G be any open subset of W. Since g is continuous, $V = g^{-1}(G)$ is open in T. Consequently, since f is continuous, $f^{-1}(V)$ is an open subset of S. Moreover, $h^{-1}(G) = f^{-1}(V)$; hence $h^{-1}(G)$ is open; i.e., h is a continuous mapping.

We consider now some equivalent conditions for the continuity of a mapping.

THEOREM 4.9. Let S and T be topological spaces, and $f: S \to T$ a mapping. If f is continuous, then both the following conditions hold. If either one of these conditions holds, then f is continuous.

(i) For every closed subset H of T, the set $f^{-1}(H)$ is a closed subset of S.

(ii) For every subset A of S, $f(\bar{A}) \subset \overline{f(A)}$.

Proof. We show first that f is continuous if and only if condition (i) holds. Suppose the condition holds, and let G be any open subset of T. Then $T - G$ is a closed subset of T; hence, by (*i*), $f^{-1}(T - G)$ is a closed subset of S. Consequently, since $f^{-1}(G) = S - f^{-1}(T - G)$, it follows that $f^{-1}(G)$ is open. Therefore, f is continuous. On the other hand, if we assume that f is continuous and let H be any closed subset of T, we may define $G = T - H$. We have, as before, $f^{-1}(G) = S - f^{-1}(T - G) = S - f^{-1}(H)$. But $f^{-1}(G)$ is open; hence $f^{-1}(H)$ is closed.

The proof of Theorem 4.9 will be completed by showing that f is continuous if and only if condition (ii) holds. Suppose first that this condition is satisfied. We shall prove that f is continuous by showing that, for any closed subset H of T, the set $f^{-1}(H)$ is closed. Let

$K = H \cap f(S)$, and note that $f^{-1}(H) = f^{-1}(K)$. Suppose that p is a limit point of $f^{-1}(K)$. It follows that $p \in \overline{f^{-1}(K)}$. By condition (ii), we see that $f(p) \in \overline{f(f^{-1}(K))} = \bar{K} \subset H$. Since $f(p) \in f(S)$, we must have $f(p) \in H \cap f(S) = K$. Thus $p \in f^{-1}(K) = f^{-1}(H)$; hence this set is closed by Theorem 3.7. Consequently, since condition (i) is satisfied, the mapping f is continuous. We prove, finally, that, if f is continuous, then condition (ii) must hold. Let A be any subset of S. We have at once $f(\bar{A}) = f(A \cup A') = f(A) \cup f(A')$. To show that condition (ii) holds, we let p be an arbitrary point of $f(\bar{A})$, and must show that $p \in \overline{f(A)}$. If $p \in f(A)$, this is immediate; hence we may assume that p is a point of $f(A')$ that is not in $f(A)$. Thus there exists a point q of $A' - A$ such that $f(q) = p$. Let U be any open set containing p. Using Theorem 4.3, we see that $f^{-1}(U)$ is a neighborhood of the point q. But q is a limit point of A; hence there exists a point x of A in $f^{-1}(U)$. The point $f(x)$ lies in $U \cap f(A)$ and is distinct from p. This means, by Definition 3.3, that p is a limit point of $f(A)$, so that $p \in \overline{f(A)}$, and condition (ii) holds.

The following fallacious argument is sometimes presented in connection with the above theorem in an attempt to establish the fact that the continuity of the mapping f implies condition (ii). Let $\{A_\alpha\}$ be the collection of all closed subsets of S, each of which contains the set A. Then $\bar{A} = \cap A_\alpha$ and $f(\cap A_\alpha) \subset \cap f(A_\alpha) \subset \overline{f(A)}$, since "$f(A_\alpha)$ is a closed set containing $f(A)$." One error lies in the claim that $f(A_\alpha)$ is a closed set. To see this, consider the next example.

Example 4.10. Let S_1 be the space consisting of three points a, b, c, whose open sets are $\emptyset$, a, $a \cup b$, $a \cup c$, and S_1. Let S_2 be the space consisting of three points d, e, f, whose open sets are $\emptyset$, d, $e \cup f$, S_2. Let $f\colon S_1 \to S_2$ be the mapping defined by $f(x) = e$, for x in S_1. This mapping is continuous, since the inverse of *any* subset of S_2 is either $\emptyset$ or S_1, and is therefore open. Now $(b \cup c)$ is a closed subset of S_1, and $f(b \cup c) = e$ which is not a closed subset of S_2. At the same time we see that the point a is an open subset of S_1, but $f(a) = e$ is not an open subset of S_2.

Definition 4.11. Let S and T be spaces and $f\colon S \to T$ a mapping. The mapping f is said to be closed if and only if the image of every closed subset of S is a closed subset of T. The mapping f is said to be open if and only if the image of every open subset of S is an open subset of T.

We have seen in Example 4.10 that a continuous mapping need not be either open or closed. However, it is natural to ask if an open mapping is necessarily closed. Consider again the spaces S_1 and S_2 of Example 4.10. The mapping $g\colon S_2 \to S_1$ defined by $g(x) = a$ for every x in S_2 is open but not closed. Likewise, the mapping $h\colon S_2 \to S_1$ defined by

$h(x) = c$ for every x in S_2 is closed but not open. In addition, each of the mappings g and h is continuous.

DEFINITION 4.12. Let S and T be spaces. Then S is said to be homeomorphic to T if and only if there exists a one-to-one open continuous mapping of S onto T. The mapping is called a homeomorphism.

The word homeomorphic comes from the Greek, and means of similar form or structure. Thus we have defined what we shall mean by the statement that two spaces are of similar structure. The homeomorphism, being a one-to-one open continuous mapping of S onto T, enables us to not only place the points of S and T into one-to-one correspondence, but also to use the same correspondence to place the open sets of S into one-to-one correspondence with the open sets of T. Closely related to homeomorphisms or, as they are sometimes called, topological mappings, is the concept of a topological property.

DEFINITION 4.13. A property of a space S is said to be a topological property if and only if every space T homeomorphic with S has the same property.

Many important problems in analysis concern the determination of the length of curves, and the determination of the areas bounded by curves. Inevitably, as the analysis of these problems became more precise, the need arose for an exact understanding of the nature and properties of curves. A special case of a curve is an arc. The topological properties common to all arcs are exactly those possessed by the closed unit interval [0, 1]. This follows immediately from what is now the accepted definition of an arc.

DEFINITION 4.14. A topological space S is said to be an arc if and only if it is homeomorphic with the closed interval [0, 1] of the space R of real numbers.

The converse problem also arises in analysis; namely, given a space with certain topological properties, is it an arc? Necessary and sufficient conditions for an affirmative answer to this question are given in Chapter 5.

Since, as above, we use the word homeomorphic to mean "topologically equal" it is important to show that this relation between spaces has properties similar to those for the equal sign in algebra; that is, that it is an equivalence relation.

THEOREM 4.15. Let S, T, W be spaces. Then

(i) S is homeomorphic to S.

(ii) If S is homeomorphic to T, then T is homeomorphic to S.

(iii) If S is homeomorphic to T and T is homeomorphic to W, then S is homeomorphic to W.

Proof. (i) The mapping $f: S \to S$ defined by $f(x) = x$ for each $x \in S$ is clearly a homeomorphism.

(ii) Let $g: S \to T$ be a homeomorphism. Then, clearly, the mapping $h: T \to S$ defined by $h(x) = g^{-1}(x)$ for each $x \in T$ is a homeomorphism.

(iii) Let $g: S \to T$ and $h: T \to W$ be homeomorphisms. We have at once that the mapping $k: S \to W$ defined by $k(x) = h[g(x)]$ for each $x \in S$ is a homeomorphism.

Under certain circumstances we can recognize that a one-to-one continuous mapping of one space onto another space is a homeomorphism. Before discussing this result, we shall consider the continuous images of compact and countably compact spaces.

THEOREM 4.16. Let S and T be spaces, S compact, and $f: S \to T$ a continuous mapping. Then $f(S)$ is compact.

Proof. The proof follows by essentially the same argument that was given for Theorem II, 9.5.

Although we shall show that the continuous image of a countably compact T_1 space is countably compact, it is not true that the continuous image of an arbitrary countably compact topological space is countably compact. We see this by means of an example.

EXAMPLE 4.17. Let S be the space of Example 3.23. Define T as the set of all positive integers with the so-called discrete topology: i.e., every subset of T is an open set. Thus T is not countably compact since it is an infinite set no subset of which has a limit point. Clearly, the mapping $f: S \to T$ defined by $f(2n) = n, f(2n - 1) = n$, for each positive integer n, is a continuous mapping of S onto T. Thus we have an example of a continuous mapping of a countably compact topological space onto a space that is not countably compact. It should be noted that the space S is not T_1.

THEOREM 4.18. Let S be a countably compact T_1 space, and T a space. Then, if $f: S \to T$ is a continuous mapping, the set $f(S)$ is countably compact.

Proof. We will show that $f(S)$ is countably compact by showing that every countable open covering of $f(S)$ contains a finite subcollection which covers $f(S)$. The result then follows by Corollary 3.22. Let $\{U_n\}$ be any countable covering of $f(S)$ by open sets. Since f is continuous, it follows that $\{f^{-1}(U_n)\}$ is a countable open covering of S. Since S is a countably compact T_1 space, this covering of S contains a finite subcollection $f^{-1}(U_{n_1}), \cdots, f^{-1}(U_{n_k})$ which covers S. The sets $U_{n_1}, \cdots, U_{n_k}$ therefore form a finite subcollection of $\{U_n\}$ which covers $f(S)$. This completes the proof.

We now give a condition in terms of compactness which, if present, implies that a one-to-one continuous mapping of one space onto another is a homeomorphism. This is followed by an example of a situation in which the condition is not fulfilled and the mapping is not a homeomorphism.

THEOREM 4.19. Let S and T be spaces and $f: S \to T$ a one-to-one continuous mapping of S onto T. If S is compact and T is Hausdorff, then f is a homeomorphism.

Proof. By Problem 41, we need only show that f is a closed mapping. Let K be any closed subset of S. By Problem 24, the set K is compact; hence $f(K)$ is compact by Theorem 4.16. Since T is a Hausdorff space, the set $f(K)$ must be closed by Theorem 3.19.

The following example, essentially due to Kuratowski, shows that a one-to-one continuous mapping of one space onto another need not be a homeomorphism.

EXAMPLE 4.20. Let R be the space of all real numbers. Define a subspace S of R to consist of the union of all points of R having the form $(3n + 2)$ and all open intervals of R of the form $(3n, 3n + 1)$, where $n = 0, 1, 2, \cdots$. Define $T = (S - \{2\}) \cup \{1\}$. We shall define three mappings: a mapping $f: S \to T$, a mapping $g: T \to S$, and a mapping $h: S \to S$. The mapping f is defined by $f(x) = x$ for $x \neq 2$, $f(2) = 1$. The mapping g is defined by $g(x) = x/2$ for $x \leq 1$, $g(x) = x/2 - 1$ for $3 < x < 4$, and $g(x) = x - 3$ for $x \geq 5$. The mapping h is defined to be the composition mapping of f and g: namely, $h(x) = g[f(x)]$ for each $x \in S$. We note that the mappings f, g, h are each one-to-one continuous onto mappings. The mapping f is not a homeomorphism since $\{2\}$ is an open subset of S, but $f(2)$ is not an open subset of T. Similarly $0 < x \leq 1$ is open in T, but the image of this set under the mapping g is not open in S. Thus g is not a homeomorphism. The mapping h is a one-to-one continuous mapping of the space S onto itself. This mapping fails to be a homeomorphism since $\{2\}$ is open in S, but $h(2)$ is not open in S.

The fact that the graph (see Definition I, 4.1) of a continuous mapping defined on a space S is homeomorphic to S has many applications. We now prove this interesting result.

THEOREM 4.21. Let S, T be spaces, $f: S \to T$ a mapping, and F the subspace of $S \times T$ which is the graph of f. Then the mapping $h: S \to F$ defined by $h(a) = (a, f(a))$ is a homeomorphism if and only if f is continuous.

Proof. Suppose f is continuous. Define $h: S \to F$ by $h(a) = (a, f(a))$ for $a \in S$. This is continuous at any point $a \in S$; for, given any open set $W = G \cap F$ of F, where G is open in $S \times T$ and contains $(a, f(a))$, there exists an open set $U \times V$ of $S \times T$, where U is open in S, V is open in T, and $(a, f(a)) \in U \times V \subset G$. Now $f^{-1}(V)$ is open and contains a; hence $U \cap f^{-1}(V)$ is a neighborhood of a in S. Furthermore $h(U \cap f^{-1}(V)) \subset (U \times V) \cap F \subset (G \cap F) = W$. Thus h is continuous at a. Since h is clearly one-to-one onto F, it remains to show that h is an open mapping. This is equivalent to showing that h^{-1} is a continuous mapping. But h^{-1} can be considered as the restriction to F of the mapping $g: S \times T \to S$ defined by $g(p, q) = p$, and clearly g is continuous.

Suppose that h is a homeomorphism. Then, given any open set G of F, $h^{-1}(G)$ is open. So, in particular, given any open set V of T, the set $(S \times V) \cap F$ is open in F, and thus $h^{-1}[(S \times V) \cap F]$ is open in S. But $h^{-1}[(S \times V) \cap F] = f^{-1}(V)$. Thus f is continuous.

PROBLEMS

29. Let S and T be metric spaces with metrics ρ_1 and ρ_2, respectively, and $f: S \to T$ a mapping. Prove that f is continuous at the point x_0 of S if and only if, given any real number $\varepsilon > 0$, there exists a real number $\delta > 0$ such that, if $x \in S$ and $\rho_1(x, x_0) < \delta$, then $\rho_2[f(x), f(x_0)] < \varepsilon$.

30. Let M_1 and M_2 be subsets of the spaces S_1 and S_2, respectively. Then the topological product $M_1 \times M_2$ is the same space as the subspace $M_1 \times M_2$ of $S_1 \times S_2$.

31. Let S be a space, K a subset of S, and M a subset of K.

(*a*) Show that, if M is open in K and K is open in S, then M is open in S.

(*b*) Show that, if M is closed in K and K is closed in S, then M is closed in S.

(*c*) Give examples to illustrate cases in which M is open in K, but M is not open in S, and also where M is closed in K, but M is not closed in S.

32. Let S and T be spaces, A a subset of S, and $f: S \to T$ a mapping. Discuss the propositions:

(i) If f is an open mapping, then the restriction of f to A is open.

(ii) If f is a closed mapping, then the restriction of f to A is closed.

In particular, is (i) true if A is open, and is (ii) true if A is closed?

33. Let S, T be spaces, A, B subsets of S, and suppose that the mappings $f: A \to T$, $g: B \to T$ are continuous. Suppose further that $f(x) = g(x)$ for $x \in A \cap B$. Consider the mapping $h: A \cup B \to T$ defined by $h(x) = f(x)$ if $x \in A$, $h(x) = g(x)$ if $x \in B$, and discuss the continuity of the mapping h. Give an example to show that the mapping h need not be continuous. Give additional hypotheses which will insure the continuity of the mapping h (compare Problem II, 79).

34. Show that metrizability is a topological property.

35. Let S be a metric space with metric ρ. Show that the mapping $\rho: S \times S \to R$ is continuous.

36. Let S be a non-empty countably compact T_1 space, and $f: S \to R$ a continuous mapping. Show that there exist points x_1 and x_2 of S such that $f(x_1) \leq f(x) \leq f(x_2)$ for every x in S.

37. Show that every subspace of a Hausdorff space is also a Hausdorff space. Show also that the property of being a Hausdorff space is a topological property.

38. Let S be the set of real numbers x such that $0 \leq x \leq 1$, and σ the collection of subsets of S consisting of all sets K such that $S - K$ is finite. Show that σ generates a topology for the set S. Thus we may regard S as a topological space. Let f be the mapping of the closed interval [0, 1] of the space R of real numbers into the space S defined by $f(x) = x$ for each $x \in [0, 1]$. Show that f is a one-to-one continuous mapping onto S, but that f is not a homeomorphism. Show further that there exists no homeomorphism of the closed interval [0, 1] of R onto S; hence S is not an arc. Show, however, that S is compact. How do these results compare with Theorem 4.19?

39. In Example 4.20 show that f, g, h are one-to-one continuous mappings onto T, S, and S, respectively.

40. Let S and T be spaces and $f: S \to T$ a one-to-one mapping of S onto T. Show that f is a homeomorphism if and only if $f(\bar{A}) = \overline{f(A)}$ for every subset A of S.

41. Let S and T be spaces and $f: S \to T$ a one-to-one continuous mapping of S onto T. Show that f is a homeomorphism if and only if it is a closed mapping.

42. Show that the property of countable compactness is a topological property.

5. Connected Sets

We have seen in Chapter 2 the usefulness of the idea of a connected set in establishing results such as the intermediate-value theorem. It is the purpose of the present section to extend this idea to general spaces, and to develop some of the basic results on connected sets which will be needed for later and important applications. The result of Theorem II, 9.7 will be used as the definition of a connected subset of a general topological space.

DEFINITION 5.1. A subset A of a space S is said to be connected if and only if there exists no continuous mapping $f: A \to R$ such that $f(A)$ consists of exactly two points.

The following theorem shows, among other things, that connectedness is a topological property. It is, of course, a generalization of the intermediate-value theorem.

THEOREM 5.2. Let S and T be spaces and $f: S \to T$ a continuous mapping. If S is connected, then $f(S)$ is connected.

Proof. Let $H = f(S)$. We suppose that H is not connected; i.e., that there exists a continuous mapping $g: H \to R$ such that $g(H) = a \cup b$, where a and b are distinct points. Consider the mapping $h: S \to R$ defined by $h(x) = g[f(x)]$ for each $x \in S$. This mapping is continuous by Theorem 4.8. Moreover, $h(S) = a \cup b$. Consequently, the set S is not connected. This proves that, if H is not connected, then S is not connected, which is another way of saying that the connectivity of S implies the connectivity of H.

We have seen that every non-empty connected subset of the space R that contains more than one point must contain an open interval, and hence be itself an infinite set. In particular, a subset of R consisting of precisely two distinct points is not connected. Taking a subset of R consisting of two distinct points as a prototype of the non-connected subsets of R, we can see a very close relationship between the definition of connected sets given in Definition 5.1 and the result on connected sets given in Theorem 5.2.

Our discussion will be facilitated by the following definition.

DEFINITION 5.3. A set K is said to be non-degenerate if and only if K contains at least two distinct points.

It is not necessarily true in a topological space that every non-degenerate finite set fails to be connected. The space of Example 1.3, consisting of three distinct points, is readily shown to be connected. The important characterization of connected sets provided by the next theorem will show us a large class of spaces in which every non-degenerate connected set is infinite.

THEOREM 5.4. Let S be a space, and A a subset of S. Then A is connected if and only if no proper subset of A is both open and closed in A.

Proof. Suppose first that H is a proper subset of A which is both open and closed in A. Let a, b be distinct points of R, and define a mapping $f: A \to R$ by $f(x) = a$ if $x \in H$, $f(x) = b$ if $x \in A - H$. Clearly, f is a continuous mapping of A onto $a \cup b$. Thus A is not connected.

To complete the proof, suppose that A is not connected: i.e., that there exists a continuous mapping $f: A \to R$ such that $f(A)$ consists of exactly two points a and b. Let $H = f^{-1}(a)$. Since f is continuous, H is both open and closed in A. The set H is a proper subset of A since b is not in $f(H)$.

Let S be a space in which every point is a closed set. It follows at once that, for every finite subset A of S, every subset of A is both open and closed in A. Thus we see that every non-degenerate finite set fails to be connected. We state this result formally as a theorem.

THEOREM 5.5. Let S be a T_1 space. Then every non-degenerate connected subset of S is infinite.

We collect in the next theorem several immediate results of Theorem 5.4.

THEOREM 5.6. In any space S, the following results hold.

(i) The empty set is connected.

(ii) Every set consisting of a single point is connected.

(iii) A set A is connected if and only if A is not the union of two disjoint non-empty sets, each of which is open in A.

(iv) A set A is connected if and only if A is not the union of two disjoint non-empty sets, each of which is closed in A.

The next theorem is a sharpening of Theorem 5.5 for the special cases in which the space S is metric.

THEOREM 5.7. Let A be a non-degenerate connected subset of a metric space S. Then A is uncountable.

Proof. Let a and b be distinct points of A. Define a mapping $f: A \to R$ by $f(x) = \rho(x, a)$ for each x in A where ρ is the metric for S. This mapping is easily seen to be continuous. Thus, by Theorem 5.2, $f(A)$ is a connected subset of R. The set $f(A)$ contains the point $f(a) = 0$ of R and the point $c = f(b)$ of R, and we know $c \neq 0$. Thus $f(A)$ must contain the interval $(0, c)$ of R. Consequently, by Problem I, 44, the set A will be proved uncountable if we can show that the open interval $(0, c)$ is uncountable. This follows at once from the fact that there exists a one-to-one mapping of any open interval onto the set R. The construction of such a mapping is left as an exercise for the student.

It is frequently convenient to phrase theorems on connected sets in terms of what are known as separated sets.

DEFINITION 5.8. Two subsets A and B of a space S are said to be separated if and only if $A \neq \emptyset$, $B \neq \emptyset$, $\bar{A} \cap B = \emptyset$, $A \cap \bar{B} = \emptyset$.

We have at once the following characterization of separated sets.

THEOREM 5.9. Two disjoint subsets of a space S are separated if and only if neither is empty and each is both open and closed in their union.

COROLLARY 5.10. A subset A of a space S is connected if and only if A is not the union of two separated sets.

COROLLARY 5.11. Let B and C be separated subsets of a space S, and A a subset of S such that $A \cap B \neq \emptyset$ and $A \cap C \neq \emptyset$. Then the sets $A \cap B$ and $A \cap C$ are separated.

COROLLARY 5.12. If A is a connected set that is contained in the union of two separated sets, then A is contained in one of these sets.

The notion of compactness involves open coverings and is a useful way of making certain statements about limit points. Similarly, it is useful to express statements concerning connectedness in terms of open coverings. To develop these ideas we shall need the concept of a simple chain.

DEFINITION 5.13. Let a and b be points of a set S, and $H_1, H_2, \cdots, H_n$ a finite collection of subsets of S. This collection is said to be a simple chain from a to b if and only if

(i) $a \in H_1$, $b \in H_n$.

(ii) $H_i \cap H_j \neq \emptyset$ if and only if $|i-j| \leq 1$, $i = 1, 2, \cdots, n$; $j = 1, 2, \cdots, n$.

Each of the sets H_i, $i = 1, 2, \cdots, n$ is called a link of the chain.

Theorem 5.4 allows us to obtain a characterization of connected spaces in terms of simple chains.

THEOREM 5.14. A space S is connected if and only if, given any two points a and b of S and any open covering $\{G_\alpha\}$ of S, there exists a finite subcollection of $\{G_\alpha\}$ which is a simple chain from a to b.

Proof. Suppose first that the space S is connected. Let a be a point of S and $\{G_\alpha\}$ an open covering of S. Define H to be the set consisting of all points b of S for which there exists a finite subcollection of $\{G_\alpha\}$ which is a simple chain from a to b. The set H is not the null set, since it contains the point a. We must prove that $H = S$. This will follow at once, by Theorem 5.4, if we show that H is both open and closed. To show that H is open, let p be any point of H. Thus there exists a simple chain of elements of $\{G_\alpha\}$ joining a to p. The link of this chain which contains p is a neighborhood of p. Moreover, every point of this neighborhood is in H, since it can be joined to a by the same simple chain that joined a to p. Consequently, H is open. To see that H is closed, we let p be any limit point of H, and G_{α_0} any element of $\{G_\alpha\}$ that contains p. Thus G_{α_0} is a neighborhood of p; hence G_{α_0} contains a point q of H. There must exist, since q belongs to H, a simple chain of elements of $\{G_\alpha\}$ from a to q. We denote this simple chain by $H_1, H_2, \cdots, H_n$. We note that $H_n \cap G_{\alpha_0} \neq \emptyset$; hence there exists a least index j such that $H_j \cap G_{\alpha_0} \neq \emptyset$. For this least index j, $H_1, H_2, \cdots, H_j, G_{\alpha_0}$ is a simple chain from a to p. Consequently, $p \in H$. Since p was an arbitrary limit point of H, it follows that H is a closed set. Thus $H = S$.

For the other part of the proof, we must show that the space S is connected if the condition of the theorem holds for every pair of points

a, b of S and every covering $\{G_\alpha\}$ of S by open sets. Suppose that S is not connected. Then, by Theorem 5.6 (iii), S is the union of two disjoint non-empty open sets G_1 and G_2. Thus G_1, G_2 is an open covering of S. Let $a \in G_1$, $b \in G_2$. It is immediate that a and b cannot be joined by a simple chain of elements of the covering since G_1 and G_2 are disjoint. Consequently, the condition of the theorem must fail if S is not connected. This completes the proof of the theorem.

COROLLARY 5.15. Let $\{G_\alpha\}$ be an open covering of a connected subset H of a space S. Then, given any two points a and b of H, there exists a simple chain of sets of $\{G_\alpha\}$ from a to b.

In order to see that this corollary is not an if and only if theorem, consider the space S of Example 1.3. The subset $b \cup c$ is not connected, since it is the union of two non-empty disjoint closed sets b and c (Theorem 5.6 iv). On the other hand, it may be easily verified that any covering of $b \cup c$ by open sets of S contains a simple chain from b to c.

The next two theorems may be considered to be applications of Corollary 5.12.

THEOREM 5.16. Let $\{G_\alpha\}$ be a collection of connected subsets of a space S. If the intersection of every pair of sets of the collection $\{G_\alpha\}$ is non-empty, then $\cup G_\alpha$ is a connected set.

Proof. Suppose that $\cup G_\alpha$ is not a connected set. Then we may write, by Corollary 5.10, $\cup G_\alpha = A \cup B$, where A and B are separated sets. Let a, b be points of A, B, respectively. Then there exist G_{α_1} and G_{α_2} such that $a \in G_{\alpha_1}$, $b \in G_{\alpha_2}$. From Corollary 5.12, it follows at once that $G_{\alpha_1} \subset A$, $G_{\alpha_2} \subset B$. This is impossible since A and B are separated sets and $G_{\alpha_1} \cap G_{\alpha_2} \neq \emptyset$.

THEOREM 5.17. Let A be a connected subset of a space S, and $A^\star$ any set such that $A \subset A^\star \subset \bar{A}$. Then $A^\star$ is connected. In particular, the closure of a connected set is connected.

Proof. If $A^\star$ is not connected, we may write $A^\star = C \cup D$, where C and D are separated sets. By Corollary 5.12, we may choose the notation in such a way that $A \subset C$. By the definition of $A^\star$, this means that every point of D is a limit point of C, a contradiction to the assumption that C and D are separated sets.

Definition 5.1 can frequently be used directly to obtain results on connected sets. The theorem that follows illustrates this type of proof.

THEOREM 5.18. Let C be a connected subset of a connected space S, and suppose that $S - C = A \cup B$, where A and B are separated sets. Then each of the sets $A \cup C$ and $B \cup C$ is connected.

Proof. Suppose that $A \cup C$ is not connected. Using Definition 5.1, there must exist a continuous mapping $f: A \cup C \to R$, such that $f(A \cup C) = a \cup c$, where a and c are distinct points. By Theorem 5.2, we may choose the notation in such a way that $f(C) = c$. We now define a mapping $g: S \to R$ as follows: $g(x) = f(x)$ if $x \in A \cup C$, $g(x) = c$ if $x \in B$. From the fact that A and B are separated sets, it follows easily that the mapping g is continuous. Moreover, $g(S) = a \cup c$. This contradicts the hypothesis that S is a connected set. Consequently, $A \cup C$ is connected. A similar proof shows that $B \cup C$ is connected.

THEOREM 5.19. Let A and B be subsets of a space S. If each of the sets A, B is closed in their union, and if $A \cup B$ and $A \cap B$ are connected, then A is connected and B is connected.

Proof. The theorem is immediate if $A \subset B$ or $B \subset A$. Otherwise, we may assume that each of the sets $(A - B)$ and $(B - A)$ is non-empty. From the fact that each of the sets A, B is closed in $A \cup B$, we have at once that $(A - B)$ and $(B - A)$ are each open in $A \cup B$. Consequently, these sets are separated, and $(A \cup B) - (A \cap B) = (A - B) \cup (B - A)$. Thus, by Theorem 5.18, $A = (A - B) \cup (A \cap B)$ and $B = (B - A) \cup (A \cap B)$ are connected.

COROLLARY 5.20. Let S be a connected space and x a point of S such that $S - x = A \cup B$, where A and B are separated. Then $A \cup x$ and $B \cup x$ are each connected. Moreover, if x is a closed subset of S, then $\bar{A} = A \cup x$ and $\bar{B} = B \cup x$.

Let S be a space. One of the results of our next theorem is the fact that S is connected if and only if the topological product $S \times S$ of S with itself is connected. A more or less common error is frequently made by students in writing a proof for this theorem. Suppose that p and q are points of S, that each of the sets $S \times q$ and $p \times S$ has been shown to be a connected subset of $S \times S$, and that these sets have been shown to have the point (p, q) in common. The error is then made in asserting that $S \times S = (S \times q) \cup (p \times S)$, which is connected by Theorem 5.12. We state a lemma, the proof of which is left to the student, that reveals the flaw in this argument.

LEMMA 5.21. Let S be a set containing at least two points, and suppose that p and q are points of S. Then

$$(S \times q) \cup (p \times S) \neq S \times S$$

THEOREM 5.22. Let S_1 and S_2 be spaces and $H = S_1 \times S_2$ their topological product. Then H is connected if and only if S_1 is connected and S_2 is connected.

Proof. Consider the so-called projection mappings $f: H \to S_1$, $g: H \to S_2$ defined by $f[(p, q)] = p$, $g[(p, q)] = q$, for each $(p, q) \in H$. It is easily seen that f and g are continuous mappings, and that $f(H) = S_1$, $g(H) = S_2$. Thus, if H is connected, it follows from Theorem 5.2 that $S_1 = f(H)$ and $S_2 = g(H)$ are connected.

We next show that, if S_1 is connected and S_2 is connected, then H is connected. Let $p \in S_1$, and define A_p as the topological product of p by S_2: namely, $A_p = p \times S_2$. The mapping $h: S_2 \to A_p$, defined by $h(q) = (p, q)$ for each $q \in S_2$ is a homeomorphism. Thus A_p is a connected set. In exactly the same way, we can show that, for a fixed point q_0 of S_2 the set $B_{q_0} = S_1 \times q_0$ is a connected set. The sets B_{q_0} and A_p each contain the point (p, q_0). Consequently, by Theorem 5.16, the set $H_p = B_{q_0} \cup A_p$ is a connected set for each $p \in S_1$. Furthermore, $\bigcup_{p \in S_1} H_p = H$, and the intersection of any two of the H_p's contains the non-empty set B_{q_0}. It follows, again by Theorem 5.16, that H is connected.

COROLLARY 5.23. The space E^2 is connected.

A theorem similar to Theorem 5.22 holds for the product of compact spaces. It is proved by similar techniques.

THEOREM 5.24. Let S_1 and S_2 be spaces and $H = S_1 \times S_2$ their topological product. Then H is compact if and only if S_1 is compact and S_2 is compact.

Proof. Suppose H is compact. Then, as in the proof of Theorem 5.22, it is seen that S_1 and S_2 are each continuous images of H. Thus S_1 and S_2 are each compact.

To complete the proof, we must show that, if S_1 and S_2 are each compact, then H is compact. Suppose S_1 and S_2 are compact, and let $\{G_\alpha\}$ be any open covering of H. Each G_α is the union of open sets of the form $U \times V$ where U and V are open in S_1 and S_2, respectively. Let $\{W_\beta\}$ be the collection of all $W_\beta = U_\beta \times V_\beta$, U_β open in S_1, V_β open in S_2, such that, for each β, W_β is contained in a G_α for some α. It is apparent that $\{W_\beta\}$ is an open covering of H and that, if a finite subcollection of $\{W_\beta\}$ covers H, then a finite subcollection of $\{G_\alpha\}$ covers H. We will show that such a finite subcollection of $\{W_\beta\}$ exists. For each $p \in S_1$, define $A_p = p \times S_2$. As shown in the proof of Theorem 5.22, A_p is homeomorphic to S_2. Thus A_p is compact for each $p \in S_1$. Since A_p is a compact subset of H, a finite subcollection $U_{\beta_1} \times V_{\beta_1}, \cdots, U_{\beta_{k_p}} \times V_{\beta_{k_p}}$ of non-empty elements of $\{W_\beta\}$ covers A_p. Define $U_p = \bigcap_{i=1}^{k_p} U_{\beta_i}$. This is an open set in S_1 containing p, since each U_{β_i}, $i = 1, \cdots, k_p$, is such a set. The open sets $U_p \times V_{\beta_i}, \cdots, U_p \times V_{\beta_{k_p}}$ cover A_p. Define

$J_p = \bigcup_{i=1}^{k_p} (U_p \times V_{\beta_i}) = U_p \times S_2$. (See Figure 7, where $k_p = 2$.) For each $p \in S_1$, J_p is an open set containing A_p. It is apparent from the construction that, if a finite number of the J_p suffice to cover H, then a finite subcollection of $\{W_\beta\}$ covers H. We now show that a finite number of the J_p cover H. First, we note that every point of H lies in A_p for some $p \in S_1$. Thus $\{J_p\}_{p \in S_1}$ is an open covering of H and thus of the subset $B_{q_0} = S_1 \times q_0$ of H defined for some fixed point q_0 of S_2. Since

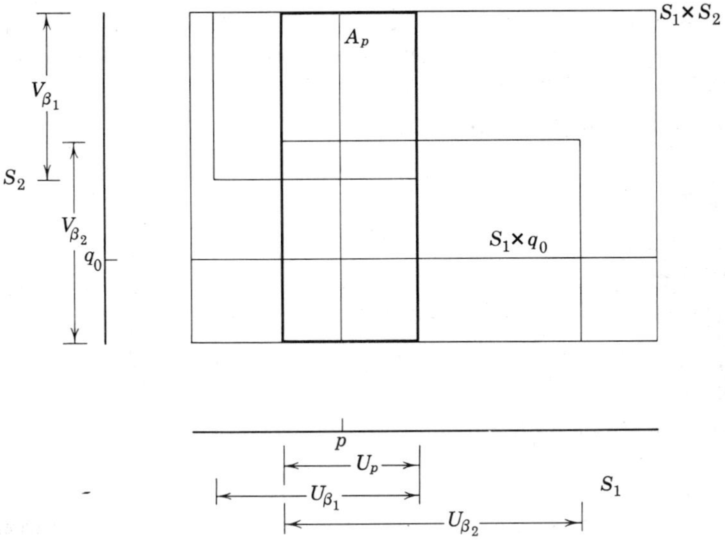

Figure 7

B_{q_0} is homeomorphic to S_1, it is compact. Thus a finite number $J_{p_1}, \cdots, J_{p_n}$ of elements of the collection $\{J_p\}_{p \in S_1}$ cover B_{q_0}. We claim that this collection $J_{p_1}, \cdots, J_{p_n}$ also covers H. Let (r, s) be any point of H. Then $(r, q_0) \in B_{q_0}$, and thus $(r, q_0) \in J_{p_i} = U_{p_i} \times S_2$ for some i, $1 \leq i \leq n$. Hence $(r, s) \in J_{p_i} = U_{p_i} \times S_2$. This completes the proof.

COROLLARY 5.25. The subset K of E^2 consisting of all points (x, y) satisfying $0 \leq x \leq 1$, $0 \leq y \leq 1$ is a compact connected subset of E^2.

PROBLEMS

43. Prove that every open interval (a, b), $a < b$ of the space R is homeomorphic with R.

44. Show that the mapping $f: A \to R$ defined in the proof of Theorem 5.4 is continuous.

45. Show that the mapping $f: A \to R$ defined in the proof of Theorem 5.7 is continuous.

46. Prove Corollary 5.12.

47. Give a proof of Theorem 5.16 based on Definition 5.1.

48. Give another proof of Theorem 5.18.

49. In the proof of Theorem 5.18 given in the text show that the mapping g is continuous.

50. Let A and B be subsets of a space S, each of which is closed in their union. Show that, if $A \cup B$ is connected while $A \cap B$ contains at most two points, then A is connected or B is connected.

51. Prove Lemma 5.21.

52. Establish the continuity of the projection mappings defined in the proof of Theorem 5.22.

53. In the proof of Theorem 5.22, show that the mapping h is a homeomorphism.

54. Modify the proof of Theorem 5.24 by taking $J_p = \bigcup_{i=1}^{k_p} (U_{\beta_i} \times V_{\beta_i})$ for each $p \in S_1$. Show that, with this new definition of J_p, the proof of Theorem 5.24 is no longer valid.

55. Prove Corollary 5.25.

Bibliographical Comment

In each of the following books in English, the reader may find supplementary material of interest on the topics of this chapter.

R. L. Moore, Foundations of Point Set Theory, *American Mathematical Society Colloquium Publications*, vol. 13, New York, 1932.

W. Sierpiński, *Introduction to General Topology*, University of Toronto Press, Toronto, 1934.

W. Sierpiński, *General Topology*, University of Toronto Press, Toronto, 1952 (Mathematical Expositions No. 7).

R. Vaidynathaswamy, *Treatise on Set Topology*, Part I, Indian Mathematical Society, Madras, 1947.

G. T. Whyburn, Analytic Topology, *American Mathematical Society Colloquium Publications*, vol. 28, New York, 1948.

R. L. Wilder, Topology of Manifolds, *American Mathematical Society Colloquium Publications*, vol. 32, New York, 1949.

In languages other than English, the following books are of interest.

P. Alexandroff and H. Hopf, *Topologie* I, Julius Springer, Berlin, 1935.

K. Kuratowski, *Topologie* I, Warsaw, 1948.

Among the many mathematical journals, the Polish journal *Fundamenta Mathematicae* has many articles in English and, particularly in its early issues, has numerous articles of interest in connection with this chapter.

CHAPTER FOUR

METRIC SPACES

1. Elementary Properties of Metrics and Metric Spaces

In Chapter 3, we introduced the concept of a metric space, proved a few theorems, and gave examples and problems to illustrate this type of space. In this chapter, we shall study metric spaces more systematically. First we shall develop concepts which will review and extend some of the material in the previous chapter. We then turn to the problem of characterizing such spaces; in particular, we find conditions which, if satisfied by the open sets of a space, will assure us that there is a metric for the space. Notice that we say "metric for the space," meaning that the spherical neighborhoods, in this metric, generate the topology of the space. The final portion of the chapter is devoted to a few of the theorems and concepts used most frequently in analysis, together with some of their applications.

Several of the ideas presented in this chapter apply to general spaces as well as to metric spaces. Although the applications are made to metric spaces, the theorems are stated in their general form, and some problems are given concerning their role in more general spaces; i.e., spaces that are not necessarily metrizable.

Among the problems on metric spaces that have been given in Chapter 3, at least two are sufficiently important to be stated as formal theorems.

THEOREM 1.1. Let S be a metrizable space. Then every space homeomorphic with S is metrizable. Thus the property of being metrizable is a topological property.

THEOREM 1.2. Let ρ be a metric for the space S. Then the mapping $\rho: S \times S \to R$ is continuous.

The student has already realized the obvious fact that, if A is a subset of a metrizable space S, then A, considered as a subspace of S, is a metrizable space. Indeed, if ρ is a metric for S, then ρ, restricted to A, is a metric for the subspace A of S.

THEOREM 1.3. Let S be a metrizable space, and A a subset of S. Then A, considered as a subspace of S, is a metrizable space, and the restriction to A of any metric ρ for the space S is a metric for the subspace A.

We have seen that the space R of all real numbers is a metric space, with a metric defined by $\rho(x, y) = |x - y|$. We shall call this particular metric for R the standard metric for R. Early in Chapter 2 the idea of a bounded subset of R was introduced. We see that, if A is a non-empty bounded subset of R, if α is any upper bound for A, and if β is any lower bound for A, then the distance, in the standard metric for R, between any two points of A is less than or equal to $\alpha - \beta$. This idea of bounded sets can easily be extended to metric spaces.

DEFINITION 1.4. A subset A of a metric space S is said to be bounded with respect to the metric ρ of S if and only if A is empty or there exists a real number α such that $\rho(x, y) \leq \alpha$ for all $x, y \in A$.

We have seen how this definition relates to the definition given in Chapter 2. In Problem 1, the student is asked to establish the equivalence of the two definitions in R.

The subset R of the space R of all real numbers is not bounded in the standard metric for R. We are going to show methods by which a new metric can be introduced for the space R such that, in this new metric, every subset of R is bounded. It is for this reason that a metric with this property is referred to as a bounded metric.

DEFINITION 1.5. Let S be a metrizable space. A metric ρ for the space S is said to be a bounded metric for S if and only if S is bounded with respect to this metric ρ. A bounded metric is frequently referred to in this book as a B metric.

One immediate consequence of producing, for every metric space S, a bounded metric, is that the space R has a bounded metric. Hence boundedness cannot be a topological property.

THEOREM 1.6. Every metrizable space S has a B metric.

Proof. Let ρ be a metric for the space S. Define the mapping $\rho_1\colon S \times S \to R$ by $\rho_1(x, y) = \rho(x, y)/[1 + \rho(x, y)]$ for $x, y \in S$. We shall show that ρ_1 is a B metric for the space S.

Clearly,

1. $\rho_1(x, y) = 0$ if and only if $x = y$.
2. $\rho_1(x, y) \geq 0$ for all $x, y \in S$.
3. $\rho_1(x, y) = \rho_1(y, x)$ for all $x, y \in S$.

It remains to be shown that the triangle property holds for ρ_1. Let $x, y, z \in S$, and suppose $x \neq z$. We have at once

$$\begin{aligned}
\rho_1(x, y) + \rho_1(y, z) &= \frac{\rho(x, y)}{1 + \rho(x, y)} + \frac{\rho(y, z)}{1 + \rho(y, z)} \\
&\geq \frac{\rho(x, y) + \rho(y, z)}{1 + \rho(x, y) + \rho(y, z)} \\
&\geq \frac{1}{1 + \dfrac{1}{\rho(x, y) + \rho(y, z)}} \\
&\geq \frac{1}{1 + \dfrac{1}{\rho(x, z)}} \\
&= \rho_1(x, z)
\end{aligned}$$

Thus, for any three points x, y, z of S, with $x \neq z$, we have $\rho_1(x, z) \leq \rho_1(x, y) + \rho_1(y, z)$. If $x = z$, this inequality is immediate.

To complete the proof, it is only necessary to verify that the topology induced by ρ_1 is the original topology of the space S. To do this, we need only observe that, for any point x of S and any real number $r > 0$, the spherical neighborhood $S_r(x)$, using ρ as metric, is precisely the same set of points as the spherical neighborhood $S_{r/(1+r)}(x)$ using ρ_1 as metric.

Corollary 1.7. Given a metric ρ for a space S, there exists a B metric ρ_1 for S such that every spherical neighborhood of S using the metric ρ is homeomorphic to a spherical neighborhood of the same point using the metric ρ_1.

Another method by which Theorem 1.6 may be proved is to define the mapping $\rho_1: S \times S \to R$ by $\rho_1(x, y) =$ the smaller of the two numbers $[\rho(x, y), 1]$. It is clear that this is a metric for the space S, but Corollary 1.7 does not then follow immediately. It was to obtain the corollary that we gave the slightly more difficult proof of the theorem.

For the space R with its standard metric, we proved in Chapter 2 that closed and bounded sets are compact, and conversely. This result cannot hold in general. For example, if we give the space R a B metric, then the subset R of R is closed and bounded, but is not compact. On the other hand, if S is a metrizable space and if ρ is any metric for S, we can prove that every compact subset of S is not only closed but is also bounded with respect to the metric ρ. This result will follow rather easily from the continuity of the metric, and we shall give it after introducing two rather useful symbols, and defining the notion of the diameter of a set.

The symbols referred to are ∞ and $-\infty$. These symbols are used in conjunction with the space R of real numbers, and obey the following rules of computation:

1. $-\infty < a < \infty$ for every $a \in R$.
2. $(\infty)a = a(\infty) = \infty$ if $0 < a \leq \infty$.
3. $(\infty)a = a(\infty) = -\infty$ if $-\infty \leq a < 0$.
4. $(-\infty)a = a(-\infty) = -\infty$ if $0 < a \leq \infty$.
5. $(-\infty)a = a(-\infty) = \infty$ if $-\infty \leq a < 0$.
6. $a/(\infty) = a/(-\infty) = 0$ if $a \in R$.
7. $\infty + a = a + \infty = \infty$ if $a > -\infty$.
8. $-\infty + a = a + (-\infty) = -\infty$ if $a < \infty$.

The concept of boundedness leads very naturally to the notion of the diameter of a set.

DEFINITION 1.8. Let A be a subset of a metric space S with metric ρ. The diameter of A, written $\delta(A)$, is defined by the equations

$\delta(A) = 0$ if $A = \emptyset$.

$\delta(A) =$ least upper bound of $\rho(x, y)$ for all $x, y \in A$ if A is non-empty and bounded.

$\delta(A) = \infty$ if A is not bounded.

The following theorem is almost immediate. Its proof is left to the student.

THEOREM 1.9. Let A, B be subsets of a metric space S with metric ρ. Then

(i) $\delta(A) = 0$ if and only if A contains at most one point.
(ii) If $A \subset B$, then $\delta(A) \leq \delta(B)$.
(iii) $\delta(\bar{A}) = \delta(A)$.
(iv) If $A \cap B \neq \emptyset$, then $\delta(A \cup B) \leq \delta(A) + \delta(B)$.
(v) A is bounded if and only if $\delta(A) < \infty$.

THEOREM 1.10. Let S be a metrizable space, ρ any metric for S, and A any compact subset of S. Then

(i) A is closed.
(ii) A is bounded.
(iii) $\delta(A) < \infty$.
(iv) If A is non-empty, then there exist points a, b of A such that $\delta(A) = \rho(a, b)$.

Proof. We know, by Theorem III, 2.7, that S is a Hausdorff space. Thus A is closed by Theorem III, 3.19. This proves (i).

To prove the remaining parts of the theorem, we assume A to be non-empty and let $H = S \times S$, $B = A \times A$. Define a mapping

$f: H \to R$ in the following way: Given a point $x \in H$, we have $x = (x_1, x_2)$, where $x_1, x_2 \in S$. We define $f(x) = \rho(x_1, x_2)$. The mapping f is continuous by Theorem 1.2. The set B is a compact subset of H by Theorem III, 5.24. The restriction of the continuous mapping f to the subspace B of H is continuous by Theorem III, 4.7. Thus $f(B)$ must be a compact subset of R by Theorem III, 4.16. It follows that $f(B)$ is closed and bounded. Define c as the least upper bound of $f(B)$. Thus $c \in f(B)$; hence there exists a point $p \in B$ such that $f(p) = c$. Let $p = (a, b)$ where $a, b \in A$. Clearly, $\rho(a, b) = \delta(A)$. This proves (iv), from which (ii) and (iii) are immediate.

COROLLARY 1.11. *Let S be a compact metrizable space. Then every metric for S is a B metric.*

It should be carefully noted that Theorem 1.10 does not state that a compact subset of S which has α for a bound in one metric will have α as a bound in all other metrics for the space S.

Another concept based upon the metric is that of the distance between two sets. If each of the sets reduces to a single point, then the distance between the sets is the distance between these points.

DEFINITION 1.12. Let A and B be subsets of a metric space with metric ρ. The distance between A and B is denoted by $\rho(A, B)$ and defined as follows:

(i) If either of the sets A, B is empty, then $\rho(A, B) = 0$.

(ii) If neither of the sets A, B is empty, then $\rho(A, B) =$ greatest lower bound of $\rho(x, y)$ for $x \in A$, $y \in B$.

It should be noted that, if A and B are non-empty subsets of a space S with metric ρ, then $\rho(x, y) \geq 0$ for every $x \in A$, $y \in B$. Thus the set of all real numbers of the form $\rho(x, y)$ for $x \in A$, $y \in B$ has a lower bound. This set, since it is not empty and has a lower bound, must have a greatest lower bound. Consequently, $\rho(A, B)$ is defined for every pair of subsets A, B of the space S.

The student should find it easy to prove the next theorem.

THEOREM 1.13. *Let A and B be subsets of a metric space S. Then*

(i) If $x \in A$, $y \in B$, then $0 \leq \rho(A, B) \leq \rho(x, y) \leq \delta(A \cup B)$.

(ii) $\delta(A \cup B) \leq \delta(A) + \rho(A, B) + \delta(B)$.

(iii) If $x \in S$, then $\rho(x, A) = 0$ if and only if $x \in \bar{A}$.

(iv) Let $H = A \times B$, and define a mapping $f: H \to R$ as follows: Given any point $x \in H$, write $x = (x_1, x_2)$, where $x_1 \in A$, $x_2 \in B$; then $f(x) = \rho(x_1, x_2)$. The mapping f is continuous.

(v) If the sets A and B are non-empty, compact, and disjoint, then $\rho(A, B) > 0$. Moreover, there exist points $a \in A$, $b \in B$ such that $\rho(a, b) = \rho(A, B)$.

To see that the word "compact" in Theorem 1.13(v) cannot be replaced by the word "closed," consider the next examples.

EXAMPLE 1.14. Let R be the space of all real numbers. Define the subspace S of R to consist of all points of R with the exception of the points 2 and 3. Let A be the set of all points of S of the form $2 - 1/n$, and B the set of all points of S of the form $3 + 1/n$, $n \in I$. It is clear that S is a metric space and that A and B are closed and bounded subsets of S. Moreover, $\rho(A, B) = 1$. However, there do not exist points $a \in A$, $b \in B$, such that $\rho(a, b) = 1$. It should be noted that neither of the sets A, B is compact.

EXAMPLE 1.15. Let S be the space defined in Example 1.14, and C the closed interval $[0, 1]$ of S. Clearly, C is compact and B is closed, while $\rho(C, B) = 2$. However, there do not exist points $c \in C$, $b \in B$, such that $\rho(c, b) = 2$.

EXAMPLE 1.16. Let the space be E^2, the Cartesian plane. Let A be the subset of E^2 consisting of all points on the x axis, and B the subset of E^2 consisting of all points on the graph of the hyperbola $y = 1/x$. Clearly, A and B are disjoint non-empty closed sets but $\rho(A, B) = 0$. Neither the set A nor the set B is compact.

The trouble that arises in Example 1.14 stems from the fact that, in the space S, the sequence $\{a_n\}$ defined by $a_n = 2 - 1/n$ "converges to a point which isn't there." In other words, we have a sequence of points $\{a_n\}$ of the space S, and we have reason to believe that this sequence should converge (it is easily seen to be a Cauchy sequence). But there is no point of the space S for it to converge to. The space S may thus be thought of as not being "complete," since it lacks a point which we would like to have in it. The next definitions will make precise the rather rough discussion of this paragraph.

DEFINITION 1.17. Let $\{a_n\}$ be a sequence of points of a metric space S. Then $\{a_n\}$ is said to be a Cauchy sequence if and only if, for each real number $\varepsilon > 0$, there exists an integer N such that $\rho(a_h, a_k) < \varepsilon$ for every pair of positive integers h, k, each greater than N.

DEFINITION 1.18. Let S be a topological space and $\{a_n\}$ a sequence of points of S. Then $\{a_n\}$ is said to converge to the point p of S if and only if, given any neighborhood U of p, there exists an integer N such that $a_n \in U$ for every $n > N$.

Using the fact that the topology of a metric space is generated by the collection of sets consisting of all spherical neighborhoods, the following theorem is immediate.

THEOREM 1.19. Let $\{a_n\}$ be a sequence of points of a metric space S and p a point of S. Then the sequence $\{a_n\}$ converges to the point p if and only if, for each $\varepsilon > 0$, there exists an integer N such that $\rho(a_n, p) < \varepsilon$ for every $n > N$.

DEFINITION 1.20. A space S is said to be complete if and only if it is metric and, given any Cauchy sequence $\{a_n\}$ of points of S, there exists a point p of S such that $\{a_n\}$ converges to p.

The idea of a complete space is not new to us. As a matter of fact we may now restate one of our most important theorems (Theorem II, 3.5) in the language of complete spaces.

THEOREM 1.21. The metric space R with its standard metric is a complete space.

THEOREM 1.22. The Cartesian plane E^2 is a complete space in its standard metric. (See Problem III, 12.)

Proof. This is immediate from Theorem 1.21 and Problem 9.

Returning to the problem of the existence of two points a, b, $a \in A$, $b \in B$ such that $\rho(a, b) = \rho(A, B)$, we give an example in which B is compact, A is closed, $A \cap B = \emptyset$, the space is complete, but a, b do not exist.

EXAMPLE 1.23. Let S be a set consisting of an object b and the set I of all positive integers. For any x, y in S, define $\rho: S \times S \to R$ as follows:

$$\rho(x, y) = 1 \quad \text{if} \quad x, y \in I, x \neq y$$

$$\rho(b, x) = \rho(x, b) = 1 + 1/x \quad \text{if} \quad x \in I$$

$$\rho(x, y) = 0 \quad \text{if} \quad x = y$$

The mapping ρ thus defined is a metric for the set S. In this metric, S is a complete space. Define the subsets A and B of S as $A = I$, $B = \{b\}$. They are closed and disjoint, and B is compact. It is also seen that $\rho(A, B) = 1$. However, there is no pair of points a, b such that $a \in A$, $b \in B$ for which $\rho(a, b) = 1$.

PROBLEMS

1. Let R be the space of all real numbers. Prove that a subset of R is bounded with respect to the standard metric of R if and only if it has an upper bound and a lower bound.

2. Let A denote the subspace of R consisting of the open interval $(-1, 1)$. Prove that the mapping $h: R \to A$ defined by $h(x) = x/(1 + |x|)$ for each $x \in R$ is a homeomorphism. Establish also the following results:

(*a*) Every two open intervals of R are homeomorphic.

(*b*) Every open interval of R is homeomorphic to R.

(*c*) Every open interval is uncountable.

(*d*) There exist spaces that are homeomorphic to proper subspaces of themselves. Could such a space be finite?

3. Give the details of the proof of the statement made in the paragraph following Corollary 1.7.

4. Prove Theorem 1.9.

5. Prove Theorem 1.13.

6. Let $\{a_n\}$ be a Cauchy sequence of points of a metric space S. Prove that $\{a_n\}$ converges to a point p of S if and only if there exists a subsequence $\{b_n\}$ of $\{a_n\}$ that converges to the point p.

7. Let S be a metric space. Then S is complete if and only if, given any Cauchy sequence $\{a_n\}$ of points of S, there exists a point $p \in S$ and a subsequence $\{b_n\}$ of $\{a_n\}$ such that $\{b_n\}$ converges to p.

8. Prove that the topological product of two metric spaces S_1, S_2 with metrics ρ_1 and ρ_2 is metrizable. Show that $\rho[(x, y), (w, z)] = [\rho_1^2(x, w) + \rho_2^2(y, z)]^{\frac{1}{2}}$ is a metric for the space $S_1 \times S_2$.

9. Prove that, if each of the spaces S_1 and S_2 of Problem 8 is complete, then $S_1 \times S_2$ is complete in the metric ρ defined in Problem 8.

10. Let S be a metric space and A a subset of S. Define a mapping $f: S \to R$ by $f(x) = \rho(x, A)$ for each $x \in S$. Prove that f is continuous.

11. Let A and B be non-empty disjoint countably compact subsets of a metric space S. Prove that $\rho(A, B) > 0$, and that there exists points $a \in A$, $b \in B$ such that $\rho(a, b) = \rho(A, B)$.

12. Let S and T be metric spaces with metrics ρ_S and ρ_T, respectively, and let $f: S \to T$ be a mapping. Then f is continuous if and only if, given any real number $\varepsilon > 0$ and any point x_0 of S, there exists a real number $\delta > 0$ such that, if $x \in S$ and $\rho_S(x, x_0) < \delta$, then $\rho_T(f(x), f(x_0)) < \varepsilon$.

13. Let S and T be metric spaces and $f: S \to T$ a mapping. Then f is continuous if and only if, given any real number $\varepsilon > 0$ and any point x_0 of S, there exists a real number $\alpha > 0$ such that, if U is any neighborhood of x_0 with $\delta(U) < \alpha$, then $\delta(f(U)) < \varepsilon$.

2. Local Connectivity and the Introduction of an *M* Metric

In our study of the space R of all real numbers, we have seen that, in the standard metric, every spherical neighborhood of a point is a connected set. Consequently, given any point $p \in R$, we can find a connected open set in R containing p and having its diameter less than any preassigned ε. We now extend these ideas to more general spaces.

DEFINITION 2.1. A space S is said to be locally connected at a point p if and only if, given any neighborhood U of p, there exists a connected neighborhood V of p such that $V \subset U$. The space S is said to be locally connected if and only if it is locally connected at each of its points.

DEFINITION 2.2. A metric ρ for a metrizable space S is said to be an M metric for S if and only if, using this metric, the spherical neighborhood $S_r(p)$ is a connected set for every real number $r > 0$ and every point $p \in S$.

We have pointed out several concepts such as boundedness, the diameter of a set, etc., which depend on the particular metric used in a metrizable space S. These concepts are not topological properties. It is important to understand, however, that they are frequently useful in establishing topological properties. In this section we shall see the part that the diameter of a set plays in establishing the existence of an M metric for any arbitrary connected and locally connected metrizable space. Our main interest in this result does not, of course, lie in the role that the diameter of a set plays in its proof, but rather in the later applications that will be made of the theorem.

The spaces R and E^2 are examples of connected and locally connected spaces having M metrics. The following theorem is easily established.

THEOREM 2.3. (*a*) Let S be a locally connected space. Then every space homeomorphic with S is locally connected. Thus local connectivity is a topological property.

(*b*) Let S be a metrizable space that has an M metric. Then every space homeomorphic with S has an M metric. Thus the property of having an M metric is a topological property.

We prove at once our main result.

THEOREM 2.4. Let S be a metrizable space. Then there is an M metric for S if and only if S is connected and locally connected.

Proof. Suppose first that S has an M metric, and denote such a metric by ρ. Clearly, S is locally connected. To show that S is connected, let $x_0 \in S$, and note that $S = \bigcup_{n \in I} S_n(x_0)$. Thus S is the union of a collection of connected sets having the point x_0 in common. It follows from Theorem III, 5.16 that S is connected.

For the other half of the proof we suppose that S is a metrizable, connected and locally connected space. We must establish the existence of an M metric for S. We start out by observing, by Theorem 1.6, that S has a B metric ρ. Define a mapping $\rho_1 \colon S \times S \to R$ by the equation

$$\rho_1(x, y) = \text{greatest lower bound of } \delta(K) \text{ for all connected subsets } K \text{ of } S \text{ that contain both } x \text{ and } y$$

The proof of the theorem will be complete if we prove the following statements:

(i) ρ_1 is a metric for the set S.

(ii) The topology induced in S by ρ_1 is the same as the given topology of S.

(iii) All spherical neighborhoods in the metric ρ_1 are connected sets.

To prove (i) we note that, for any pair of points x, y of S, there exists a bounded connected subset of S containing both x and y: namely, the set S itself. Thus $\rho_1(x, y)$ is defined for every pair of points x, y of S. Clearly, for any such points, $\rho_1(x, y) \geq 0$, and $\rho_1(x, y) = \rho_1(y, x)$. We note also that, for any set K containing x and y, $\delta(K) \geq \rho(x, y)$. Consequently, $\rho_1(x, y) \geq \rho(x, y)$. This and the definition of ρ_1 gives immediately the result that $\rho_1(x, y) = 0$ if and only if $x = y$.

The proof that ρ_1 is a metric for the set S will be complete when we show that the triangle property holds. To do this, let x, y, z be three points of S, and let $\varepsilon > 0$ be an arbitrary real number. From the definition of ρ_1, it follows that there exist connected subsets K_1 and K_2 of S such that K_1 contains $x \cup y$, K_2 contains $y \cup z$, and

$$\delta(K_1) \leq \rho_1(x, y) + \varepsilon$$

$$\delta(K_2) \leq \rho_1(y, z) + \varepsilon$$

Consequently,

$$\rho_1(x, z) \leq \delta(K_1 \cup K_2) \leq \delta(K_1) + \delta(K_2) \leq \rho_1(x, y) + \rho_1(y, z) + 2\varepsilon$$

Since this holds for every $\varepsilon > 0$, it follows that

$$\rho_1(x, z) \leq \rho_1(x, y) + \rho_1(y, z)$$

Thus the triangle property holds for ρ_1; hence (i) is proved.

We now have two kinds of spherical neighborhoods in S, those using ρ as metric [which will be denoted by $S_r(x)$] and those using ρ_1 as metric [which will be denoted by $S_r^1(x)$]. We have seen that, for every pair of points a, b of S, $\rho_1(a, b) \geq \rho(a, b)$. Consequently, $S_r^1(x) \subset S_r(x)$ for every $x \in S$ and every $r \in R$ such that $r > 0$. It follows that every set open in the original topology of S is also open in the topology induced by ρ_1. On the other hand, let $S_r^1(x)$ be an arbitrary spherical neighborhood in the ρ_1 topology. Since S is locally connected in the topology induced by ρ, there must exist a connected neighborhood V of x contained in $S_{r/3}(x)$; hence, in the metric ρ, $\delta(V) < r$. Thus $V \subset S_r^1(x)$. Thus every set open in the ρ_1 topology is also open in the ρ topology. This proves (ii). Thus ρ_1 is a metric for the space S.

To prove (iii), let $S_r^1(x)$ be an arbitrary spherical neighborhood in the ρ_1 topology, and let $y \in S_r^1(x)$. Thus $\rho_1(x, y) = \alpha < r$. It follows that there exists a connected subset $K(y)$ of S containing x and y and of diameter less than $(\alpha + r)/2$. It is clear that $S_r^1(x) = \bigcup_{y \in S_r^1(x)} K(y)$. Consequently, since $x \in K(y)$, for every $y \in S_r^1(x)$, the set $S_r^1(x)$ is connected by Theorem III, 5.16. This completes the proof of Theorem 2.4.

COROLLARY 2.5. Let S be a metrizable space. Then S is connected and locally connected if and only if S has a metric that is both a B metric and an M metric.

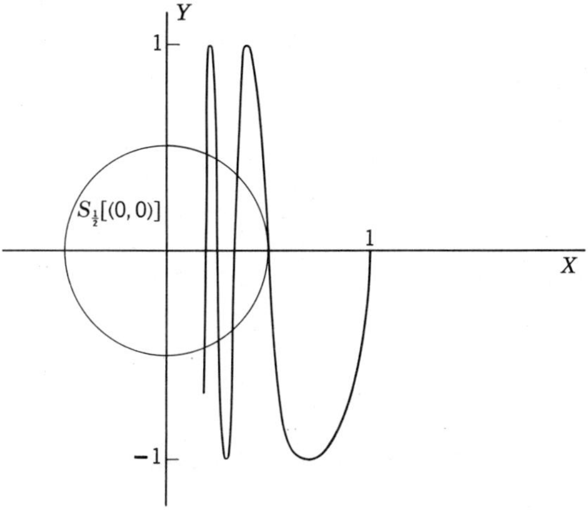

Figure 8

As previously observed, the space E^2 is connected, locally connected, and metrizable. We now give an example of a subset of E^2 that is connected but not locally connected. It is, of course, metrizable, since it is a subset of E^2. However, by Theorem 2.4, we see that it can have no M metric.

EXAMPLE 2.6. (See Figure 8.) Let F be the subspace of E^2 which consists of the graph of the mapping $f: J \to R$, where J is the closed unit interval $[0, 1]$ of R and f is defined by

$$f(x) = \sin [\pi/x], \qquad 0 < x \leq 1$$
$$f(0) = 0$$

It is established in treatises on advanced calculus* that the function $g(t) = \sin t$ is continuous for every $t \in R$. We know that the function $t = \pi/x$ is continuous at every point $x \neq 0$. Thus, by Theorem II, 8.7,

* See, for example, Franklin, *A Treatise on Advanced Calculus*, John Wiley & Sons, New York, 1940.

the function $f(x) = \sin[\pi/x]$ defined on the set A, consisting of all x such that $0 < x \leq 1$, is continuous. Consequently, since A is connected, the graph G of f restricted to $0 < x \leq 1$ is connected. But G consists exactly of every point of F except the point (0, 0), and this point is a limit point of G. By Theorem III, 5.17, the set F is connected.

We shall now show that the set F is not locally connected at the point (0, 0). In the standard metric for E^2, let $S_{1/2}[(0, 0)] = W$; i.e. W consists of all points of E^2 inside the circle with center at the origin and radius $\frac{1}{2}$. In the subspace F of E^2, let V be the neighborhood of (0, 0) defined as the intersection of W and F. We shall prove that F is not locally connected at (0, 0) by showing that there is no connected neighborhood of the origin contained in V. To this end, let U be any neighborhood of (0, 0) contained in V. We note that, for every integer n, the point $(1/n, 0)$ is a point of F. Let N be an integer sufficiently large so that $(1/n, 0) \in U$ for every n such that $n > N$. Consider the mapping $g\colon U \to R$ defined by $g(p, q) = p$, for $(p, q) \in U$. We note that this mapping is continuous. The set $g(U)$ contains the point (0, 0) and the point $(1/(N+1), 0)$, but does not contain the point $(2/(2N+3), 0)$. Since $0 < 2/(2N+3) < 1/(N+1)$, the set $g(U)$ is not connected. Consequently, since g is continuous, U is not a connected set. We have thus shown that the set F is not locally connected at the point (0, 0).

It should also be observed, in connection with this example, that the set F cannot be homeomorphic with the set J. This is immediate since J is locally connected, F is not locally connected, and local connectivity is a topological property. In particular, the set F is not an arc.

PROBLEMS

14. Prove that a space is locally connected if and only if it has a basis consisting of connected sets.

15. Prove that every connected subset of R is locally connected.

16. Prove that a space S is locally connected if and only if, given any open set G of S and any point $p \in G$, then the union of all connected subsets of G that contain p is an open set. Show also that, if the space S is compact, then S is locally connected if, given any point $p \in S$ and any open set V containing p, then there exists an open set W containing p that lies in a connected subset of V.

17. Consider the subset K of E^2 which is the union of two sets F and H. The set F is described in Example 2.6, the set H consists of all points $(0, y)$, $-1 \leq y \leq 1$. Prove that the subset K of E^2 is compact and connected, but not locally connected at any point of H. Show also that F is not compact, and conclude that the mapping $f\colon J \to R$ defined in Example 2.6 is not continuous.

18. (See Figure 9.) Consider the subset K of E^2 which is the union of the two sets L and M, where

L consists of all points (x, y) such that $0 \leq x \leq 1$, $y = 0$,

M consists of all points (x, y) such that $0 \leq x \leq 1$, $0 \leq y \leq 1$, $x^2 + y^2 = 1$.

Let J be the closed unit interval $[0, 1]$ of R.

(*a*) Show that K is connected and locally connected.

(*b*) Show that J and K are homeomorphic.

(*c*) Using the standard metric ρ of E^2, determine the distance from the point $(0, 1)$ to an arbitrary point $(a, 0)$ of L in the M metric constructed from ρ as in the proof of Theorem 2.4.

(*d*) Show also that $S_r(p) \cap K$, for $p \in K$ is not necessarily connected [the spherical neighborhood $S_r(p)$ is here taken in the sense of the standard metric ρ of E^2].

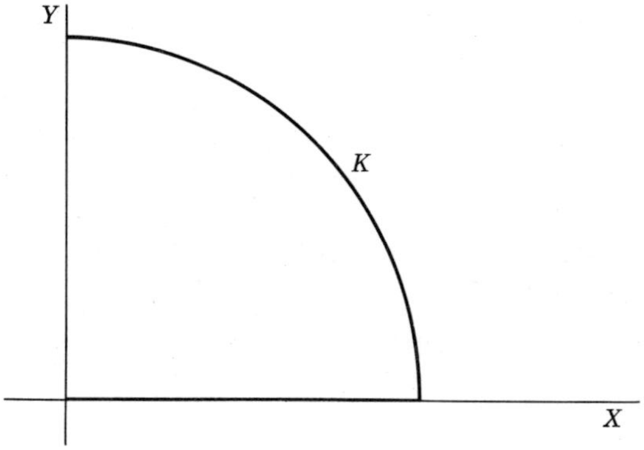

Figure 9

3. Sequences

In contrast with our work in the space R of real numbers, we have, thus far, made little or no use of sequences in general topological spaces. To see some of the things that may happen in such spaces, as opposed to the situation in R, we consider some examples.

EXAMPLE 3.1. Let S be the space of Example III, 1.3; i.e. S consists of three distinct points a, b, c, and a non-empty subset of S is open if and only if it contains the point a. Let $\{x_n\}$ be the sequence of points of S defined by $x_n = a$ for every n. Clearly, this sequence converges both to the point a and to the point b. Thus, without some restriction upon a topological space, we cannot prove that there is at most one point to which a given sequence converges.

The space considered in Example 3.1 is not a Hausdorff space. In Hausdorff spaces the desired result is true as we shall now show.

THEOREM 3.2. Let $\{x_n\}$ be a sequence of points of a Hausdorff space S. If this sequence converges to a point a and also to a point b, then $a = b$.

Proof. Suppose that the points a and b are distinct. Since S is a Hausdorff space, there must exist disjoint neighborhoods U, V of a, b, respectively. From the convergence of the sequence, we see that there exist integers N_1, N_2 such that $n > N_1$ implies $x_n \in U$ while $n > N_2$ implies $x_n \in V$. Let $s = N_1 + N_2 + 1$. Then $x_s \in U \cap V = \emptyset$, which is a contradiction. Consequently, $a = b$.

EXAMPLE 3.3. Let S be a set consisting of all real numbers x such that $0 \leq x \leq 1$. Define a subset K of S to be open if and only if K is the null set or $S - K$ is countable. It is easily seen that, with this topology, S is a topological space which is T_1 but not Hausdorff. Consider the subset L of S consisting of all x such that $\frac{1}{2} \leq x \leq 1$. Since every non-empty open subset U of S intersects L in at least two points (indeed, the intersection is uncountable), it follows that every point of S is a limit point of L; i.e. $\bar{L} = S$. Thus L is not a closed set in the space S. Even more peculiar things happen in this space! The point $\frac{1}{3}$ is, as we have seen, a limit point of L. However, no sequence $\{x_n\}$ of points of L may converge to the point $\frac{1}{3}$. This can be seen by noticing that $S - \bigcup_{n \in I} x_n$, where $\{x_n\}$ is any sequence of points of L, is a neighborhood of the point $\frac{1}{3}$.

Although we see from Example 3.3 that a limit point of a set need not be the limit of a sequence of distinct points of the set, we do have the following result. The simple proof should be supplied by the student.

THEOREM 3.4. Let A be a subset of a space S, and $\{x_n\}$ a sequence of distinct points of A which converges to a point p of S. Then p is a limit point of A.

In our study of the real line, we originally gave an if and only if condition for a mapping to be continuous in terms of sequences (Definition II, 8.1). The next example shows that this condition does not necessarily imply continuity in a general topological space.

EXAMPLE 3.5. Let S be the space described in Example 3.3. It is readily seen that a sequence $\{x_n\}$ of points of S converges to a point x of S if and only if there exists an integer N such that $x_n = x$ for every $n > N$. Let J denote the subspace of R consisting of all x such that $0 \leq x \leq 1$, and consider the mapping f: $S \rightarrow J$ defined by $f(x) = x$ for every x in S. Clearly the statement "if a sequence $\{x_n\}$ of points of S converges to a point x of S, then the sequence $\{f(x_n)\}$ converges to $f(x)$" is true. Yet the mapping f is not continuous, since the set $f^{-1}(K)$, where

K is the open subset of J consisting of all x such that $\frac{1}{4} < x < \frac{3}{4}$, is not open in S.

We may, however, salvage half our condition for continuity in terms of sequences for general spaces.

THEOREM 3.6. Let S and T be spaces, $f: S \to T$ a continuous mapping, and $\{x_n\}$ a sequence of points of S that converges to a point x of S. Then the sequence $\{f(x_n)\}$ converges to the point $f(x)$.

The usual restrictions placed on a space S to insure that the converses of Theorems 3.4 and 3.6 hold are that the space be both Hausdorff and first countable. To define the concept of a space S being first countable, it is convenient to introduce the idea of a basis for a space S at a point p of S.

DEFINITION 3.7. A collection σ of neighborhoods of a point p in a space S is said to be a basis at p if and only if, given any neighborhood U of p in S, there exists a $V \in \sigma$ such that $V \subset U$.

DEFINITION 3.8. A space S is said to be first countable if and only if, for each point p of S, there exists a countable basis at p.

DEFINITION 3.9. A topological space S will be called a I space if and only if it is first countable, and each of its points is a closed set.

THEOREM 3.10. Every metric space S is a I space.

Proof. We know that every point of S is closed and hence need show only that S is first countable. This is immediate from the fact (which the student should prove) that, for each point $p \in S$, the collection of neighborhoods $\{U_n(p)\}$ defined by $U_n(p) = S_{1/n}(p)$ is a countable basis at p.

In any I space S, we can find a countable basis at each point which will satisfy rather convenient conditions.

THEOREM 3.11. Let S be a I space and $p \in S$. Then there exists a countable basis $\{U_n\}$ at p such that

(i) $U_{n+1} \subset U_n$ for every $n \in I$.
(ii) $p = \bigcap_{n \in I} U_n$.

Proof. Let $\{V_n\}$ be a countable basis at p. Define $U_1 = V_1$. If U_k has been defined, let $U_{k+1} = U_k \cap V_{k+1}$. Since the intersection of two neighborhoods of a point p is a neighborhood of p, we have defined, by induction, a sequence $\{U_n\}$ of neighborhoods of p. Since $U_n \subset V_n$, for every $n \in I$, the sequence $\{U_n\}$ is a basis at p. Condition (i) is automatic from the definition of the U_n. To prove (ii), note that $p \in \bigcap_{n \in I} U_n$, since every U_n is a neighborhood of p. Let q be any point of $S - p$. Since

q is a closed set, we can find a neighborhood W of p not containing q. Since U_n is a basis at p, there exists an integer N such that $U_N \subset W$. Thus q is not in U_N; hence (ii) holds.

THEOREM 3.12. Let A be a subset of a I space S. Then a point p of S is a limit point of A if and only if there exists a sequence of distinct points of A that converges to p.

Proof. By virtue of Theorem 3.4, we need only show that, if p is a limit point of A, then there exists a sequence of distinct points of A that converges to p. Let $\{U_n\}$ be a countable basis at p, as described in Theorem 3.11. Since p is a limit point of A, there exists, for each n, a point x_n of $U_n \cap A$ distinct from p. From the fact that $\{U_n\}$ is a basis at p, we see at once that the sequence $\{x_n\}$ converges to p. Let $H = \bigcup_{n \in I} x_n$. If H is an infinite set, we can find a subsequence $\{x_n'\}$ of $\{x_n\}$ consisting of distinct points of A. Since $\{x_n'\}$ converges to p, the theorem is proved for this case. On the other hand, if H is a finite set, then H is closed since S is a I space. There must thus exist an integer N such that U_N is contained in the open set $S - H$. But this is impossible, since $x_N \in H$.

Recalling Theorem 3.2, and making use of the obvious fact that every point of a Hausdorff space S is a closed subset of S, we have the following interesting corollary.

COROLLARY 3.13. Every countably compact subset of a first countable Hausdorff space S is a closed subset of S.

In the proof of Theorem 3.12, we made no use of property (ii) of the $\{U_n\}$ or of the fact that each point of S is a closed subset of S until we reached the point in the proof where it became necessary to find a sequence of *distinct* points of A converging to the point p. Thus we have another corollary.

COROLLARY 3.14. Let A be a subset of a first countable space S. Then a point p of S is a limit point of A if and only if there exists a sequence of points of A distinct from p which converges to the point p.

In connection with Corollary 3.14, the student should restudy the first countable space S described in Example 3.1.

THEOREM 3.15. Let S and T be spaces, S first countable, and $f: S \to T$ a mapping. Then f is continuous if and only if, given any point $x \in S$ and any sequence $\{x_n\}$ of points of S that converges to x, then the sequence $\{f(x_n)\}$ converges to $f(x)$.

Proof. We have proved half of this theorem as Theorem 3.6. For the other half, we must show that the mapping f is continuous if the given condition holds. We shall make use of Theorem III, 4.9 and show that f is continuous by proving that for *every* subset A of S it is true that $f(\bar{A}) \subset \overline{f(A)}$: i.e., that the image of the closure of A is contained in the closure of its image. To do this let A be *any* subset of S. We may assume that A is not the null set. Let $q \in f(\bar{A})$. Then there exists a point $p \in \bar{A}$ such that $f(p) = q$. We may then find a sequence $\{p_n\}$ of points of A which converges to the point p. (If $p \in A$, let $p_n = p$ for every n; otherwise use Corollary 3.14.) The sequence of points $\{f(p_n)\}$ must converge to the point $q = f(p)$, by the condition of the theorem. Since $f(p_n) \in f(A)$ for every n, we must have $q \in \overline{f(A)}$. Consequently, $f(\bar{A}) \subset \overline{f(A)}$; hence f is continuous.

In order to use sequences in a space S in much the same way as they are used in the space R of all real numbers, we see that it is sufficient to have S a I space. The limitations imposed by this restriction may be avoided if the concept of a sequence is generalized so as to permit uncountably many points in a sequence. The necessary definitions and some of the elementary and important theorems are listed in a problem at the end of this chapter to be worked out as a short paper by the student. Our work is so arranged that none of the results given in this text depend upon the results of this problem, although the techniques involved could have been used in many places. One important fact which we do not obtain as a result of this omission is that a topological space S is a Hausdorff space if and only if each of its generalized sequences that converges to a point converges to one and only one point. Since analysts are much interested in the uniqueness of limits of sequences, this explains in part the large amount of attention that is paid to Hausdorff spaces.

Besides the method of generalizing the idea of a sequence to allow it to contain uncountably many elements, there is another direction in which the concept of a sequence may be generalized: namely, by considering sequences of subsets of a topological space S. We conclude this section by introducing a few of the definitions and elementary results on sequences of sets.

DEFINITION 3.16. Let $\{A_n\}$ be a sequence of subsets of the space S. The limit superior of this sequence is denoted by $\overline{\text{Lim}}\, A_n$ and consists of all points x of S with the property that every neighborhood of x contains points from infinitely many sets of the sequence. The limit inferior of the sequence is denoted by $\underline{\text{Lim}}\, A_n$ and consists of all points x of S with the property that, given any neighborhood U of x there exists an integer N such that U contains a point of A_n for every $n > N$. This last statement is frequently abbreviated by saying that every neighborhood of x contains points from "almost all the sets A_n."

The student should have no difficulty in establishing the following two fundamental theorems.

THEOREM 3.17. Let $\{A_n\}$ be a sequence of subsets of a space S. Then each of the sets $\underline{\text{Lim}}\, A_n$ and $\overline{\text{Lim}}\, A_n$ is a closed subset of S.

THEOREM 3.18. Let $\{A_n\}$ be a sequence of subsets of a space S, and $\{A_{n_i}\}$ any subsequence of $\{A_n\}$. Then $\underline{\text{Lim}}\, A_n \subset \underline{\text{Lim}}\, A_{n_i} \subset \overline{\text{Lim}}\, A_{n_i} \subset \overline{\text{Lim}}\, A_n$.

The study of sequences of real numbers was greatly simplified by the fact that every bounded sequence of real numbers has a convergent subsequence. It is a remarkable fact (see Problem 93) that in a T_1 space S which has a countable basis, every sequence of sets $\{A_n\}$ has a subsequence $\{A_{n_i}\}$ such that $\underline{\text{Lim}}\, A_{n_i} = \overline{\text{Lim}}\, A_{n_i}$.

DEFINITION 3.19. A sequence of sets $\{A_n\}$ is said to be convergent if and only if $\underline{\text{Lim}}\, A_n = \overline{\text{Lim}}\, A_n$. The sequence $\{A_n\}$ is then said to converge to the set $K = \underline{\text{Lim}}\, A_n = \overline{\text{Lim}}\, A_n$, and we write $\text{Lim}\, A_n = K$.

It should be noted that, when we say that a sequence of sets converges, we mean simply that its limit inferior and limit superior are the same set of points. This set may be the null set as we now show.

EXAMPLE 3.20. Let R be the space of all real numbers. For each integer n, define A_n as the closed interval $[2n, 2n + 1]$. This sequence of sets converges to the null set.

EXAMPLE 3.21. Let R be the space of all real numbers, and $\{a_n\}$ the sequence consisting of all rational numbers in R. For each $n \in I$, define A_n as the subset of R consisting of the single point a_n. The sequence $\{A_n\}$ is not convergent since $\overline{\text{Lim}}\, A_n = R$, $\underline{\text{Lim}}\, A_n = \emptyset$.

EXAMPLE 3.22. The space S of this example is a subspace of E^2 consisting of the union of two sets F and G. The set F has been described in Example 2.6. The set G consists of the closed interval $[-1, 1]$ of the y axis of E^2. It is easily seen that the space S is connected, but fails to be locally connected at every point of G. For each $n \in I$, define A_n as the subset of S consisting of all points (x, y) of S such that x lies in the closed interval $[2/(4n + 1), 2/(4n - 1)]$. It is easily seen that each of the sets A_n is an arc, that A_i and A_j are disjoint for $i \neq j$, and that the sequence $\{A_n\}$ converges to G.

One or two of the problems that follow take up material not discussed in the text. For the purpose of these problems the next definition will be needed.

DEFINITION 3.23. Let M be a subset of a metric space S, a and b points of M, and $\varepsilon > 0$ a real number. By an ε chain of points from a to b in M, we shall mean a finite collection of points $x_1, x_2, \cdots, x_n$ such that $a = x_1$, $b = x_n$, $\rho(x_i, x_{i+1}) < \varepsilon$ for $i = 1, 2, \cdots, n-1$. The set M is said to be well-chained if and only if, given any pair of points a, b of M and any $\varepsilon > 0$, there exists an ε chain of points from a to b in M.

PROBLEMS

19. Prove Theorem 3.4.

20. Give the details of the unproved statements made in Example 3.5.

21. Prove Theorem 3.6.

22. Let S be a I space, $p \in S$, and $\{U_n\}$ a countable basis at the point p. Prove that condition (ii) of Theorem 3.11 holds for the basis $\{U_n\}$.

23. Prove Corollary 3.13.

24. Prove that every compact Hausdorff space is regular.

25. Let $\{A_n\}$ be a sequence of subsets of a space S, and K a subset of S. Prove that $\{A_n\}$ converges to K if and only if every subsequence of $\{A_n\}$ converges to K.

26. Prove Theorem 3.17.

27. Prove Theorem 3.18.

28. Assume $\{A_n\}$ a sequence of subsets of a space S, and define $K = \underline{\text{Lim}}\, A_n$, $L = \overline{\text{Lim}}\, A_n$. Prove that, for any point $p \in L - K$, there exists a neighborhood U of p and a subsequence $\{A_{n_i}\}$ of $\{A_n\}$ such that $A_{n_i} \cap U = \emptyset$ for every i. Conclude that U is disjoint with the limit superior of the subsequence.

29. Let R be the space of real numbers and $\{a_n\}$ a sequence of points of R. For each $n \in I$, define A_n as the set consisting of the single point a_n. Let K, L denote, respectively, the limit inferior and limit superior of the sequence $\{A_n\}$. Prove that the sequence $\{a_n\}$ converges to a point p if and only if $K = L = p$.

30. Establish the results of Example 3.21.

31. Establish the results of Example 3.22.

32. Let $\{A_n\}$ be a sequence of sets in a T_1 space S with limit superior L, and suppose that $\overline{\bigcup_{n \in I} A_n}$ is countably compact. Prove that, given any open set G containing L, there exists an integer N such that $A_n \subset G$ for every $n > N$.

33. Let M be a subset of a metric space S, and A a subset of M. Given $\varepsilon > 0$, define M_A as the set of all points x of M that can be joined to at least one point of A by an ε chain of points of M. Prove that M_A is both open and closed in M.

34. Prove that every connected subset of a metric space is well-chained.

35. Give an example of a well-chained countable subset M of the space R, no two points of which lie in a connected subset of M.

36. Repeat Problem 35, substituting the word "uncountable" for the word "countable."

37. Let S be a metric space and p a point of S. Show that the collection of spherical neighborhoods $\{S_{1/n}(p)\}$ is a countable basis at p.

38. Let A be a subset of a space S and $\{A_n\}$ a sequence of subsets of S such that $A_n = A$ for every $n \in I$. Show that this sequence converges to the closure of the set A in the space S.

39. For two points a and b of a subset M of a metric space S, compare the following statements:

(*a*) a and b may be joined by an ε chain of points of M.

(*b*) Given any covering of M by open sets, there exists a simple chain of sets of this covering joining a and b.

(*c*) Given any covering of M by ε spheres with centers in M, there exists a simple chain of sets of this covering joining a and b.

4. Second Countable Spaces

A second countable space is one that has a countable basis for its topology. The spaces R and E^2 are examples of such spaces. The space T, where T is the set of all real numbers and in whose topology every subset of T is open, is not a second countable space.

Our interest in second countable spaces is due primarily to three results which depend on this notion. The first (Theorem 4.12) states that a subset of a second countable T_1 space is compact if and only if it is countably compact. The second and closely related result (Theorem 4.16) states that a subset of a metric space is compact if and only if it is countably compact. The third (Theorem 4.25) states that every regular second countable T_1 space is completely normal. The significance of this last result is that it is a major step in the proof (given in Section 6) that every regular second countable T_1 space is metrizable. [Note that the space T, where T is the set of all real numbers and in whose topology every subset of T is open, is metrizable (set $\rho(x, y) = 1$, if $x \neq y$, $\rho(x, y) = 0$, if $x = y$) but not second countable.]

DEFINITION 4.1. A space S is said to be second countable if and only if there exists a countable basis for S.

Closely associated with second countability is the idea of a subset of a space being separable. This will be defined in terms of being dense in a set.

DEFINITION 4.2. A subset A of a set B is said to be dense in B if and only if $B \subset \bar{A}$.

DEFINITION 4.3. A subset A of a space S is said to be separable if and only if there exists a countable subset H of A which is dense in A.

The next theorem is immediate from the definitions.

THEOREM 4.4. (*a*) Every second countable space is first countable.
(*b*) Every second countable space is separable.
(*c*) Every subspace of a second countable space is second countable.

EXAMPLE 4.5. Let S be the space of Example III, 3.15. We have seen that the space S is Hausdorff, but not regular. Thus, in particular, S is not metrizable. The space S is not second countable. To see this, suppose that S is second countable. Then the subspace L of S is second countable, which is a contradiction, since each point of L is an open subset of L, and the set L contains uncountably many distinct points. Thus S is not second countable.

Consider this space further. Let H be the set of all points (x, y) of S, such that $y > 0$ and both x and y are rational. We note that H is a countable set which is dense in S, so that the space S is separable. Clearly, the subspace L of S is not separable.

Finally, the student may show that S is first countable. Thus S is an example of a separable first countable space that is not second countable.

THEOREM 4.6. Every separable metric space S is second countable.

Proof. Let H be a countable subset of S which is dense in S. Define Σ as the collection of subsets of S consisting of all spherical neighborhoods in S that have centers at points of H and rational radii. We shall show that Σ is a basis for S. This will prove the theorem, since the collection Σ is known to be countable. Let $q \in S$, and let U be any neighborhood of q. There exists a real number $\varepsilon > 0$ such that $S_\varepsilon(q) \subset U$. Let r be any rational number satisfying $0 < r < \varepsilon/2$. Since H is dense in S, there exists a point $p \in H$ such that $\rho(q, p) < r$. The spherical neighborhood $S_r(p)$ is an element of Σ that contains q and is contained in U. Thus S is second countable.

We recall that every real number is the limit of a convergent sequence of rational numbers. Differently stated, the set of all rational numbers is dense in the space R.

THEOREM 4.7. The space R of all real numbers is second countable.

THEOREM 4.8. The topological product of two separable spaces is separable.

THEOREM 4.9. The topological product of two second countable spaces is second countable.

THEOREM 4.10. The space E^2 is second countable.

The next theorem, known as the Lindelöf theorem, shows that, in considering open coverings of a second countable space, it is sufficient to consider only countable open coverings. From this result, and the connection between countable compactness and countable coverings established in Theorem III, 3.27, the equivalence of compactness and countable compactness in second countable T_1 spaces is evident.

THEOREM 4.11. Let A be a subset of a second countable space S, and $\{G_\alpha\}$ a collection of open sets covering A. Then some countable subcollection of $\{G_\alpha\}$ also covers A.

Proof. Let $\{V_i\}$ be the non-empty sets of a countable basis for the space S. Given $x \in A$, there exists an index α such that $x \in G_\alpha$. By the definition of a basis, we see that there exists an integer i such that $x \in V_i \subset G_\alpha$. Denote by $\{V_j\}$ the subcollection of $\{V_i\}$ defined as follows: A set V_i is a V_j if and only if there exists an index α such that $V_i \subset G_\alpha$. For each j, denote by G_j one of the G_α that contains V_j. Clearly the countable subcollection $\{G_j\}$ of $\{G_\alpha\}$ covers A.

THEOREM 4.12. A subset of a second countable T_1 space is compact if and only if it is countably compact.

Proof. This follows at once from the Lindelöf theorem and Theorem III, 3.27.

The proof that countable compactness and compactness are equivalent in a metric space depends on the theorem just established. This may seem unreasonable since, as already remarked, a metric space is not necessarily second countable; yet the theorem we intend to use depends on second countability. No logical difficulty is involved, however, since the key step is the proof that a countably compact subset of a metric space, regarded as a space, is second countable. This proof is achieved by the use of the notion of an ε net.

DEFINITION 4.13. Let F be a finite subset of a metric space S, and $\varepsilon > 0$ a real number. Then F is said to be an ε net for the space S if and only if, given any point p of S, there exists a point q of F such that $\rho(p, q) < \varepsilon$.

THEOREM 4.14. Every countably compact metric space S has an ε net for every $\varepsilon > 0$.

Proof. Suppose the theorem false, i.e., that there exists an $\varepsilon > 0$

for which S has no ε net. Let p_1 be an arbitrary point of S, and define $U_1 = S_\varepsilon(p_1)$. Since the subset of S consisting of the single point p_1 is not an ε net for S, there must exist a point $p_2 \in S$ which is not in U_1. Thus $\rho(p_1, p_2) \geq \varepsilon$. Let $U_2 = S_\varepsilon(p_2)$. Since the subset of F consisting of the points p_1 and p_2 is not an ε net for S, there must exist a point $p_3 \in S$ which is not in $U_1 \cup U_2$. We have at once that $\rho(p_i, p_j) \geq \varepsilon$ for $i \neq j$, $i = 1, 2, 3$; $j = 1, 2, 3$. We continue this process. Suppose that we have found n points $p_1, p_2, \cdots, p_n$ in S such that $\rho(p_i, p_j) \geq \varepsilon$ for $i \neq j$, $i = 1, 2, \cdots, n$; $j = 1, 2, \cdots, n$, and define $U_i = S_\varepsilon(p_i)$, for $i = 1, 2, \cdots n$. The finite set $H = \bigcup_{i=1}^{n} p_i$ is not an ε net for S; hence there exists a point p_{n+1} not in $\bigcup_{i=1}^{n} U_i$. We have at once $\rho(p_i, p_j) \geq \varepsilon$ for $i \neq j$, $i = 1, 2, \cdots, (n+1)$; $j = 1, 2, \cdots, (n+1)$. We have thus defined, by mathematical induction, a sequence $\{p_n\}$ of distinct points of the countably compact set S such that $\rho(p_i, p_j) \geq \varepsilon$ for $i \neq j$. Define $K = \bigcup_{n \in I} p_n$. By countable compactness, the set K has a limit point q in K. Thus $S_{\varepsilon/3}(q)$ must contain infinitely many points of the set K. This contradicts the fact that $\rho(p_i, p_j) \geq \varepsilon$ for $i \neq j$. We conclude that the space S has an ε net.

THEOREM 4.15. Every countably compact metric space S is separable.

Proof. The conclusion is an easy consequence of Theorem 4.14. For each $n \in I$, let F_n be a finite subset of S which is a $1/n$ net for S. Let $F = \bigcup_{n \in I} F_n$. The set F is countable, since it is the union of a countable collection of countable sets. Thus there exists a one-to-one mapping $f: I \to F$ of I onto F. For each $n \in I$, let $p_n = f(n)$. The proof of the theorem will be complete if we show that $S \subset \overline{\bigcup_{n \in I} p_n}$. To do this, let $p \in S$, and let $\delta > 0$ be any real number. Pick an integer N sufficiently large so that $1/N < \delta$. Using the fact that F_N is a $1/N$ net for S, we can find a point $w \in F_N$ such that $\rho(p, w) < 1/N < \delta$. Define $k = f^{-1}(w)$, and note that $w = p_k$. We have thus proved that, for every $\delta > 0$, the spherical neighborhood $S_\delta(p)$ contains a point of $\bigcup_{n \in I} p_n$. Therefore $p \in \overline{\bigcup_{n \in I} p_n}$, and the space S is separable.

THEOREM 4.16. A subset A of a metric space S is compact if and only if it is countably compact.

Proof. If A is compact, then A is countably compact (Theorem III, 3.21). If A is countably compact, then A, considered as a subspace of the metric space S, is a countably compact metric space. Thus A is separable, by Theorem 4.15. Hence A is second countable, by Theorem 4.6, and it follows at once, from Theorem 4.12, that A is compact. Since the topology for the space A is the subspace topology for A in S, A is a compact subset of S.

The next results involve two new concepts: normality and complete normality.

DEFINITION 4.17. A space S is said to be normal if and only if, given any two disjoint closed subsets F_1 and F_2 of S, there exist disjoint open subsets G_1 and G_2 of S containing F_1 and F_2, respectively.

DEFINITION 4.18. A space S is said to be completely normal if and only if, given any two separated subsets F_1 and F_2 of S, there exist disjoint open subsets G_1 and G_2 of S containing F_1 and F_2, respectively.

THEOREM 4.19. Every normal T_1 space is regular.

THEOREM 4.20. Every completely normal space is normal.

The student should be able to show that every subspace of a completely normal space is completely normal.

THEOREM 4.21. Every metric space S is completely normal.

Proof. Let A and B be two separated subsets of S, and p a point of A. We know that p is not a point of $\bar{B}$. Consequently, there exists a real number $r_p > 0$, such that the spherical neighborhood $S_{r_p}(p)$ contains no point of B. Denote by U_p the spherical neighborhood with center p and radius $\frac{1}{2}r_p$.

Similarly, for each $q \in B$, we find a spherical neighborhood $S_{r_q}(q)$ containing no point of A, and denote by V_q the spherical neighborhood with center q and radius $\frac{1}{2}r_q$. Define $U = \bigcup_{p \in A} U_p$, $V = \bigcup_{q \in B} V_q$, and note that U and V are open sets containing A and B, respectively. Suppose there exists a point $p \in U \cap V$. Then there must exist points $a \in A$, $b \in B$ such that $\rho(p, a) < \frac{1}{2}r_a$, $\rho(p, b) < \frac{1}{2}r_b$. Consequently, $\rho(a, b) < \frac{1}{2}(r_a + r_b)$. If $r_b \leq r_a$, it follows that $b \in S_{r_a}(a)$, whereas if $r_a \leq r_b$, it follows that $a \in S_{r_b}(b)$. Thus we have a contradiction in either case, and the proof of the theorem is complete.

COROLLARY 4.22. Every metric space is normal.

COROLLARY 4.23. The space R is normal.

COROLLARY 4.24. The space E^2 is normal.

The next theorem shows us an important class of spaces which are known to be completely normal. This theorem is used in Section 6 in the proof that each space of this type is metrizable.

THEOREM 4.25. Every regular second countable T_1 space S is completely normal.

Proof. Let A and B be separated subsets of S. For each point $p \in A$, we can find a neighborhood W_p of p which does not intersect the set $\bar{B}$. By regularity, there exists a neighborhood U_p such that $p \in U_p \subset \bar{U}_p \subset W_p$. The collection of all the sets $\{U_p\}$ for $p \in A$ is an open covering of A. By the Lindelöf theorem, there must exist a countable subcollection $\{U_i\}$ of $\{U_p\}$ which also covers A. We note that $\bar{U}_i \cap \bar{B} = \emptyset$ for each $i \in I$.

By exactly the same process, we can find a countable covering $\{V_i\}$ of B by open sets such that $\bar{V}_i \cap \bar{A} = \emptyset$ for each $i \in I$. Define $W_1 = U_1$, $W_{n+1} = U_{n+1} - [U_{n+1} \cap \bigcup_{i=1}^{n} \bar{V}_i]$, $Z_n = V_n - [V_n \cap \bigcup_{i=1}^{n} \bar{U}_i]$. It then follows from Problem 56 that U and V where $U = \bigcup_{n \in I} W_n$, $V = \bigcup_{n \in I} Z_n$ are disjoint open sets containing A and B, respectively.

COROLLARY 4.26. Every regular second countable Hausdorff space is completely normal.

To clarify the relationship among completely normal, normal, regular, Hausdorff and T_1 spaces, we state the following result. From the definitions involved, the proof follows immediately.

THEOREM 4.27. Let S be a T_1 space. Then the following statements concerning S are true.

(i) If S is completely normal, then S is normal.

(ii) If S is normal, then S is regular.

(iii) If S is regular, then S is Hausdorff.

(iv) If S is Hausdorff, then S is T_1.

We have already given examples (III, 3.15 and IV, 3.3) that show the converses of statements (iii) and (iv) are false. An example showing that the converse of statement (ii) is false can be obtained by slightly modifying Example III, 3.15. Take the sets $U_r(p)$ as in that example, if $p \in S - L$. If $p \in L$, then take $U_r(p)$ as the union of the point p and the set $S_r(q)$, where q is the point (x, r) if p is the point $(x, 0)$. Incidentally, this is also an example of a regular Hausdorff space that is not second countable.

The space T of Example VII, 3.5 is normal and Hausdorff, but not completely normal since it has a subspace that is not normal. Thus the converse of (i) is also false.

We conclude this section with two definitions which will be useful in the problems.

DEFINITION 4.28. Let A be a non-empty subset of a metric space S, and $\varepsilon > 0$ a real number. The spherical ε neighborhood of A is denoted by $S_\varepsilon(A)$, and consists of all points x of S such that $\rho(x, A) < \varepsilon$.

DEFINITION 4.29. A point p of a space S is said to be a condensation point of a subset A of S if and only if every neighborhood of p contains uncountably many distinct points of A.

PROBLEMS

40. Prove that the set of positive integers I with the topology defined in Example III, 3.23 is second countable. Why does Theorem 4.12 not apply to this space?

41. Prove Theorem 4.4.

42. Fill in the missing details of Example 4.5.

43. Prove Theorem 4.8.

44. Prove Theorem 4.9.

45. Prove that every subspace of a completely normal space is completely normal.

46. Prove or disprove that a space S is completely normal if and only if every subspace of S is normal.

47. Prove that, for any non-empty subset A of a metric space S, and any real number $\varepsilon > 0$, the spherical neighborhood $S_\varepsilon(A)$ is an open set.

48. Let M be the subspace of E^2 consisting of the point $a = (0, -1)$, the point $b = (0, 1)$, and all points (x, y) of E^2, such that $x = 1/n$ for some $n \in I$ and $-1 \leq y \leq 1$. Prove the following facts about the space M.

   (a) No connected subset of M contains both the point a and the point b.

   (b) The space M cannot be written as the union of two separated sets A and B containing a and b, respectively.

49. Let S be a second countable space and A the subset of S consisting of all points of S that are not condensation points of S. Prove that A is countable.

50. Prove that every second countable space that contains uncountably many distinct points contains at most a countable number of points which are not points of condensation.

51. Let S be a regular second countable T_1 space and M a subset of S, with the property that every infinite subset of M has at least one limit point in S. Prove that $\bar{M}$ is countably compact.

52. Prove that every subspace of a second countable space is second countable.

53. Show, by considering Example III, 3.15, that it is not true that every subspace of a separable space is separable.

54. Show that normality is a topological property.

55. Prove that every compact Hausdorff space is normal.

56. Show that the sets U and V defined in the proof of Theorem 4.25 are disjoint open sets such that $U \supset A$, $V \supset B$.

5. Metrizability*

In this section we shall consider conditions which, when satisfied by the topology of a space S, will insure that S is metrizable. Since every metric space satisfies these conditions, it follows that a space is metrizable if and only if these conditions are fulfilled.

To begin with, we define the concept of a K basis for a space S. The letter K comes from the German word Kugel, meaning sphere, and is appropriate here since, as will be seen, the sets of a K basis have properties like those of spherical neighborhoods in metric spaces. Our object is to prove that a topological space S is metrizable if and only if S has a K basis.

Closely related to the question of metrizability is the following question: "If S and T are each normal Hausdorff spaces, what additional restrictions are needed to insure that $S \times T$ is normal?" In particular, it is unknown whether $S \times J$ is normal if S is a normal Hausdorff space and J is the closed unit interval. To see the relationship with metrizability, notice that, if S is metrizable, then each of the spaces S and $S \times S$ is a normal Hausdorff space. This follows from the fact that $S \times S$ is metrizable if S is metrizable.

The concept of a K basis which we are about to introduce is a special kind of development (see Bibliographical Comment). We can show that a space having this kind of development is metrizable, but it is unknown whether a normal space possessing a development is metrizable. The definition of a K basis will now be given.

DEFINITION 5.1. Let S be a space, and σ a basis for S. Suppose that there is associated with each positive integer n, and each point p of S, a unique non-empty element of σ which we denote by $V_n(p)$, and every non-empty element of σ is a $V_n(p)$ for some n and some p. Then σ is said to be a K basis for S if and only if the following conditions hold:

(*a*) For each $p \in S$, the collection $\{V_n(p)\}_{n \in I}$ is a countable basis at p.

(*b*) $p = \bigcap_{n \in I} V_n(p)$.

(*c*) Given $n \in I$, for every pair of points p, q of S, $q \in V_n(p)$ implies $p \in V_n(q)$.

(*d*) Given $n \in I$, if $q \in V_{n+1}(p)$ and $r \in V_{n+1}(q)$, then $r \in V_n(p)$.

If this definition looks a bit mysterious, consider a metric space S, and define, for each $p \in S$ and each positive integer n, the set $V_n(p)$ to be the spherical neighborhood $S_{1/2^n}(p)$. Let σ be the collection of subsets of S, consisting of all the sets $V_n(p)$ just defined. It is clear that σ is a

* This section, which is, except for Definitions 5.5 and 5.6, independent of the remainder of the text, may be omitted if desired.

basis for S. Considering the conditions of Definition 5.1, we see that (a) and (b) are immediate. Condition (c) follows since $\rho(p, q) = \rho(q, p)$, whereas (d) comes at once from the triangle property of the metric; i.e., $\rho(p, r) \leq \rho(p, q) + \rho(q, r) < 1/2^{n+1} + 1/2^{n+1} = 1/2^n$. Thus we have established the following theorem:

THEOREM 5.2. Every metric space has a K basis.

We have seen in our previous work that the triangle property of the metric is important in establishing properties of metric spaces. Property (d) of Definition 5.1 is the corresponding property for spaces having a K basis. We use it here to prove an important lemma.

LEMMA 5.3. Let S be a space with a K basis. Then, given any point $p \in S$ and any integer $n \in I$, we must have $\overline{V_{n+1}(p)} \subset V_n(p)$.

Proof. Let $r \in \overline{V_{n+1}(p)}$. Then there must exist a point q in the intersection of $V_{n+1}(p)$ and $V_{n+1}(r)$. Thus $q \in V_{n+1}(r)$; hence $r \in V_{n+1}(q)$. But $q \in V_{n+1}(p)$; hence, by Definition 5.1d, $r \in V_n(p)$, and the lemma is proved.

THEOREM 5.4. Every space S that has a K basis is a regular Hausdorff space.

Proof. We know that S is a T_1 space by condition (a) and (b) of Definition 5.1. Thus, to prove that S is a regular Hausdorff space, we need only show that S is regular. That is, we must show that, given any point p of S and any neighborhood U of p, there exists a neighborhood V of p such that $p \in V \subset \bar{V} \subset U$. We need only remark that, by Definition 5.1, there exists an integer N such that $V_N(p) \subset U$. The neighborhood $V = V_{N+1}(p)$ thus satisfies our requirements, by Lemma 5.3.

We consider now a space S which has a K basis. How do we show that S is metrizable? If we think of the sets $V_n(p)$ of the K basis for S as being akin to spherical neighborhoods of radius $1/2^n$ we can attempt to define the distance between two points p and q in terms of inequalities. For instance, if q is in $V_n(p)$ we can think of p and q as being closer together than $1/2^n$. To sharpen these inequalities we might say that, if there is a point r of S such that $q \in V_m(r)$ and $r \in V_n(p)$, then q and p are closer together than $1/2^n + 1/2^m$. Here we are, of course, using ideas closely related to simple chains. If we could assign a number to be the distance between p and q by such inequalities, there would remain the difficulties of showing that a metric for the set S had been defined which would generate the given topology of S. The general scheme of our proof follows this informal discussion fairly closely. In particular, we make extensive use of chains and of numbers which are sums of numbers of the form $1/2^n$, where n is a positive integer. These numbers are the dyadic fractions.

DEFINITION 5.5. A real number τ such that $0 < \tau < 1$ is said to be a dyadic fraction if and only if it can be expressed in the form

$$\tau = \sum_{j=1}^{m} \frac{1}{2^{n_j}}$$

where n_j is a positive integer for $j = 1, 2, \cdots, m$, and $n_1 < n_2 < \cdots < n_m$.

It will be seen in the problems that the expression for a dyadic fraction is unique. This means that a given dyadic fraction determines uniquely, and is determined uniquely, by a given integer m and set of integers $n_1 < n_2 < \cdots < n_m$. This uniqueness of the representation of dyadic fractions will be of the utmost importance in our proofs.

Using the notion of chains, we now establish that every space that has a K basis has what we shall call a dyadic basis. In effect, it gives us a sharpening of the inequalities obtainable from the $V_n(p)$ alone, these inequalities being the ones referred to in the informal description of the construction of a metric.

DEFINITION 5.6. A space S is said to have a dyadic basis if and only if there exists, for each $p \in S$ and each dyadic fraction τ satisfying $0 < \tau \leq \frac{1}{2}$ a neighborhood $U_\tau(p)$ such that the following conditions hold:

(i) The collection $\{U_\tau(p)\}$ is a countable basis at p.

(ii) $U_{\tau_1}(p) \subset U_{\tau_2}(p)$ if $\tau_1 < \tau_2$.

(iii) If $p \in U_\tau(q)$, then $q \in U_\tau(p)$.

The collection $\{U_\tau(p)\}$ is said to be a dyadic basis.

It should be clear that, in a metric space S, the collection of all spherical neighborhoods of radius τ, for all dyadic fractions τ satisfying $0 < \tau \leq \frac{1}{2}$, forms a dyadic basis.

THEOREM 5.7. Every metrizable space S has a dyadic basis.

In order to lay the groundwork for the existence of a dyadic basis, we first define what we mean by a τ chain, and then establish, in several lemmas, some important properties of these chains.

DEFINITION 5.8. Let S be a space having a K basis σ, and let τ be any dyadic fraction satisfying $0 < \tau < 1$. Then we may express τ uniquely in the form $\tau = \sum_{j=1}^{m} \frac{1}{2^{n_j}}$, where $n_1 < n_2 < \cdots < n_m$. By a τ chain of points from p to q in S, we mean a finite collection of points $p_1 = p$,

$p_2, \cdots, p_m, p_{m+1} = q$ of S, such that $p_i \in V_{n_{i-1}}(p_{i-1})$ for $i = 2, 3, \cdots, m+1$. Here, of course, the sets $V_{n_i}(p_i)$, for $i = 1, 2, \cdots, m$, are elements of the basis σ.

It should be carefully noted that a τ chain of points from p to q is not necessarily a τ chain of points from q to p.

LEMMA 5.9. Let τ be a dyadic fraction of the form $(2h + 1)/2^{n+1}$, and p, q be points of S. Define $\tau_3 = h/2^n$. Then there exists a τ chain from p to q if and only if there exists a point y such that $q \in V_{n+1}(y)$ and there exists a τ_3 chain of points from p to y.

Proof. Suppose first that such a point y exists, and let $p_1 = p$, p_2, $\cdots, p_f = y$ be the τ_3 chain joining p to y. Since $q \in V_{n+1}(y)$, the points $p_1 = p, p_2, \cdots, p_f, q$ clearly form a τ chain from p to q.

If, on the other hand, there exists a τ chain $p_1 = p$, $p_2, \cdots, p_{f-1}$, $p_f = q$, we may define $y = p_{f-1}$. Now $\tau = \tau_3 + 1/2^{n+1}$; hence $q \in V_{n+1}(y)$. Clearly the points $p_1 = p$, $p_2, \cdots, p_{f-1} = y$ form the required τ_3 chain from p to y.

LEMMA 5.10. Let $\tau_1 = k/2^m$, $\tau_2 = (k + 1)/2^m$ be dyadic fractions such that $(k + 1)/2^m \leq \frac{1}{2}$, and let p, q be points of S. Suppose that there exists a point z of S and a τ_1 chain of points from p to z. Then, if $q \in V_m(z)$, there is a τ_2 chain of points from p to q.

Proof. The proof we give is by mathematical induction on the integer m. The smallest value of m for which the statement has meaning is $m = 2$, and for this case we must have $k = 1$. We must thus show that, if there exists a point z of S and $1/2^2$ chain of points from p to z [i.e., if $y \in V_2(p)$] and if $q \in V_2(y)$, then there exists a $\frac{1}{2}$ chain of points from p to q; i.e., $q \in V_1(p)$. This is immediate from Definition 5.1*d*.

Suppose now that the lemma has been proved for $m = 1, 2, \cdots, n$. We shall show that, under this assumption, it must also be true for $m = n + 1$. We suppose that there exists a $k/2^n$ chain from p to z and that $q \in V_{n+1}(z)$. If k is even, it is immediate that there exists a $(k + 1)/2^{n+1}$ chain from p to q: namely, the chain from p to z with the point q adjoined to it.

There remains for consideration the case when k is odd: i.e., $k = 2h + 1$. We know that there exists a $(2h + 1)/2^{n+1}$ chain from p to z, and that $q \in V_{n+1}(z)$, and must show that there exists a $(2h + 2)/2^{n+1} = (h + 1)/2^n$ chain from p to q. By Lemma 5.9, there exists a point y such that $z \in V_{n+1}(y)$, and there exists an $h/2^n$ chain from p to y. From the hypothesis that $q \in V_{n+1}(z)$ and Definition 5.1*d*, we see that $q \in V_n(y)$. It follows from our inductive assumption that there must exist an $(h + 1)/2^n$ chain from p to q. This completes the proof of Lemma 5.10.

We note that, for any integer m, $q \in V_m(q)$. Thus we have an immediate application of Lemma 5.10.

LEMMA 5.11. Let $\tau_1 = k/2^m$, $\tau_2 = (k+1)/2^m$ be dyadic fractions such that $(k+1)/2^m \leq \frac{1}{2}$, and let p, q be points of S. If there exists a τ_1 chain from p to q then there must exist a τ_2 chain from p to q.

Our next lemma is a sharpening of the result of Lemma 5.11.

LEMMA 5.12. Let τ_1 and τ_2 be dyadic fractions satisfying $\tau_1 < \tau_2 \leq \frac{1}{2}$, and p, q points of S. Then, if there exists a τ_1 chain from p to q, there must also exist a τ_2 chain from p to q.

Proof. Since τ_1 and τ_2 are dyadic fractions, there must exist integers h, m, t, such that $\tau_1 = h/2^m$, $\tau_2 = t/2^m$, and $h < t$. There exists an integer j such that $h + j + 1 = t$. Consequently, $h/2^m < (h+1)/2^m < (h+2)/2^m < \cdots < (h+j)/2^m < t/2^m$. The proof is then easily completed by successive applications of Lemma 5.11.

THEOREM 5.13. Every space S that has a K basis has a dyadic basis at each of its points.

Proof. For each point p of S and each dyadic fraction τ satisfying $0 < \tau \leq \frac{1}{2}$, we define a set $U_\tau(p)$ in the following way. A point q of S is an element of $U_\tau(p)$ if and only if there exists a point r of S satisfying the following conditions:

(*a*) There is a τ chain from p to r in S.

(*b*) There is a τ chain from q to r in S.

We shall show that, for each p, the collection of sets $\{U_\tau(p)\}$ is a dyadic basis at p. To do this we must show that conditions (i), (ii), (iii) of Definition 5.6 are satisfied. Now (iii) follows at once from the definition of the set $U_\tau(p)$, whereas (ii) is immediate from Lemma 5.12. We shall show that (i) holds.

We first show that the set $U_\tau(p)$ is open. To do this, let $q \in U_\tau(p)$, and let r be a point of S satisfying conditions (*a*) and (*b*). Let $q_1 = q$, $q_2, \cdots, q_{m+1} = r$ be a τ chain from q to r in S. Then $q_2 \in V_{n_1}(q)$; hence $q \in V_{n_1}(q_2)$. It is clear that any point x of the neighborhood V_{n_1} of q can be joined to r by a τ chain of points of S. Thus this neighborhood of q is contained in $U_\tau(p)$. Therefore, $U_\tau(p)$ is an open set.

To complete the proof of (i), it is sufficient to show that $U_{1/2^{n+1}}(p) \subset V_n(p)$. Let $q \in U_{1/2^{n+1}}(p)$. Then there exists a point $r \in S$ such that $r \in V_{n+1}(q)$ and $r \in V_{n+1}(p)$. Thus, by Definition 5.1, $q \in V_{n+1}(r)$ and hence $q \in V_n(p)$.

We now use the dyadic basis which we have constructed to define a metric for the set S, and then show that this metric is a metric for the space S. We first define a certain auxiliary mapping of $S \times S$ into R

and then use this mapping to define the actual metric. The following lemma introduces this auxiliary mapping, and establishes its fundamental properties.

LEMMA 5.14. Let S be a space that has a K basis, and define a mapping $f: S \times S \to R$ in the following way: (i) $f(p, q) = 0$ if $p = q$, (ii) if $p \neq q$, then $f(p, q)$ is the least upper bound of all τ such that q is not in $U_\tau(p)$. This mapping f has the following properties:

(*a*) $f(p, q) = f(q, p)$ for each point (p, q) of $S \times S$.

(*b*) $f(p, q) = 0$ implies $p = q$.

(*c*) The mapping f is continuous.

Proof. Condition (*a*) is immediate from the definition of the mapping f, and Definition 5.6(iii). For condition (*b*), suppose $f(p, q) = 0$. This means that $q \in U_\tau(p)$ for every dyadic fraction τ. Consequently, $q = p$.

We now establish the continuity of f at an arbitrary point (p, q) of $S \times S$. Note that there are two cases to be considered: namely, $f(p, q) = 0$, $0 < f(p, q) \leq \frac{1}{2}$.

Suppose first that $f(p, q) = 0$: i.e., that $p = q$. To establish the continuity of f at this point (p, p), it is sufficient to show that, given any integer m, there exists a neighborhood W of p such that $r \in W$, $s \in W$ implies $|f(r, s) - f(p, p)| < 1/2^m$. Such a neighborhood of p is immediately available, and we take $W = V_m(p)$, observing that $r \in V_m(p)$, $s \in V_m(p)$ implies $r \in U_{1/2^m}(s)$.

For the remaining case, we pick integers k and m such that

$$k/2^m < f(p, q) \leq (k + 1)/2^m \tag{1}$$

Let (r, s) be any point of $S \times S$ satisfying $r \in V_{m+1}(p)$, $s \in V_{m+1}(q)$. We shall show that

$$(2k - 1)/2^{m+1} < f(r, s) < (2k + 3)/2^{m+1} \tag{2}$$

It follows, from (1) and (2), that $|f(p, q) - f(r, s)| < 5/2^{m+1}$, which proves the continuity of f at the point (p, q).

To prove (2), suppose it false. In the first place, we might have $f(r, s) \leq (2k - 1)/2^{m+1}$, i.e., $r \in U_{(2k-1)/2^{m+1}}(s)$. Recalling that $r \in V_{m+1}(p)$, $s \in V_{m+1}(q)$, and observing that $(2k - 1)/2^{m+1} = k/2^m - 1/2^{m+1}$, we see easily that $q \in U_{k/2^m}(p)$, contradicting (1). A similar contradiction follows if we assume that $f(r, s) \geq (2k + 3)/2^{m+1}$. Thus (2) is proved, and the continuity of f is established.

We are now ready to prove our main result.

THEOREM 5.15. A topological space S is metrizable if and only if S has a K basis.

Proof. We have seen in Theorem 5.2 that every metrizable space has a K basis. To establish the other half of the theorem, we suppose that the given space S has a K basis, and shall make use of the continuous mapping $f: S \times S \rightarrow R$ defined in Lemma 5.14. We note, from the definition of this mapping, that $f(p, q) = f(q, p)$ for each point (p, q) of $S \times S$.

We now define a mapping $\rho: S \times S \rightarrow R$ in the following way: For $(p, q) \in S \times S$, $\rho(p, q)$ is the least upper bound of $|f(p, r) - f(q, r)|$ for all $r \in S$. We show first that ρ is a metric for the set S. This means that we must establish the following properties:

(*a*) $\rho(p, q) \geq 0$ for every $(p, q) \in S \times S$.

(*b*) $\rho(p, q) = \rho(q, p)$ for every $(p, q) \in S \times S$.

(*c*) $\rho(p, q) = 0$ if and only if $p = q$.

(*d*) $\rho(p, r) \leq \rho(p, q) + \rho(q, r)$ for every three points p, q, r of S.

Properties (*a*) and (*b*) are immediate from the definition of ρ. Consider (*c*). If $p = q$, then $\rho(p, q) = 0$ from the definition of ρ. For the converse, suppose $\rho(p, q) = 0$. Then we must have $f(p, r) - f(q, r) = 0$ for every $r \in S$. In particular, for $r = q$, we have $f(p, q) = 0$. Thus $p = q$, by Lemma 5.14. Property (*d*) follows easily from the following inequality for absolute values:

$$| f(p, s) - f(r, s) | \leq | f(p, s) - f(q, s) | + | f(q, s) - f(r, s) |$$

Consequently, ρ is a metric for the set S.

To complete the proof of Theorem 5.15, we show that ρ is a metric for the space S: that is, that the topology induced by the metric ρ is the same as the original topology of S. For $p \in S$, let $S_r(p)$ denote the spherical neighborhood using the metric ρ. Let m be an integer sufficiently large so that $5/2^{m+1} < r$. We know that $s \in V_{m+1}(s)$. Thus, by the proof of Lemma 5.14, if $q \in V_{m+1}(p)$, then $|f(p, s) - f(q, s)| < 5/2^{m+1}$; i.e., $\rho(p, q) < r$. Stated differently, $V_{m+1}(p) \subset S_r(p)$. Thus every set open in the topology generated by ρ is open in S.

Similarly, given a point p and an integer m, there will always exist an r such that $S_r(p) \subset V_m(p)$; namely $r = 1/2^{m+1}$. To show this, we shall prove that, for any point q not in $V_m(p)$, q is not in $S_{1/2^{m+1}}(p)$. Since q is not in $V_m(p)$, it cannot be in $V_{m+1}(p)$, by Lemma 5.3. Consequently, q cannot be joined to any point of S by a $1/2^{m+1}$ chain. Therefore, q is not in $U_{1/2^{m+1}}(p)$; hence $f(p, q) \geq 1/2^{m+1}$. Consequently, by Problem 61, $\rho(p, q) \geq 1/2^{m+1}$; thus q is not in $S_{1/2^{m+1}}(p)$. Therefore, every set open in S is open in the topology generated by ρ. It follows that ρ is a metric for the space S, and the proof of Theorem 5.15 is complete.

PROBLEMS

57. Show that every space that has a K basis is a T_1 space.

58. Prove that the expression for a dyadic fraction is unique.

59. What is the relationship between rational numbers of the form $h/2^m$, where h and m are integers, and dyadic fractions?

60. For the collection $\{U_\tau(p)\}$ of Theorem 5.13, show that $\overline{U_{\tau_1}(p)} \subset U_{\tau_2}(p)$ if $\tau_1 < \tau_2$.

61. Using the notation of Theorem 5.14, show that $\rho(p, q) \geq f(p, q)$ for every element (p, q) of $S \times S$.

62. Let S be a space. Then S is said to have a K_1 basis if and only if there is associated with each point p of S and each positive integer n a subset $V_n(p)$, and these subsets satisfy the following conditions:

(1) The collection $\{V_n(p)\}$ for fixed p is a basis at p.

(2) $\bigcap_{n \in I} V_n(p) = p$.

(3) Given m, $n \in I$, there exists $k \in I$ such that $V_k(p) \subset V_m(p) \cap V_n(p)$ for every point p of S.

(4) Given $n \in I$, there exists $m \in I$ such that $p \in V_m(r)$, $q \in V_m(r)$ implies $q \in V_n(p)$.

The collection of all the sets $\{V_n(p)\}$ for $p \in S$, $n \in I$ is called a K_1 basis for S. Prove that S has a K basis if and only if it has a K_1 basis.

63. Let S be a space and Σ a non-empty index set. Then S is said to be a uniform space if and only if there is associated with each point p of S and each element α of Σ a subset $V_\alpha(p)$, and these subsets satisfy the following conditions:

(1) The collection $\{V_\alpha(p)\}$ for fixed p is a basis at p.

(2) $\bigcap_{\alpha \in \Sigma} V_\alpha(p) = p$.

(3) Given α, $\beta \in \Sigma$, there exists $\lambda \in \Sigma$ such that $V_\lambda(p) \subset V_\alpha(p) \cap V_\beta(p)$ for every $p \in S$.

(4) Given $\alpha \in \Sigma$, there exists $\beta \in \Sigma$ such that $p \in V_\beta(r)$, $q \in V_\beta(r)$ implies $q \in V_\alpha(p)$.

Show that every uniform space is a regular Hausdorff space.

6. The Metrization of Regular, Second Countable T_1 Spaces

We shall establish in this section the important result that every regular, second countable T_1 space is metrizable. It follows from Theorem 4.25 that every such space is normal. As a first step in the argument establishing the metrizability of such a space, we give a result that characterizes normal spaces by means of mappings.

Throughout this section, the letter J will denote the closed unit interval [0, 1] considered as a subspace of the space R of all real numbers.

THEOREM 6.1. A space S is normal if and only if, given any two disjoint closed subsets A and B of S, there exists a continuous mapping $f: S \to J$, such that $f(x) = 0$ if $x \in A$, and $f(x) = 1$ if $x \in B$.

Proof. Suppose first that the condition holds, i.e., that such a mapping $f: S \to J$ has been defined for two given disjoint closed subsets of S. Let J_1 and J_2 be the open subsets of the space J consisting of the intervals $0 \leq x < \frac{1}{4}$ and $\frac{3}{4} < x \leq 1$, respectively. Define $U = f^{-1}(J_1)$, $V = f^{-1}(J_2)$. Clearly, U and V are disjoint open subsets of S containing A and B, respectively. Thus S is normal.

To prove the converse, suppose that S is normal and that the two disjoint closed subsets A and B of S are given. By normality, there exist disjoint open sets U and Z of S containing A and B, respectively. The set Z, being open, can contain no limit point of U. Consequently, we must have $A \subset U \subset \bar{U} \subset S - B$. We shall keep this set U fixed throughout the proof.

Our next step will be to define for each dyadic fraction τ (Definition 5.5) satisfying $0 < \tau < 1$ an open set U_τ. The collection of all the open sets $\{U_\tau\}$ will satisfy the following condition:

1. If $\tau_1 < \tau_2$, then $A \subset U \subset \bar{U} \subset U_{\tau_1} \subset \bar{U}_{\tau_1} \subset U_{\tau_2} \subset \bar{U}_{\tau_2} \subset S - B$.

To define this collection of sets, we begin by observing that $\bar{U}$ and B are disjoint closed subsets of S. Precisely as we defined the set U, we can define a set $U_{1/2}$ such that $\bar{U} \subset U_{1/2} \subset \bar{U}_{1/2} \subset S - B$. This definition of the set $U_{1/2}$ is the first step in our inductive definition of the collection of sets $\{U_\tau\}$. We observe that every dyadic fraction may be written in the form $h/2^m$, where h and m are positive integers such that $h < 2^m$. We have defined $U_{h/2^m}$ for all integers h and m such that $m = 1$ and $0 < h/2^m < 1$. We suppose now that $U_{h/2^m}$ has been defined for all integers h and m such that $m = 1, 2, \cdots, k$ and $0 < h/2^m < 1$ so as to satisfy condition (1) and shall show how to define $U_{h/2^m}$ for all τ of the form $h/2^{k+1}$, $0 < h/2^{k+1} < 1$ so as to satisfy condition (1). Since $2/2^{k+1} = 1/2^k$, the set $U_{2/2^{k+1}}$ has already been defined. The sets $\bar{U}$ and $S - U_{2/2^{k+1}}$ are disjoint closed subsets of S. We may thus find an open set, which we denote by $U_{1/2^{k+1}}$ such that $\bar{U} \subset U_{1/2^{k+1}} \subset \bar{U}_{1/2^{k+1}} \subset U_{2/2^{k+1}}$. In exactly the same manner we may define an open set $U_{(2^{k+1}-1)/2^{k+1}}$ to satisfy $\bar{U}_{(2^{k+1}-2)/2^{k+1}} \subset U_{(2^{k+1}-1)/2^{k+1}} \subset \bar{U}_{(2^{k+1}-1)/2^{k+1}} \subset S - B$. We may now define $U_{h/2^{k+1}}$ for the remaining values of h: namely, the odd integers h such that $1 < h < (2^{k+1} - 1)$. This is easily done, by defining $U_{h/2^{k+1}}$ as an open set satisfying $\bar{U}_{(h-1)/2^{k+1}} \subset U_{h/2^{k+1}} \subset \bar{U}_{h/2^{k+1}} \subset U_{(h+1)/2^{k+1}}$. Thus, by mathematical induction, we have defined an open set U_τ for every dyadic fraction τ satisying $0 < \tau < 1$. Condition (1) is fulfilled.

The mapping $f: S \to J$ required by the theorem may now be defined in the following way. If $x \in U_\tau$ for every τ, then we set $f(x) = 0$. Otherwise, we define $f(x)$ as the least upper bound of all τ such that x is not in U_τ. Clearly, $f(x) = 0$ for $x \in A$ and $f(x) = 1$ for $x \in B$.

To complete the proof, we must establish the continuity of the mapping just defined; i.e., we must show that f is continuous at each point of S. We leave to the student the details of the proof for the cases where $f(p) = 0$ and $f(p) = 1$, and consider only the case where $0 < f(p) < 1$. For any real number $\varepsilon > 0$, we can find two dyadic fractions τ_1 and τ_2 satisfying $f(p) - \varepsilon < \tau_1 < f(p) < \tau_2 < f(p) + \varepsilon$. The set $U_{\tau_2} - \bar{U}_{\tau_1}$ is a neighborhood W of p having the property that, for any point q of W, $|f(p) - f(q)| < \varepsilon$. Thus f is continuous at the point p, and the proof of the theorem is complete.

The concept of complete regularity is related to mappings into J such as we have just constructed.

Definition 6.2. A space S is said to be completely regular if and only if, given any point p of S and any closed subset M of S not containing p, then there exists a continuous mapping $f: S \to J$ such that $f(p) = 0$ and $f(x) = 1$ for every x in M.

We now state a corollary to Theorem 6.1 in terms of complete regularity.

Corollary 6.3. Every normal T_1 space is completely regular. Also, every completely regular space is regular.

Example VII, 3.5 shows that there exist completely regular T_1 spaces which are not normal. It is also true that there exist regular T_1 spaces which are not completely regular. Thus the concept of complete regularity fits between that of normality and that of regularity for T_1 spaces.

Theorem 6.4. Every regular, second countable, T_1 space S is metrizable.

Proof. Let $\{V_n\}$ be a countable basis for the space S. Given any integer k and any point p of V_k, we know, since S is regular, that there exists a neighborhood U of p such that $p \in U \subset \bar{U} \subset V_k$. Thus, since $\{V_n\}$ is a basis for S, there must exist an integer j such that $p \in V_j \subset \bar{V}_j \subset V_k$. We shall say that (V_r, V_s) is a *pair* if and only if V_r and V_s are elements of the basis $\{V_n\}$ satisfying $\bar{V}_r \subset V_s$. We have seen that, given any integer k, there exists an integer j such that (V_j, V_k) is a pair.

The collection of all such pairs is countable by Theorem I, 7.3; hence we may index this collection with the set of positive integers. Let $P_n = (V_j, V_k)$ be the nth pair under this indexing. For this pair P_n, the sets $\bar{V}_j$ and $S - V_k$ are disjoint closed subsets of S. Since S is normal, there must exist, by Theorem 6.1, a continuous mapping $f_n: S \to J$ having the property that $f_n(x) = 0$ for $x \in \bar{V}_j$ and $f_n(x) = 1$ for $x \in S - V_k$.

The sequence of functions $\{f_n(x)\}$ leads naturally to the definition of a metric for the set S. We define a mapping $\rho: S \times S \to R$ by the equation

$$\rho(x, y) = \sum_{n \in I} |f_n(x) - f_n(y)|/2^n$$

It should be noted that this infinite series of non-negative real numbers is convergent since its nth term cannot exceed the nth term of the convergent geometric series $\sum_{n \in I} 1/2^n$. Thus, we have at once the following two properties for $\rho(x, y)$:

1. $0 \leq \rho(x, y) < \infty$.
2. $\rho(x, y) = \rho(y, x)$ for any pair of points x, y of S.

For any three points x, y, z of S, it is immediate that

$$|f_n(x) - f_n(z)| \leq |f_n(x) - f_n(y)| + |f_n(y) - f_n(z)|$$

Thus we have the triangle property:

3. $\rho(x, z) \leq \rho(x, y) + \rho(y, z)$ for $x, y, z \in S$.

To complete the proof that ρ is a metric for the *set* S, we need only show that the following property holds:

4. $\rho(x, y) = 0$ if and only if $x = y$.

Clearly, if $x = y$, then $\rho(x, y) = 0$. To prove the converse, suppose that $x \neq y$. Then there exists an integer k such that $x \in V_k$, $y \in S - V_k$. For this integer k, there exists an integer j such that (V_j, V_k) is a pair, and $x \in V_j$. Thus, for some integer N, $P_N = (V_j, V_k)$. It is immediate that $f_N(x) = 0$, $f_N(y) = 1$; hence $\rho(x, y) \geq 1/2^N > 0$. Thus (4) holds, and ρ is a metric for the set S.

It remains to be shown that ρ is a metric for the space S: i.e., that the topology induced by ρ is the same as the original topology of the space S. We show first that, for $p \in S$, and $\delta > 0$, the spherical neighborhood $S_\delta(p)$ contains a neighborhood of p in the topology of S. We begin by picking an integer N sufficiently large so that $1/2^N < \delta/2$. We see at once that, for any $x, y \in S$,

$$\sum_{i=N+1}^{\infty} \frac{|f_i(x) - f_i(y)|}{2^i} < \frac{\delta}{2}$$

Now each of the functions $f_i(x)$ for $i = 1, 2, \cdots, N$ is continuous at the point p. Thus, for each i, there exists a neighborhood U_i of p such that $q \in U_i$ implies $|f_i(p) - f_i(q)|/2^i < \delta/2N$. Define $U = \bigcap_{i=1}^{N} U_i$. Then U is an open set containing p. For any point $q \in U$, we must have $|f_i(p) - f_i(q)|/2^i < \delta/2N$ for $i = 1, 2, \cdots, N$; hence

$$\sum_{i=1}^{N} \frac{|f_i(p) - f_i(q)|}{2^i} < \frac{\delta}{2}$$

From these inequalities we have $\rho(p, q) < \delta$ for $q \in U$; i.e., $p \in U \subset S_\delta(p)$. If $q \in S_\delta(p)$, then, from the triangle property, there exists $\alpha > 0$ such that $S_\alpha(q) \subset S_\delta(p)$. From the above argument, there then exists a neighborhood V of q such that $q \in V \subset S_\alpha(q) \subset S_\delta(p)$. Thus $S_\alpha(p)$ is open in the original topology.

To complete the proof that ρ is a metric for the space S, it will suffice to show that, given any point $p \in S$ and any neighborhood U of p, there exists a real number $\alpha > 0$ such that $S_\alpha(p) \subset U$. To do this, we shall show that there exists $\alpha > 0$ such that $q \in S - U$ implies $\rho(p, q) \geq \alpha$. We know that there exists a pair $P_h = (V_j, V_k)$ such that $p \in V_j \subset \bar{V}_j \subset V_k \subset U$. Thus, for $q \in S - U$, $\rho(p, q) \geq |f_h(p) - f_h(q)|/2^h = 1/2^h$, and we define $\alpha = 1/2^h$. We have shown that $q \in S - U$ implies q not in $S_\alpha(p)$. Thus $S_\alpha(p) \subset U$; hence U is open in the topology induced by ρ.

PROBLEMS

64. Show that every uniform space (see Problem 63) and, consequently, every metric space is completely regular. [*Hint*. Establish the existence of a countable subcollection $\{V_n(p)\}$ of $\{V_\alpha(p)\}$ such that

(1) If $q \in V_n(p)$, then $p \in V_n(q)$;

(2) Given $V_{n+1}(p)$. If $q \in S$ and there exists a point $r \in V_{n+1}(p)$ such that $q \in V_{n+1}(r)$, then $q \in V_n(p)$;

(3) $V_1(p) \subset S - M$, where M is the closed set disjoint with p;

and apply the methods of this section.]

65. Give another proof that every metric space is completely regular by making explicit use of the metric.

66. Show that every countably compact, second countable Hausdorff space is metrizable.

7. Extension of Mappings

Our discussion of the extension of mappings begins with an informal discussion of the function $f(x) = 2^x$, where x is a real number. Exactly how is the mapping $f: R \to R$, where $f(x) = 2^x$ defined; i.e., precisely what do we mean by the symbol 2^x, where x is a real number? We recall that, for any positive integer n,

$$2^n = \underbrace{2 \cdot 2 \cdot \cdots \cdot 2}_{n \text{ terms}}$$

$$2^0 = 1$$

$$2^{-n} = 1/2^n$$

Similarly, for any integer $q > 0$, $2^{1/q}$ is defined to be the unique positive root of the equation $z^q = 2$. For any positive rational number p/q, we define $2^{p/q} = (2^{1/q})^p$. If x is a negative rational, then, by definition, $2^x = 1/2^{-x}$.

Let S be the subset of R consisting of all rational numbers. We have thus far defined a mapping $g: S \to R$ such that $g(x) = 2^x$ for $x \in S$. It is certainly true that $f(x) = g(x)$ for $x \in S$, but the question remains as to what we mean by $f(x) = 2^x$ when x is an irrational. One way in which 2^x might be defined for x irrational would be to form a sequence $\{a_n\}$ of rationals which converges to the point x, and consider the sequence $\{2^{a_n}\}$. If it were proved that this sequence converged to a real number, then we could define 2^x as the limit to which the sequence $\{2^{a_n}\}$ converged. This would still not be a satisfactory definition unless we could show that the limit of the sequence $\{2^{a_n}\}$ is independent of the choice of the sequence $\{a_n\}$ converging to x. In other words, we would have to prove that, for any two sequences $\{a_n\}$ and $\{b_n\}$, each of which converges to x, the sequences $\{2^{a_n}\}$ and $\{2^{b_n}\}$ both converged and had the same limit. Even after proving all these things, we would still not know, without further investigation, whether or not the mapping $f: R \to R$ defined by $f(x) = 2^x$ was continuous.

Equivalent to this method of defining $f(x) = 2^x$ for x irrational is that of showing that the mapping g defined above is continuous, and that there exists one and only one continuous mapping $f: R \to R$ such that the restriction of f to S is g. The image of the real number x under this uniquely determined mapping f is then, by definition, 2^x. One advantage of this second method of definition is that the continuity of the mapping f is obtained at once. Since we started with 2^x defined for rational numbers x, and then extended our definition to include numbers x which are irrational, we think of f as an extension of g. More precisely, we have the following definition.

Definition 7.1. Let S and T be sets, A and B subsets of S such that $A \subset B$, and $g: A \to T$ a mapping. Then a mapping $f: B \to T$ is said to be an extension of g (from A to B) if and only if $f(x) = g(x)$ for every $x \in A$: that is, if and only if the restriction of f to A is g. If S and T are spaces, and g is continuous, then a mapping $f: B \to T$ is said to be a continuous extension of g from A to B if and only if f is continuous and is an extension of g from A to B.

We have not indicated any proof that the function $g(x) = 2^x$ defined on the space of all rational numbers S is continuous on S nor that it possesses a continuous extension to the space of all real numbers R. Before giving the details of such a proof, we give an example. Let K be the set of all points x of R such that $x \neq 2$, and consider the mapping $g: K \to R$ defined by $g(x) = 0$ if $x < 2$, $g(x) = x$ if $x > 2$. This mapping is easily seen to be continuous. To see that no extension f of g from K to R could be continuous, note that the sequences $\{f(a_n)\}$ and $\{f(b_n)\}$

where $a_n = 2 - 1/n$, $b_n = 2 + 1/n$ for $n \in I$ converge to 0 and 2, respectively, while each of the sequences $\{a_n\}$ and $\{b_n\}$ converges to 2. Closely related to this phenomenon is the fact that

$$\delta(g(U \cap K)) > 2$$

for every neighborhood U of 2 in the space R. The greatest lower bound of these diameters is 2, and this bound is known as the oscillation of the mapping g at the point 2.

DEFINITION 7.2. Let A be a subset of a topological space S, T a metric space, and $f: A \to T$ a mapping. The oscillation of f at a point p of S, written $\omega(p)$, is defined to be the greatest lower bound of $\delta(f(U \cap A))$ for all neighborhoods U of p in the space S.

It is important that the student understand that the oscillation of a mapping f is defined at every point of the space, including the points where the mapping f itself is not defined.

We list in the next theorem results that follow easily from Definitions 7.1 and 7.2.

THEOREM 7.3. Let A and B be subsets of a space S such that $A \subset B$. Let T be a metric space and $g: A \to T$ a mapping. Suppose that $f: B \to T$ is an extension of the mapping g from A to B. Then

(*a*) The mapping g is continuous at a point p of A if and only if the oscillation of g at the point p is zero.

(*b*) If $\omega_f(p)$ and $\omega_g(p)$ denote, respectively, the oscillations of f and g at a point p of S, then $\omega_f(p) \geq \omega_g(p)$.

(*c*) If there exists a point p of B for which $\omega_g(p) > 0$, then there exists no continuous extension of g from A to B.

We return to our consideration of the function 2^x. Let S be the subspace of R consisting of all the rationals. We recall that we have defined a mapping $g: S \to R$, by $g(x) = 2^x$, for each $x \in S$. In order to have any hope of proving that this mapping has a continuous extension from S to R, we must show, as a consequence of Theorem 7.3*c*, that the oscillation of g is zero at each point x of R. The proof that this is the case can be based on very elementary considerations. We note that, for $n \in I$, the first two terms of the binomial expansion of $(1 + 1/n)^n$ are $1 + 1$. Thus $(1 + 1/n)^n \geq 2$, and it follows at once that, for $n \in I$,

$$1 \leq 2^{1/n} \leq 1 + 1/n$$

Let x be a real number, $\varepsilon > 0$ a positive number, and N an integer greater than the larger of the two numbers x and 1. Since x is a real number, there must exist integers p, q, such that $q > 2^N/\varepsilon$ (hence $2^N/q < \varepsilon$) and $p/q < x < (p+1)/q$. Then, for any pair of rational numbers r_1 and r_2 satisfying $p/q < r_i < (p+1)/q$, for $i = 1, 2$, we must have

$$|2^{r_1} - 2^{r_2}| \le |2^{(p+1)/q} - 2^{p/q}|$$
$$= 2^{p/q}|1 - 2^{1/q}| \le 2^N|1 - 2^{1/q}| < 2^N/q < \varepsilon$$

This proves that the oscillation of g at the point x is less than ε. Consequently, since $\varepsilon > 0$ is arbitrary, the oscillation at the point x is zero. This shows not only that g is continuous at each rational point of R but also that the necessary condition of Theorem 7.3*c* is satisfied.

To obtain a sufficient condition for the continuous extension of a continuous mapping, we introduce and study the concept of equivalent Cauchy sequences.

DEFINITION 7.4. Let S be a metric space with metric ρ. Then two Cauchy sequences $\{x_n\}$ and $\{y_n\}$ of points of S are equivalent if and only if, given $\varepsilon > 0$, there exists an integer N such that $n > N$ implies $\rho(x_n, y_n) < \varepsilon$.

The following three theorems are easily established.

THEOREM 7.5. Let S be a metric space, then

(i) Any Cauchy sequence of S is equivalent to itself.

(ii) If $\{x_n\}$ and $\{y_n\}$ are Cauchy sequences of S, then $\{x_n\}$ is equivalent to $\{y_n\}$ if and only if $\{y_n\}$ is equivalent to $\{x_n\}$.

(iii) Let $\{x_n\}$, $\{y_n\}$, $\{z_n\}$ be Cauchy sequences of S. If $\{x_n\}$ is equivalent to $\{y_n\}$ and $\{y_n\}$ is equivalent to $\{z_n\}$, then $\{x_n\}$ is equivalent to $\{z_n\}$.

THEOREM 7.6. Let S be a metric space and $p \in S$. If the Cauchy sequence $\{x_n\}$ of S converges to the point p, then a sequence of points $\{y_n\}$ of S converges to the point p if and only if $\{x_n\}$ and $\{y_n\}$ are equivalent Cauchy sequences.

THEOREM 7.7. Let S and T be metric spaces, A a subset of S, and $f: A \to T$ a mapping. If the oscillation of f at a point p of S is zero, and $\{x_n\}$ and $\{y_n\}$ are sequences of points of A, each converging to p, then $\{f(x_n)\}$ and $\{f(y_n)\}$ are equivalent Cauchy sequences.

Our main theorem on continuous extensions now follows quite easily.

THEOREM 7.8. Let S and T be metric spaces, T complete. Suppose that A is a subset of S such that $\bar{A} = S$, and that $g: A \to T$ is a continuous mapping. Then there exists a continuous extension f of g from A to S if and only if $\omega(p)$, the oscillation of g, is zero at every point of S. Furthermore, there is at most one continuous extension of g from A to S.

Proof. It follows from Theorem 7.3 that, if a continuous extension exists, then $\omega(p) = 0$ for $p \in S$. To see that there is at most one continuous extension, suppose that f and h are continuous extensions of g from A to S. Then $f(x) = h(x)$ for $x \in A$. For any point $p \in S - A$, we know that $p \in \bar{A} - A$; hence there exists a sequence of points $\{a_n\}$ of A which converges to the point p. Since f and h are continuous, the sequences $\{f(a_n)\}$ and $\{h(a_n)\}$ converge to the points $f(p)$ and $h(p)$, respectively. But $f(a_n) = h(a_n)$ for every n; hence $f(p) = h(p)$. Consequently, $f(x) = h(x)$ for every $x \in S$.

It remains to establish, under the assumption that $\omega(p) = 0$ for $p \in S$, the existence of a continuous extension f of g from A to S. Let x be any point of $S - A$. Then there exists a sequence $\{a_n\}$ of points of A which converges to x. But $\omega(x) = 0$; thus, by Theorem 7.7, the sequence $\{g(a_n)\}$ is a Cauchy sequence of points of the complete space T. There must thus exist a point y of T such that $\{g(a_n)\}$ converges to y. Moreover, since $\omega(x) = 0$, the image of any sequence of points of A which converges to x is a sequence of points of T which converges to y. This means that the point y of T which we have found is independent of the sequence of points of A which we chose converging to x. We define $f(x) = y$. This defines f for each point x of $S - A$. For $x \in A$, we define $f(x) = g(x)$. Thus we have defined a mapping $f\colon S \to T$ which is an extension of the mapping g from A to S.

We now show that the mapping f just defined is continuous on S. By virtue of Theorem 7.3*a*, this will follow if we show that the oscillation of f at each point of S is zero. Knowing that the oscillation of g at each point of S is zero, it follows that we need only show that, if p is any point of S, and U is any neighborhood of p, then $\delta(f(U)) = \delta(g(U \cap A))$. This is immediate, from the facts that $g(U \cap A) \subset f(U) \subset \overline{g(U \cap A)}$ and the diameter of any subset of a metric space equals the diameter of its closure. This completes the proof of Theorem 7.8.

This theorem applies in the case of the function $g(x) = 2^x$ for all rational numbers x, since the oscillation of this function g is zero for each point of R. Thus there exists one and only one continuous mapping $f\colon R \to R$ which is an extension of this g. We define 2^x, for every irrational number x, by the equation $2^x = f(x)$.

EXAMPLE 7.9. Consider the function $y = (x^2 + 2x + 1)/(x + 1)$, defined for all real numbers $x \neq -1$. Clearly, for all such x, $y = x + 1$. The oscillation of this function is zero for all real numbers x. Hence, by convention, analysts usually agree that the unique continuous extension, $y = x + 1$, of this function is to be used to define y for all real values of x. This convention simplifies the details of many discussions.

The next example show that the hypothesis of completeness is essential in Theorem 7.8.

EXAMPLE 7.10. Let S be the subspace of R consisting of all rational numbers, and note that S is a metric space satisfying $\bar{S} = R$. The mapping $g: S \to S$ defined by $g(x) = x$ for each $x \in S$ is continuous, and the oscillation of this mapping is zero at every point of R. Thus all the hypotheses of Theorem 7.8 are satisfied except the completeness of S. The student should be able to show that the mapping g possesses no continuous extension from S to R.

PROBLEMS

67. Let q be a positive integer. Show that there is one and only one positive real number z such that $z^q = 2$.

68. Let $\{x_n\}$ and $\{y_n\}$ be two Cauchy sequences of points of a metric space S with metric ρ. Prove the following statements:

(i) $\{x_n\}$ and $\{y_n\}$ are equivalent Cauchy sequences if and only if the sequence $\{\rho(x_n, y_n)\}$ of real numbers converges to zero.

(ii) $\{x_n\}$ and $\{y_n\}$ are equivalent Cauchy sequences if and only if the sequence $\{z_n\}$ of points of S, where $z_{2n} = x_n$ for every n and $z_{2n-1} = y_n$ for every n, is a Cauchy sequence.

69. Let $\{x_n\}$ and $\{y_n\}$ be two Cauchy sequences of a metric space S with metric ρ. Prove that the sequence $\{\rho(x_n, y_n)\}$ of real numbers is a Cauchy sequence.

70. Let R_+ denote the subspace of the space R consisting of all x such that $x > 0$. Show that there exists one and only one continuous mapping $f: R_+ \to R$ which has the properties

(i) $f(xy) = f(x) + f(y), \qquad x, y \in R_+$.

(ii) $f(10) = 1$.

(compare this function with $\log_{10} x$.)

71. Prove Theorem 7.3.

72. Prove Theorem 7.5.

73. Prove Theorem 7.6.

74. Prove Theorem 7.7.

8. Function Spaces

To introduce the idea of a function space, we discuss briefly the solution of two problems in analysis. Consider first the problem of finding a root of the equation

$$x^5 + 3x + 4 = 0$$

A standard approach to this problem is to use Newton's method of approximating a root. Let x_n be the nth such approximation. The sequence $\{x_n\}$, thus formed by Newton's method, usually turns out to be a Cauchy sequence of real numbers, and thus converges to a real number. Furthermore, this real number is a root of the given equation. To describe this process slightly differently, let $f\colon R \to R$ be the continuous mapping defined by $f(x) = x^5 + 3x + 4$ for $x \in R$. To say that the approximations x_n "improve with n" is a crude way of making the statement that the sequence $\{f(x_n)\}$ converges to zero. If the sequence $\{x_n\}$ is a Cauchy sequence, it must converge to a real number x_0; hence $\{f(x_n)\}$ converges to $f(x_0)$, since $f(x)$ is continuous. But the approximations "improve with n," hence $f(x_0) = 0$ and x_0 is a root.

The second problem we consider is that of finding a function $y(x)$ that satisfies the first-order differential equation

$$\frac{dy}{dx} = g(x, y)$$

where g is a function of the pair of real numbers (x, y). One method of solving this problem is to find a process analogous to Newton's method which will give rise to a sequence $\{y_n\}$ of approximate solutions (Picard's method, for example). Of course, we want the approximations to "improve with n"; i.e., we want the sequence of functions $\left\{\frac{dy_n(x)}{dx} - g(x, y_n(x))\right\}$ to converge to the function that is identically zero. Also, we would like to have the sequence $\{y_n\}$ converge to a solution of the differential equation.

To give the concept of convergence a meaning, we give a topology to the set of functions $y(x)$ for which the differential equation makes sense. Let X denote the topological space thus defined. There are many useful different topologies which can be assigned, but usually the resultant space is a metric space. Since its elements or points are functions, we commonly speak of it as a function space.

If we now consider the mapping $f\colon X \to X$ defined by $f(y) = \frac{dy(x)}{dx} - g(x, y(x))$ for each $y \in X$, and suppose that this mapping is continuous, then we see that our second problem has exactly the same structure as our first one. As in the first problem, we want our approximation process to be such that $\{f(y_n)\}$ converges to the identically zero function of X. And we want $\{y_n\}$ to be a Cauchy sequence of X which converges.

The discussion of these two problems in analysis gives an idea of the manner in which some of the concepts and theorems that we have given and will give can be applied. We wish to emphasize the economy of effort which can be achieved by establishing results that can be applied in many situations. For instance, in the interest of economy of effort, analysts have devoted a great deal of study to Banach spaces (and their generalizations, the linear spaces), since the spaces associated with many problems in analysis are function spaces of these types. Of course, to gain this economy, the similarity of problems or the fact that they fall into the same general pattern must first be recognized.

In this and the next section, we shall consider two particular function spaces and establish their completeness. One is the space of continuous functions on the closed unit interval. The other is the space of equivalent Cauchy sequences of a metric space.

THEOREM 8.1. Let J be the subspace of R consisting of all x in the closed unit interval $[0, 1]$, and denote by B the set of all continuous mappings $f: J \to R$. Then the mapping $\rho: B \times B \to R$ defined by

$$\rho(f, g) = \underset{x \in J}{\text{least upper bound}} \mid f(x) - g(x) \mid$$

is a metric for the set B.

Proof. Given two elements f, g, of B, we know, since J is compact, that there exists a real number M such that $| f(x) - g(x) | \leq M$ for every x in J. Thus $\rho(f, g)$ exists, and in fact, there exists a point $x_0 \in J$ such that $\rho(f, g) = | f(x_0) - g(x_0) |$.

Clearly,

(i) $\rho(f, g) = \rho(g, f)$ for every $f, g \in B$.

(ii) $\rho(f, g) \geq 0$.

(iii) $\rho(f, g) = 0$ if and only if $f(x) = g(x)$ for every $x \in J$; i.e., if and only if f and g are the same mapping.

The triangle property for the metric remains to be checked. Let $f, g, h \in B$. We know that there exists a point $x_0 \in J$ such that $\rho(f, h) = | f(x_0) - h(x_0) |$. Since $\rho(f, g) \geq | f(x_0) - g(x_0) |$ and $\rho(g, h) \geq | g(x_0) - h(x_0) |$, we have at once $\rho(f, g) + \rho(g, h) \geq | f(x_0) - g(x_0) | + | g(x_0) - h(x_0) | \geq | f(x_0) - h(x_0) | = \rho(f, h)$. Therefore, ρ is a metric for the set B.

DEFINITION 8.2. Let J be the subspace of R consisting of all real numbers in the closed unit interval $[0, 1]$. Then $B(J)$ denotes the metric space which has as its elements all continuous mappings $f: J \to R$, and whose metric is the one given in Theorem 8.1.

THEOREM 8.3. The metric space $B(J)$ is complete.

Proof. Let $\{f_n\}$ be a Cauchy sequence of points of $B(J)$. We shall show that there exists a point f of $B(J)$ to which this Cauchy sequence converges.

From the definition of a Cauchy sequence, we know that, given $\varepsilon > 0$, there exists an integer N such that $m, n > N$ implies $\rho(f_m, f_n) < \varepsilon/2$.

Consequently, for each $x \in J$, the sequence $\{f_n(x)\}$ is a Cauchy sequence of points of the complete space R. Consequently, for each $x \in J$, there exists a unique point of R to which the Cauchy sequence $\{f_n(x)\}$ converges. We denote this point by $f(x)$. Thus we have defined a mapping $f: J \to R$. This mapping is well defined, since, for each $x \in J$, the limit of the Cauchy sequence $\{f_n(x)\}$ is unique.

Before considering the convergence of the sequence $\{f_n\}$ to f, we must show that f is a point of the space $B(J)$: i.e., that f is continuous on J. To do this, we first show that, given $\varepsilon > 0$, there exists an integer N such that $n > N$ implies $|f_n(x) - f(x)| < \varepsilon$ for all $x \in J$. This is quite easy. Given $x \in J$, the sequence $\{f_n(x)\}$ is known to converge to the point $f(x)$. Consequently, there exists an integer M such that $m > M$ implies $|f_m(x) - f(x)| < \varepsilon/2$. Let N be the integer given in the previous paragraph, $n > N$, and m any integer larger than $M + N$. Then

$$|f_n(x) - f(x)| \leq |f_n(x) - f_m(x)| + |f_m(x) - f(x)| < \varepsilon \tag{1}$$

which is the inequality we desired to establish.

To establish the continuity of f, we let x_0 be a point of J, and $\varepsilon > 0$ a real number. By what we have just proved, there exists an integer K such that $|f_n(x) - f(x)| < \varepsilon/3$ for every $x \in J$ and $n > K$. In particular, $|f_{K+1}(x) - f(x)| < \varepsilon/3$. But f_{K+1} is a continuous mapping of J into R. There must thus exist a $\delta > 0$ such that $x_1 \in J$ and $|x_1 - x_0| < \delta$ implies $|f_{K+1}(x_1) - f_{K+1}(x_0)| < \varepsilon/3$. It follows that, for any point x_1 of J satisfying $|x_1 - x_0| < \delta$, we must have $|f(x_1) - f(x_0)| \leq |f(x_1) - f_{K+1}(x_1)| + |f_{K+1}(x_1) - f_{K+1}(x_0)| + |f_{K+1}(x_0) - f(x_0)| < \varepsilon$. Consequently, the mapping f is continuous; hence f is a point of the space $B(J)$.

To complete the proof, we need only show that the sequence $\{f_n\}$ of points of $B(J)$ converges to the point f of $B(J)$. This is immediate from (1), since, given $\varepsilon > 0$, there must exist an integer N such that $n > N$ implies $|f_n(x) - f(x)| < \varepsilon$, for all x in J. Thus, for $n > N$, $\rho(f_n, f) < \varepsilon$, since J is compact.

There are several observations to be made in connection with a Cauchy sequence $\{f_n\}$ of points of $B(J)$. Since a Cauchy sequence is bounded, we may easily deduce that there exists a real number M such that $|f_n(x)| \leq M$ for $x \in J$ and $n \in I$. This means that we have found a bound M that works for every f_n, and it is for this reason that we say that the sequence $\{f_n\}$ is uniformly bounded.

It is not true that every sequence of elements of $B(J)$ is uniformly bounded as we see at once by considering the sequence $\{h_n\}$ defined by $h_n(x) = n$, for $x \in J$.

Definition 8.4. Let S be a space and T a metric space with metric ρ. Denote by $\{f_\alpha\}$ a collection of mappings of S into T. This collection is said to be uniformly bounded if and only if there exists a real number M and a point p of T such that $\rho(f_\alpha(s), p) \leq M$ for every $s \in S$ and for all α.

Another property of the mappings f_n of a Cauchy sequence $\{f_n\}$ of points of $B(J)$ is that not only is each continuous at every point of J but also, given $\varepsilon > 0$ and a point $x_0 \in J$, there exists a real number $\delta > 0$ such that $x \in J$ and $|x - x_0| < \delta$ implies $|f_n(x) - f_n(x_0)| < \varepsilon$. That is, the choice of δ can be made independent of the index n.

Definition 8.5. Let S and T be metric spaces with metrics ρ_S and ρ_T, respectively, and denote by $\{f_\alpha\}$ a collection of mappings of S into T. This collection is said to be equicontinuous if and only if, given any $\varepsilon > 0$ and any point $p \in S$, there exists $\delta > 0$ such that, if $q \in S$ and $\rho_S(p, q) < \delta$, then $\rho_T(f_\alpha(p), f_\alpha(q)) < \varepsilon$ for every index α.

We may summarize the two facts that we have just pointed out about Cauchy sequences of $B(J)$ in the next theorem.

Theorem 8.6. Every Cauchy sequence of points of $B(J)$ is a uniformly bounded and equicontinuous collection of mappings.

It will be recalled that, in our proof of the completeness of $B(J)$, we defined the function $f(x)$ by observing that, for each fixed $x \in J$, $\{f_n(x)\}$ is a Cauchy sequence of points of R.

Definition 8.7. Let S and T be spaces, and $\{f_n\}$ a sequence of mappings of S into T. This sequence is said to converge pointwise to a mapping $f: S \to T$ if and only if, for each point x of S, the sequence $\{f_n(x)\}$ of points of T converges to the point $f(x)$ of T.

When the image space T is a metric space, we can describe the concept of the sequence $\{f_n(x)\}$ converging at the same rate for every point x of S and thus obtain the notion of uniform convergence.

Definition 8.8. Let S be a space and T a metric space. A sequence of mappings $\{f_n\}$ of S into T is said to converge uniformly to a mapping $f: S \to T$ if and only if, given any $\varepsilon > 0$, there exists an integer N such that $n > N$ implies $\rho(f_n(x), f(x)) < \varepsilon$ for all $x \in S$.

The proof of the following theorem follows very closely a portion (fourth paragraph) of the proof of the completeness of $B(J)$. This occurs since convergence in $B(J)$ is uniform convergence in the sense above.

Theorem 8.9. Let S be a space, and T a metric space. If $\{f_n\}$ is a sequence of continuous mappings of S into T which converges uniformly to a mapping f of S into T, then f is continuous.

Still another important property is exhibited by the elements of $B(J)$. If f is an element of this space, then not only is f continuous but also, given $\varepsilon > 0$, there exists a $\delta > 0$ such that $x_1, x_2 \in J$ and $|x_1 - x_2| < \delta$ implies $|f(x_1) - f(x_2)| < \varepsilon$. That is, the choice of δ depends only on ε and not on the points x_1 and x_2 of J under consideration. This property of f is known as uniform continuity.

DEFINITION 8.10. Let S and T be metric spaces with metrics ρ_S and ρ_T, respectively. Then a mapping $f: S \to T$ is said to be uniformly continuous on S if and only if, given any $\varepsilon > 0$, there exists a $\delta > 0$ such that $s_1, s_2 \in S$ and $\rho_S(s_1, s_2) < \delta$ implies $\rho_T(f(s_1), f(s_2)) < \varepsilon$.

The concept of uniform continuity is related to that of the oscillation of a function.

THEOREM 8.11. Let S and T be metric spaces with metrics ρ_S and ρ_T respectively, and let K be a subset of S. If the mapping $f: K \to T$ is uniformly continuous, then the oscillation $\omega(p)$ of f is zero at every point of S.

Proof. For each point p of K, we have $\omega(p) = 0$, since f is continuous at p. For each point p of $S - \bar{K}$, we must have $0 \leq \omega(p) \leq \delta[f((S - \bar{K}) \cap K)] = 0$, since K is disjoint with $S - \bar{K}$.

It remains to consider the case $p \in \bar{K} - K$. Let $\varepsilon > 0$ be given. Then by uniform continuity, there exists $\sigma > 0$ such that

$$0 \leq \omega(p) \leq \delta[f(S_{\sigma/2}(p) \cap K)] < \varepsilon$$

where $S_{\sigma/2}(p)$ is the spherical neighborhood with center p and radius $\sigma/2$ in the metric ρ_S. To see that this is the case, suppose $q, r \in S_{\sigma/2}(p) \cap K$. Then $\rho_S(q, r) \leq \rho_S(q, p) + \rho_S(p, r) < \sigma/2 + \sigma/2 = \sigma$; hence $\rho_T(f(q), f(r)) < \varepsilon$. Since $0 \leq \omega(p) < \varepsilon$ for every $\varepsilon > 0$, it follows that $\omega(p) = 0$ for $p \in \bar{K} - K$. But $S = (S - \bar{K}) \cup (\bar{K} - K) \cup K$. Thus $\omega(p) = 0$ for every point p of S.

COROLLARY 8.12. Let S and T be metric spaces, and suppose that T is complete. Then, if A is a subset of S and $g: A \to T$ is a uniformly continuous mapping, there exists a unique continuous extension of g from A to $\bar{A}$.

We have seen in our study of the space R that, if $f: R \to R$ is a continuous mapping, then f preserves convergent sequences; i.e., if $\{x_n\}$ is a sequence of points of R that converges to x_0, then the sequence $\{f(x_n)\}$ converges to $f(x_0)$. Uniformly continuous mappings have the property that they preserve Cauchy sequences whether the Cauchy sequences converge or not.

THEOREM 8.13. Let S and T be metric spaces with metrics ρ_S and ρ_T respectively, K a subset of S, and $f: K \to T$ a mapping which is uniformly continuous on K. Then, if $\{k_n\}$ is any Cauchy sequence of points of K, $\{f(k_n)\}$ is a Cauchy sequence of points of T.

Proof. Let $\{k_n\}$ be a Cauchy sequence of points of K, and suppose that $\varepsilon > 0$ has been given. From the uniform continuity of f, we see that there exists a $\delta > 0$ such that $x_1, x_2 \in K$ and $\rho_S(x_1, x_2) < \delta$ implies $\rho_T(f(x_1), f(x_2)) < \varepsilon$. For this $\delta > 0$, there exists an integer N such that $m, n > N$ implies $\rho_S(k_m, k_n) < \delta$. Thus $m, n > N$ implies $\rho_T(f(k_m), f(k_n)) < \varepsilon$. Consequently, $\{f(k_n)\}$ is a Cauchy sequence of points of T.

EXAMPLE 8.14. Let S be the subspace of R, consisting of all points x of R such that $x > 0$, and K the subset of S, consisting of all points of S such that $0 < x \leq 1$. Consider the continuous mapping $f: K \to R$ defined by $f(x) = 1/x$. This mapping is not uniformly continuous, in view of Theorem 8.13, since the sequence $\{k_n\}$ of points of K defined by $k_n = 1/n$ is a Cauchy sequence, but the sequence $\{f(k_n)\}$ is not a Cauchy sequence. We note also that the oscillation of f is zero at every point of S, which shows us that having the oscillation of f zero at each point of S is a less stringent condition than that f be uniformly continuous on K. Thus the converse of Theorem 8.11 is false.

In the introductory remarks to this section we emphasized that an economy of effort can be achieved by establishing results which can be applied in many situations. We also stated that economy can be achieved if the similarity between two problems is recognized. In this case the method of solution for one of the problems may apply to the other also. We close this section by illustrating how the same method of proof applies to two apparently diverse theorems, one concerning uniform continuity, the other concerning uniform local connectivity.

DEFINITION 8.15. Let K be a subset of a metric space S. Then K is said to be uniformly locally connected if and only if, given any $\varepsilon > 0$, there exists a $\delta > 0$ such that, if x and y are points of K and $\rho(x, y) < \delta$, then x and y lie in a connected subset of K of diameter less than ε.

Clearly, any subset K of a metric space S is locally connected at each of its points if it is uniformly locally connected.

THEOREM 8.16. Let S and T be metric spaces with metrics ρ_S and ρ_T, respectively, and let K be a compact subset of S. Then

(i) If $f: K \to T$ is any continuous mapping of K into T, then f is uniformly continuous.

(ii) If K is locally connected, then K is uniformly locally connected.

Proof. Let $\varepsilon > 0$ be given. We know that there exists, for each point $p \in K$, a real number $\delta_p > 0$ defined as follows: If f is continuous and q is any point of $K \cap S_{\delta_p}(p)$, then $\rho_T(f(p), f(q)) < \varepsilon/2$. On the other hand, if K is locally connected, and q is any point of $K \cap S_{\delta_p}(p)$, then p and q lie in a connected subset of K of diameter less than $\varepsilon/2$. The collection of spherical neighborhoods $\{S_{\delta_p/2}(p)\}$ for all $p \in K$ is an open covering of the compact set K. Consequently, a finite number of these sets, say, $S_{\delta_{p_1}/2}(p_1), \cdots, S_{\delta_{p_n}/2}(p_n)$, cover K. Let 2δ be the greatest lower bound of the finite set of positive numbers $\delta_{p_1}, \delta_{p_2}, \cdots, \delta_{p_n}$. Clearly, $\delta > 0$. Let x, y be arbitrary points of K satisfying $\rho_S(x, y) < \delta$. We know that there exists an integer i such that $1 \leq i \leq n$, and $\rho_S(p_i, x) < \delta \leq \delta_{p_i}/2$. Consequently, $\rho_S(p_i, y) \leq \rho_S(p_i, x) + \rho_S(x, y) < \delta_{p_i}$. If K is locally connected, then each of the points x, y of K lies with p_i in a connected subset of K of diameter less than $\varepsilon/2$. The union of these two connected sets is a connected subset of K of diameter less than ε. Therefore, K is uniformly locally connected. If, on the other hand, f is continuous, then $\rho_T(f(x), f(y)) \leq \rho_T(f(x), f(p_i)) + \rho_T(f(p_i), f(y)) < \varepsilon/2 + \varepsilon/2 = \varepsilon$. Thus f is uniformly continuous.

PROBLEMS

75. Prove Theorem 8.6.

76. For any compact metric space S, define the function space $B(S)$ analogous to $B(J)$. Prove that $B(S)$ is complete.

77. Let $\{f_\alpha\}$ be a collection of uniformly bounded equicontinuous functions, each belonging to $B(J)$. Prove that every sequence of elements of $\{f_\alpha\}$ has a convergent subsequence [convergence in $B(J)$]. If $K = \bigcup_\alpha f_\alpha$, show that the closure of K in $B(J)$ is compact. (This is essentially the Ascoli theorem of analysis. See, for example, McShane, *Integration Theory*, Princeton University Press, 1944).

78. Let $\{f_n\}$ be a sequence of points of $B(J)$ defined as follows for each $n \in I$:

$$f_n(x) = 1 \quad \text{for} \quad 0 \leq x \leq 1 - 1/n$$

$$f_n(x) = -n(x - 1) \quad \text{for} \quad (1 - 1/n) \leq x \leq 1$$

Show that $\{f_n\}$ is not a Cauchy sequence of $B(J)$. Use this result to prove that no spherical neighborhood of a point of $B(J)$ has the property that its closure in $B(J)$ is compact. Also conclude that $B(J)$ is not compact.

79. Let D and E be subsets of the metric spaces S and T, respectively, where T is complete. Suppose that there exists a one-to-one uniformly continuous mapping $g\colon D \to T$ such that $g(D) = E$, and the mapping $h\colon E \to S$ defined by

$h(x) = g^{-1}(x)$, for $x \in E$, is also uniformly continuous. Prove that there is a unique continuous extension f of g from D to $\bar{D}$. Let $E^\star = f(\bar{D})$. Thus $E \subset E^\star \subset \bar{E}$. Also prove that the mapping $k\colon \bar{D} \to E^\star$ defined by $k(x) = f(x)$, for $x \in \bar{D}$, is a homeomorphism.

80. Write a short, rigorous discussion of the application of Newton's method to finding real roots of polynomials with real coefficients. Decide, in particular, if it always works.

9. The Completion of Metric Spaces

We now turn our attention to the set of all Cauchy sequences of a given metric space S. Every such sequence is a mapping of the set of all positive integers into S. Thus the set of all Cauchy sequences of S may be regarded as a collection of mappings.

THEOREM 9.1. Let S be a metric space with metric ρ, and C the set of all Cauchy sequences of S. Then the mapping $\rho_C\colon C \times C \to R$ defined by

$$\rho_C(k_1, k_2) = \lim \rho(x_n, y_n)$$

where k_1 and k_2 are the Cauchy sequences $\{x_n\}$ and $\{y_n\}$ of S, satisfies the following conditions:

(i) $\rho_C(k_1, k_2) \geq 0$.

(ii) $\rho_C(k_1, k_2) = 0$ if and only if k_1 and k_2 are equivalent Cauchy sequences.

(iii) $\rho_C(k_1, k_2) = \rho_C(k_2, k_1)$.

(iv) For any Cauchy sequence $k_3 = \{z_n\}$ of S, $\rho_C(k_1, k_3) \leq \rho_C(k_1, k_2) + \rho_C(k_2, k_3)$.

Proof. Let k_1, k_2, k_3 denote, throughout this proof, three Cauchy sequences $\{x_n\}$, $\{y_n\}$, $\{z_n\}$ of S, respectively. From Problem 69, $\{\rho(x_n, y_n)\}$ is a Cauchy sequence of real numbers; hence $\lim \rho(x_n, y_n)$ exists. Furthermore, for every n, $\rho(x_n, y_n) \geq 0$ and $\rho(x_n, y_n) = \rho(y_n, x_n)$. Consequently, (i) and (iii) must hold. Property (ii) follows from Problem 68. To prove (iv), note that, for every n, $\rho(x_n, z_n) \leq \rho(x_n, y_n) + \rho(y_n, z_n)$. Thus, $\rho_C(k_1, k_3) = \lim \rho(x_n, z_n) \leq \lim (\rho(x_n, y_n) + \rho(y_n, z_n)) = \lim \rho(x_n, y_n) + \lim \rho(y_n, z_n) = \rho_C(k_1, k_2) + \rho_C(k_2, k_3)$, and (iv) holds.

It is clear that ρ_C fails to be a metric for C only because $\rho_C(k_1, k_2) = 0$ does not imply that the sequence k_1 is identical with the sequence k_2. This condition does imply, however, that these sequences are equivalent, and this is the motivation for our next definition.

DEFINITION 9.2. Let C denote the set of all Cauchy sequences of a metric space S with metric ρ. Then $C^\star(S)$ denotes the collection of all non-empty subsets of C such that two elements, k_1 and k_2 of C, belong to the same element of $C^\star(S)$ if and only if $\rho_C(k_1, k_2) = 0$, where ρ_C is the mapping defined in Theorem 9.1.

Clearly the points of $C^\star(S)$ are sets of the decomposition of C into equivalence classes.

THEOREM 9.3. Let S be a metric space. Then the mapping $\rho^\star: C^\star(S) \times C^\star(S) \to R$ defined by $\rho^\star(k_1^\star, k_2^\star) = \rho_C(k_1, k_2)$, where k_1 and k_2 are any elements of the sets of equivalent Cauchy sequences $k_1^\star$ and $k_2^\star$, respectively, is a metric for $C^\star(S)$.

Proof. We show first that $\rho^\star$ is well defined. To do this, it must be demonstrated that, if $t_1, t_2 \in k_1^\star$ and $t_3, t_4 \in k_2^\star$, then $\rho_C(t_1, t_3) = \rho_C(t_2, t_4)$; hence $\rho^\star(k_1^\star, k_2^\star)$ is independent of the choice of elements of $k_1^\star$ and $k_2^\star$ by which we choose to define it. We have at once $\rho_C(t_1, t_2) = 0$, $\rho_C(t_3, t_4) = 0$. Hence,

$$\rho_C(t_1, t_3) \leq \rho_C(t_1, t_2) + \rho_C(t_2, t_4) + \rho_C(t_4, t_3) = \rho_C(t_2, t_4)$$

Similarly, $\rho_C(t_2, t_4) \leq \rho_C(t_1, t_3)$. Thus $\rho_C(t_1, t_3) = \rho_C(t_2, t_4)$, and accordingly $\rho^\star$ is well defined.

It is clear that $\rho^\star(k_1^\star, k_2^\star) = 0$ if and only if $k_1^\star = k_2^\star$, where $k_1^\star, k_2^\star \in C^\star(S)$. This completes the proof of the theorem, in view of the results of Theorem 9.1.

Whenever we are discussing a metric space S, and refer to the metric space $C^\star(S)$, we shall always mean the space $C^\star(S)$ with the particular metric defined in Theorem 9.3.

DEFINITION 9.4. Let S be a metric space. Then the metric space $C^\star(S)$ is the set $C^\star(S)$ with the metric of Theorem 9.3.

THEOREM 9.5. Let S be a metric space with metric ρ. Then the metric space $C^\star(S)$ is complete.

Proof. Let $\{k_n^\star\}$ be a Cauchy sequence of elements of $C^\star(S)$. For each $n \in I$, denote by k_n a Cauchy sequence of S such that $k_n \in k_n^\star$. Thus we have the following array of sequences, where the ith element of the nth sequence k_n is denoted by $y_{n,i}$:

$$\begin{array}{l} k_1: y_{1,1}, y_{1,2}, y_{1,3}, \cdot \quad \cdot \quad \cdot \\ k_2: y_{2,1}, y_{2,2}, y_{2,3}, \cdot \quad \cdot \quad \cdot \\ k_3: y_{3,1}, y_{3,2}, y_{3,3}, \cdot \quad \cdot \quad \cdot \\ \cdot \quad \cdot \quad \cdot \quad \cdot \\ \cdot \quad \cdot \quad \cdot \quad \cdot \\ \cdot \quad \cdot \quad \cdot \quad \cdot \end{array}$$

Consider the mth sequence k_m. There exists an integer $M(m)$ such that $y_{m,M(m)}$ and $y_{m,i}$ satisfy $\rho(y_{m,M(m)}, y_{m,i}) < 1/m$ for every $i \geq M(m)$. Define h_m to be the Cauchy sequence all of whose terms are equal, each term being $y_{m,M(m)}$. We arrange these sequences in an array as follows:

$$\begin{array}{l} h_1\colon y_{1,M(1)}, y_{1,M(1)}, y_{1,M(1)}, \cdots \\ h_2\colon y_{2,M(2)}, y_{2,M(2)}, y_{2,M(2)}, \cdots \\ h_3\colon y_{3,M(3)}, y_{3,M(3)}, y_{3,M(3)}, \cdots \\ \cdot \quad \cdot \quad \cdot \quad \cdot \\ \cdot \quad \cdot \quad \cdot \quad \cdot \\ \cdot \quad \cdot \quad \cdot \quad \cdot \end{array}$$

Let $h_m^\star$ denote, for each m, the set of all Cauchy sequences of S that are equivalent to h_m. By construction, we have $\rho^\star(k_m^\star, h_m^\star) < 1/m$. Thus $\lim \rho^\star(k_m^\star, h_m^\star) = 0$. Hence $\{k_m^\star\}$ and $\{h_m^\star\}$ are equivalent Cauchy sequences of $C^\star(S)$.

We shall show that the sequence $\{h_m^\star\}$ converges to a point of $C^\star(S)$. It then follows by Theorem 7.6 that $\{k_m^\star\}$ also converges to this point, and this fact establishes the completeness of the space $C^\star(S)$.

Since $\{h_m^\star\}$ is a Cauchy sequence of $C^\star(S)$, it follows that, given $\varepsilon > 0$, there exists an integer L such that $n, m > L$ imply $\rho_C(h_m, h_n) < \varepsilon$. But this means $\rho(y_{m,M(m)}, y_{n,M(n)}) < \varepsilon$. Thus the sequence $\{y_{n,M(n)}\}$, the diagonal sequence of the second array, is a Cauchy sequence of S. Let this diagonal sequence be denoted by h. We also see that $\rho_C(h_n, h) < \varepsilon$ for $n > L$. Thus $n > L$ implies $\rho^\star(h_n^\star, h^\star) < \varepsilon$, where $h^\star$ denotes the set of all Cauchy sequences of S equivalent to h. Consequently, $\{h_n^\star\}$ converges to $h^\star$ in $C^\star(S)$, and the proof of Theorem 9.5 is complete.

The method of the proof just given is frequently given the name of the Cantor diagonal process. The student should compare this proof with that of the proof that the set of real numbers is not countable. The proof of the Ascoli theorem (Problem 77), as usually given, is also of this same general type.

We now introduce the concept of an isometry between two metric spaces in order to lay the groundwork for an important application of Theorem 9.5.

Definition 9.6. Let S and T be metric spaces with metrics ρ_S and ρ_T, respectively. Then S is said to be isometric to T if and only if there exists a homeomorphism h of S onto T such that, for any pair of points x, y of S, we have $\rho_S(x, y) = \rho_T(h(x), h(y))$. The mapping h is then said to be an isometry.

The word isometric means "of the same metric." Two metric spaces that are isometric are indistinguishable as far as their metric and topological properties are concerned. Thus we say that they are equivalent, and we want the following easily proved theorem which justifies this use of the term equivalent. It establishes "isometric to" as an equivalence relation for metric spaces.

THEOREM 9.7. Let S, T, W be metric spaces. Then

(i) S is isometric to itself.

(ii) If S is isometric to T, then T is isometric to S.

(iii) If S is isometric to T, and T is isometric to W, then S is isometric to W.

From our work with $C^\star(S)$, it easily follows that S is isometric to a subset of $C^\star(S)$ which is dense in $C^\star(S)$. The isometry is given in the following theorem.

THEOREM 9.8. Let S be a metric space with metric ρ. Then the mapping $f: S \to C^\star(S)$ defined by $f(x) = x^\star$, where $x^\star$ denotes the set of all Cauchy sequences of S equivalent to the Cauchy sequence $\{a_n\}$ of S such that $a_n = x$, for every $n \in I$, is an isometry of S onto the subspace $f(S)$ of $C^\star(S)$. Here the metric of $f(S)$ is that of $C^\star(S)$ restricted to $f(S)$. Furthermore, $\overline{f(S)} = C^\star(S)$.

In one development of the real numbers, the natural numbers or positive integers are given by the so-called Peano axioms. From these axioms the set of all integers is constructed, and from this set the metric space S of all rational numbers is built up. For this part of such a development, the reader is referred to Landau's book, *Grundlagen der Analysis*. The real numbers are then constructed exactly as $C^\star(S)$ was built up; that is, each real number is an equivalence class of Cauchy sequences of rational numbers. Then, taking advantage of the fact that Theorem 9.8 holds and that, in the abstract, the metric spaces S and $f(S)$ (using the notation of Theorem 9.8) are indistinguishable, the rationals are considered as a subset of the reals. For example, the rational number 1/4 and any Cauchy sequence equivalent to the sequence 2/10, 25/100, 250/1000, 2500/10,000, $\cdots$ which is often written $0.250000\cdots$ are considered to be the same. Following this convention we call the space R of real numbers in its standard metric the completion of the metric space of rationals, since the closure of the set of rationals is the reals, and the metric space R of all real numbers is complete. For a reference on this part of the procedure the reader should see the first chapter of *Theory and Application of Infinite Series* by Konrad Knopp.

DEFINITION 9.9. Let S be a metric space. A metric space T is said to be a completion of S if and only if T is complete, $S \subset T$, $\bar{S} = T$, and the metric for T, restricted to S, is the metric for S.

Using the notation of Theorem 9.8 and the convention that S and $f(S)$ are to be identified as the same metric space, we have the following consequence of Theorem 9.8 and Definition 9.9.

THEOREM 9.10. Let S be a metric space. Then there exists a metric space T which is a completion of S. Furthermore, every point of T is the limit of a Cauchy sequence of points of S.

In case the metric of the space S is an M metric, then a completion T of S has some special properties.

THEOREM 9.11. Let S be a metric space such that its metric is an M metric. Then the metric of any completion T of S is an M metric.

Proof. Let S be a metric space with an M metric ρ, and T a completion of S. Denote the metric of T by $\rho^\star$. Let $S_r(p)$ and $S_r^\star(q)$ denote, respectively, the spherical neighborhood of radius r of a point p of S in the metric ρ, and the spherical neighborhood of radius r of a point q of T in the metric $\rho^\star$.

For any $r > 0$ and any $q \in T$, we must prove that $S_r^\star(q)$ is connected. The proof will be divided into two cases: (i) $q \in S$, (ii) $q \in T - S$. We consider first case (i). Suppose that $q \in S$ and $r > 0$. Clearly, $S_r(q) = S_r^\star(q) \cap S$. We show that every point t of $S_r^\star(q) - S_r(q)$ is a limit point of $S_r(q)$. It is evident that $t \in T - S$; hence t is a limit point of S since S is dense in T. But $S = (S - S_r(q)) \cup S_r(q)$, so that t must be a limit point of at least one of the sets $S - S_r(q)$ and $S_r(q)$. The point t cannot be a limit point of $S - S_r(q)$ since this set is contained in the closed subset $T - S_r^\star(q)$ of T, and t is not in $T - S_r^\star(q)$. Therefore, t is a limit point of $S_r(q)$. We have thus proved that $S_r(q) \subset S_r^\star(q) \subset \overline{S_r(q)}$, where this closure is taken in the space T. Consequently, by Theorem III, 5.17, $S_r^\star(q)$ is connected.

For our discussion of case (ii), suppose that $q \in T - S$, and consider the set $S_r^\star(q)$, $r > 0$. For each integer $n > 4/r$, there exists a point q_n of S such that $\rho^\star(q, q_n) < 1/n$. Define $\beta_n = r - \rho^\star(q, q_n)$. Furthermore, $q \in S_{\beta_n}^\star(q_n) \subset S_r^\star(q)$. To see that $q \in S_{\beta_n}^\star(q_n)$, notice that

$$\rho^\star(q, q_n) < 1/n < r/4 < r - r/4 < r - 1/n < r - \rho^\star(q, q_n) = \beta_n$$

To see that $S_{\beta_n}^\star(q_n) \subset S_r^\star(q)$, suppose that $t \in S_{\beta_n}^\star(q_n)$; then

$$\rho^\star(t, q_n) < r - \rho^\star(q, q_n)$$

that is, $\rho^\star(t, q) \leq \rho^\star(t, q_n) + \rho^\star(q, q_n) < r$. Furthermore, for each integer n, the set $S_{\beta_n}^\star(q_n)$ is connected, by case (i), since $q_n \in S$.

We shall now show that $S_r^\star(q) = \bigcup_{\substack{n \in I \\ n > 4/r}} S_{\beta_n}^\star(q_n)$. Since $S_{\beta_n}^\star(q_n) \subset S_r^\star(q)$ for every $n > 4/r$, it follows that $S_r^\star(q)$ contains the union. To show the inclusion the other way, we need only show that, given $t \in S_r^\star(q)$, there exists an integer N such that $t \in S_{\beta_N}^\star(q_N)$. The least integer N such that $1/N < (r - \rho^\star(q, t))/2$ and $N > 4/r$ will serve the purpose. For, with this N,

$$\begin{aligned} \rho^\star(t, q_N) &\leq \rho^\star(t, q) + \rho^\star(q, q_N) \\ &< \rho^\star(t, q) + (r - p^\star(q, t))/2 \\ &= (r + \rho^\star(q, t))/2, \end{aligned}$$

and since

$$\begin{aligned} \beta_N &= r - \rho^\star(q, q_N) \\ &\geq r - 1/N \\ &> r - (r - \rho^\star(q, t))/2 \\ &= (r + \rho^\star(q, t))/2 \end{aligned}$$

it follows that

$$\rho^\star(t, q_N) < \beta_N$$

i.e.,

$$t \in S_{\beta_N}^\star(q_N)$$

We have thus shown that, for every $r > 0$ and every point $q \in T$, the set $S_r^\star(q)$ is the union of a collection of connected sets, each of which contains the point q. It follows from Theorem III, 5.16 that $S_r^\star(q)$ is a connected set. This proves that $\rho^\star$ is an M metric for the space T.

In the proof just given, we showed that all spherical neighborhoods in the $\rho^\star$ metric of T are connected. With a small additional step, it can be shown that the intersection of any such spherical neighborhood of T with the space S is a connected set.

COROLLARY 9.12. For every $r > 0$ and every $q \in T$, the set $S_r^\star(q) \cap S$ is connected.

Proof. If $q \in S$, then $S_r^\star(q) \cap S = S_r(q)$; hence this set is connected. If $q \in T - S$, then

$$\begin{aligned} S \cap S_r^\star(q) &= \Big[\bigcup_{\substack{n \in I \\ n > 4/r}} S_{\beta_n}^\star(q_n) \Big] \cap S \\ &= \bigcup_{\substack{n \in I \\ n > 4/r}} S_{\beta_n}(q_n) \end{aligned}$$

Now each of the sets $S_{\beta_n}(q_n)$ of this union is connected. We shall show that these sets have the point q_M in common, where M is the least integer greater than $4/r$. Now

$$\begin{aligned}\rho(q_M, q_n) = \rho^\star(q_M, q_n) &\leq \rho^\star(q_M, q) + \rho^\star(q, q_n)\\ &< 1/M + 1/n\\ &< r/4 + r/4\\ &= r/2\end{aligned}$$

and

$$\begin{aligned}\beta_n &= r - \rho^\star(q_n, q)\\ &> r - r/4\\ &= 3r/4\end{aligned}$$

Thus $q_M \in S_{\beta_n}(q_n)$ for every $n > 4/r$, and the proof that the set $S \cap S_r^\star(q)$ is connected is complete by Theorem III, 5.16.

The result of this corollary can be extended to the following theorem which includes the corollary as a special case.

THEOREM 9.13. Let S be a metric space having the property that its metric is an M metric, and let T be any completion of S. Then, if A is any open connected subset of T, the set $A \cap S$ is connected.

Proof. We know that the metric of T is an M metric, and that the intersection of any spherical neighborhood of a point of T with S is a connected set. Let A be any open connected subset of T. If $A \cap S = \emptyset$, then $A \cap S$ is a connected set. We may thus suppose that $A \cap S \neq \emptyset$. Let p_0 be a point of $A \cap S$. We shall show that, for any point q of $A \cap S$, there exists a connected subset H_q of $A \cap S$ containing the point q and the point p_0. This will complete the proof of the theorem, since we may use again Theorem III, 5.16 and observe that $A \cap S = \bigcup_{q \in A \cap S} H_q$. To demonstrate the existence of the set H_q, note that A is an open subset of T. Thus we may write $A = \bigcup_{\alpha} V_\alpha$, where each V_α is a spherical neighborhood, in the space T, of some point of T that lies in A. The collection $\{V_\alpha\}$ is an open covering of A. Therefore, since A is connected, it follows, from Theorem III, 5.14, that there exists a simple chain $V_{\alpha_1}, V_{\alpha_2}, \cdots, V_{\alpha_n}$ of sets of the collection $\{V_\alpha\}$ joining the point p_0 and the point q. Using Corollary 9.12, we see that the set $V_{\alpha_j} \cap S$ is connected for $j = 1, 2, \cdots, n$. Furthermore, $V_{\alpha_i} \cap V_{\alpha_{i+1}}$ is a non-empty open set for $i = 1, 2, \cdots, n-1$. Since S is dense in T, there must thus exist for each $i = 1, 2, \cdots, n-1$, a point $p_i \in (V_{\alpha_i} \cap V_{\alpha_{i+1}}) \cap S = (S \cap V_{\alpha_i}) \cap (S \cap V_{\alpha_{i+1}})$. Consequently, the set H_q defined by $H_q = \bigcup_{i=1}^{n} (V_{\alpha_i} \cap S)$ is a connected subset of $A \cap S$ containing the point p_0 and the point q. This completes the proof of Theorem 9.13.

PROBLEMS

81. Show that every compact metric space is complete.

82. Let S be a metric space and T a completion of S. Show that S is separable if and only if T is separable.

83. Let S be a metric space and T a completion of S. Show that S is second countable if and only if T is second countable.

84. Let A be a subset of a metric space S. Prove that, given any point $p \in \bar{A}$ and any real number $r > 0$, the spherical neighborhood $S_r(p)$ can be expressed as the union of a countable collection of spherical neighborhoods of points of A.

85. Let S be a separable metric space. Show that there exists a subset T of $B(J)$, such that S is isometric to the metric space T, whose metric is the metric of $B(J)$ restricted to T. (See Sierpiński, *Fundamenta Mathematicae*, vol. 33, q. 115, or Sierpiński, *General Topology*, Toronto, 1952.) Show also that the closure of T in $B(J)$ is a completion of S, where again the metric is that of $B(J)$ restricted to this closure.

10. The Euclidean Spaces

The topological product of two spaces has already been defined. Obviously, given three spaces, S, T, and W, we can form the spaces $S \times T$, $(S \times T) \times W$, $S \times (T \times W)$ and various other combinations. The next theorem, which is easily proved, gives some insight into the relationship between these spaces.

THEOREM 10.1. Let S, T, W, S', T' be spaces. Then

(*a*) $S \times T$ and $T \times S$ are homeomorphic.

(*b*) $(S \times T) \times W$ and $S \times (T \times W)$ are homeomorphic.

(*c*) If S' and T' are homeomorphic to S and T, respectively, then $S \times T$ and $S' \times T'$ are homeomorphic.

With respect to (*b*) of this theorem, recall the analogous statement of algebra $(a + b) + c = a + (b + c)$, the so-called associative law. In algebra we drop the parentheses, since it does not matter in which order we do the addition. Similarly, in topology, we feel free to write $S \times T \times W$ since, whichever way we introduce parentheses, the result is the same. From (*a*) and (*c*) together with (*b*), it follows that $S \times T \times W$, $S \times W \times T$, $T \times S \times W$, $T \times W \times S$, $W \times S \times T$, and $W \times T \times S$ are all the same: i.e., homeomorphic. Thus the phrase "the topological product of S, T, and W" denotes the "same" space no matter in what order the spaces S, T, and W are arranged. It then follows that the topological product of n spaces, where n is a positive integer greater than unity, is defined without specifying (by means of parentheses) the order in which the product is to be built up from the product of two spaces, since we agree to identify as the same any two spaces that are homeomorphic. In spite of these remarks, it is frequently useful to use parentheses to indicate clearly how the product is being considered.

In an earlier chapter, it resulted from a problem that the topological product of two metric spaces was metrizable. Indeed, the following results are easily shown.

THEOREM 10.2. Let S and T be metric spaces with metrics ρ_S and ρ_T, respectively. Then the mapping $\rho: (S \times T) \times (S \times T) \rightarrow R$ defined by $\rho((a, b), (c, d)) = \rho_S(a, c) + \rho_T(b, d)$, for any two points (a, b) and (c, d) of $S \times T$, is a metric for the space $S \times T$.

THEOREM 10.3. Let S and T be metric spaces with metrics ρ_S and ρ_T, respectively. Then the mapping $\rho: (S \times T) \times (S \times T) \rightarrow R$ defined by $\rho((a, b), (c, d)) = \sqrt{(\rho_S(a, c))^2 + (\rho_T(b, d))^2}$, for any two points (a, b) and (c, d) of $S \times T$, is a metric for the space $S \times T$.

DEFINITION 10.4. Let S and T be metric spaces with metrics ρ_S and ρ_T, respectively. Then the metric space $S \times T$ is defined to be the space $S \times T$ with the metric given by Theorem 10.3.

DEFINITION 10.5. The symbol E^1 denotes the metric space R of all real numbers. The symbol E^2 is used to denote the metric space $E^1 \times E^1$, and, in general, for any positive integer $n > 1$, the symbol E^n is used to denote the metric space $E^{n-1} \times E^1$. The metric space E^n is often referred to as Euclidean n space.

The space E^3 is easily recognized as the familiar space of solid analytical geometry.

The student should find it fairly easy to establish the following theorem, which describes some of the important properties of E^n.

THEOREM 10.6. For any positive integer n, the Euclidean n space E^n has the following properties:

1. E^n is connected and locally connected.
2. The metric of E^n is an M metric.
3. E^n is separable, complete and second countable.
4. Given any spherical neighborhood $S_\varepsilon(p)$ of a point of E^n, the set $\overline{S_\varepsilon(p)}$ is compact, and consists of all points of E^n whose distance from p is at most ε.
5. E^n is isometric to the metric space $E^j \times E^k$ for any two positive integers j, k satisfying $j + k = n$.

PROBLEMS

86. Prove Theorem 10.1.

87. Prove Theorem 10.2.

88. Prove Theorem 10.3.

89. Prove Theorem 10.6.

90. Write a short paper, developing the theory of generalized sequences, based on the following definitions and including, as a minimum requirement, proofs of the theorems stated below.

Definition 1. Let Q be a non-empty set on which a binary relation F is defined (see Section I, 5). Then Q is said to be a Moore-Smith set with respect to F if and only if the following conditions hold:

(i) (Transitivity): Given $q_1, q_2, q_3 \in Q$ such that $q_1 F q_2$ and $q_2 F q_3$. Then $q_1 F q_3$.

(ii) (The composition property): Given $q_1, q_2 \in Q$, there exists $q_3 \in Q$ such that $q_3 F q_1$ and $q_3 F q_2$.

The symbol $q F p$ is usually read "q follows p."

The important role of the composition property will become clear when a few of the theorems concerning Moore-Smith sets have been established. Let us make only the brief comment that in many applications the Moore-Smith set Q under consideration will have the property that for some pair of elements p, q, both $p F q$ and $q F p$ are false. The composition property is then exactly what is needed to assure the existence of an element that follows both p and q.

Definition 2. A mapping f of a Moore-Smith set Q into a set S is known as a Moore-Smith or M.S. sequence. It is often written as $\{x_q\}$ where $f(q) = x_q$. If S is a space and $x_0 \in S$, then a M.S. sequence $\{x_q\}$ of points of S is said to converge to x_0 if and only if, given any neighborhood U of x_0, there exists an element q_0 of Q such that $q F q_0$ implies $x_q \in U$. Notationally, this is written as $\lim_{q,F} f(q) = x_0$, or $\lim_{q,F} x_q = x_0$, or $\lim_q x_q = x_0$, or $\{x_q\} \to x_0$.

Examples of M.S. Sequences from Analysis.

1. $\lim_{n \to \infty} x_n = x$. Interpret Q as the set of positive integers, where $n F m$ if and only if $n \geq m$.

2. $\lim_{x \to x_0} f(x) = b$. Interpret Q as the set of real numbers with the exception of x_0, and $x F y$ if and only if $|x - x_0| < |y - x_0|$.

Example of a Moore-Smith Set from Topology.

3. Let S be a topological space and Q the topology of S. If $U, V \in Q$, then $U F V$ if and only if $U \subset V$. Many of our theorems on topological spaces relied heavily on the fact that the composition property holds for Q. This is an immediate consequence of Axiom 5 for topological spaces.

Theorem 1. Let $\{x_q\}$ be a M.S. sequence of a Hausdorff space S. If $\lim_q x_q = a$, and $\lim_q x_q = b$, then $a = b$.

Theorem 2. Let S be a space. If every convergent M.S. sequence of points of S has a unique limit point in S, then S is a Hausdorff space.

Theorem 3. Let S be a metric space and $\{x_q\}$ a M.S. sequence of S. If $\{x_q\} \to a \in S$, then, given $\varepsilon > 0$, there exists an element q_ε of Q such that $q_1 \, F \, q_\varepsilon$ and $q_2 \, F \, q_\varepsilon$ implies $\rho(x_{q_1}, x_{q_2}) < \varepsilon$. (That is, the sequence is a generalized Cauchy sequence.) If S is also complete, then a M.S. sequence converges if and only if it is a generalized Cauchy sequence.

Theorem 4. Let S and T be Hausdorff spaces and $f: S \to T$ a mapping. Then f is continuous on S if and only if, given any $x \in S$ and any M.S. sequence $\{x_q\}$ of S that converges to x, the M.S. sequence $\{f(x_q)\}$ of T converges to $f(x)$.

Definition 3. Let P, Q be Moore-Smith sets; S a metric space; and $f: P \times Q \to S, g: Q \to S$ mappings, and suppose that $\lim_p f(p, q) = g(q)$. Then $\lim_p f(p, q) = g(q)$ is said to be uniform in q if and only if $\varepsilon > 0$ implies the existence of an element p_ε of P such that $p \in P$ and $p \, F \, p_\varepsilon$ implies $\rho(f(p, q), g(q)) < \varepsilon$. (That is, p_ε does not depend on q.)

Theorem 5. Let P, Q be Moore-Smith sets, S a complete metric space, and $f: P \times Q \to S$ a mapping. If $\lim_p f(p, q)$ exists for every $q \in Q$, uniformly in q, and if $\lim_q f(p, q)$ exists for every p, then $\lim_q (\lim_p f(p, q))$, $\lim_p (\lim_q f(p, q))$, and $\lim_{p,q} f(p, q)$ each exist and all are equal. (Note that, with respect to the limit $\lim_{p,q} f(p, q)$, it is understood that (p_1, q_1) follows (p_2, q_2) in $P \times Q$ if and only if p_1 follows p_2 in P and q_1 follows q_2 in Q.)

If you wish to include further material in your paper, then:

Look up some problems on iterated limits in analysis and give some applications of our Theorem 5. You might also discuss infinite series (of real or complex numbers) where the sequence of partial sums is taken to be a M.S. sequence. In still another direction, you might compare the notion of M.S. sequences with that of *nets* (see J. L. Kelley, Convergence in Topology, *Duke Mathematical Journal*, vol. 17, pp. 277–283, 1950).

91. Write a short paper, developing a few facts concerning the Hilbert cube and Hilbert space. Base your work on the following definitions, and include, as a minimum requirement, proofs of the theorems stated below.

Definition 1. The set of all sequences $\{x_n\}$ of real numbers for which $\sum_{n=1}^{\infty} x_n^2$ converges, and where the distance between two such sequences $\{x_n\}$ and $\{y_n\}$ is defined to be $\left[\sum_{n=1}^{\infty} (x_n - y_n)^2\right]^{1/2}$ is denoted by E^ω and known as Hilbert space.

Definition 2. The subspace of E^ω consisting of all sequences $\{x_n\}$ such that $|x_n| \leq 1/n$ is denoted by J^ω and known as the Hilbert cube.

Theorem 1. Hilbert space is a metric space.

Theorem 2. The Hilbert cube is compact and contains no non-empty open subset of Hilbert space.

Theorem 3. Let W be the set of all sequences $\{x_n\}$ of real numbers such that $0 \leq x_n \leq 1$ for $n \in I$. Then, if $x = \{x_n\}$ and $y = \{y_n\}$, denote points of W, the mapping $f\colon W \times W \to R$ defined by

$$f(x, y) = \sum_{n=1}^{\infty} 2^{-n} \,|\, x_n - y_n|$$

is a metric for W.

Theorem 4. The set W with metric f as defined in Theorem 3 is homeomorphic to J^ω.

Theorem 5. Every regular Hausdorff space having a countable basis is homeomorphic with a subspace of E^ω and hence metrizable.

92. Write a short paper, developing the elements of category theory. Base your work on the following definitions, and include, as a minimum requirement, proofs of the theorems stated below.

Definition 1. Let X and Y be subsets of a space S. Then

1. X is *dense* in Y if and only if $Y \subset \overline{X \cap Y}$.
2. X is *nowhere dense* in S if and only if, for no $p \in \bar{X}$, does there exist an open subset U of S such that $p \in U \subset \bar{X}$.
3. X is of *category* I in S if and only if there exists a collection of subsets $\{V_n\}_{n \in I}$ of S such that
   (*a*) $X = \bigcup_{n \in I} V_n$.
   (*b*) For each $n \in I$, V_n is nowhere dense in S.
4. X is of *category* II in S if and only if X is not of category I in S.
5. X is of category I at a point $x \in S$ if and only if there is a neighborhood V of x in S such that $V \cap X$ is of category I in S; otherwise X is of category II at x.
6. By $D(X)$ is denoted the set of all points of S at which X is of category II.
7. By X^o is denoted the union of all open sets of S contained in X. The set X^o is called the *interior* of X in S.

Theorem 1. Let X, Y be subsets of a Hausdorff space S. Then

1. $D(X) = \emptyset$ if and only if X is of category I.
2. $D(X \cup Y) = D(X) \cup D(Y)$.
3. $D[D(X)] = D(X)$.
4. $D(X)$ is closed.
5. If $X \subset Y$, then $D(X) \subset D(Y)$.
6. $D(X) = \overline{[D(X)]^o}$.
7. If Y is open, then $Y \cap D(X) = Y \cap D(Y \cap X)$.
8. $D(X \cap Y) \subset [D(X) \cap D(Y)]$.

Theorem 2. If S is a complete metric space, then S is of category II at every point of S.

Theorem 3 (*Baire Category Theorem*). Let S be a complete metric space, $\{U_n\}_{n \in I}$ a collection of subsets of S. If $S = \bigcup_{n \in I} U_n$, then there exists $\varepsilon > 0$, $p \in S$, and an integer N such that $S_\varepsilon(p) \subset \bar{U}_N$.

Also deduce, as a consequence of Theorem 3 and using the ideas outlined by Kuratowski (see *Topologie* I, p. 327), the following result.

Theorem 4. There exists a continuous real-valued function defined on the closed unit interval which has a derivative at no point of the closed unit interval.

93. Let S be a second countable T_1 space. Prove that every sequence $\{A_n\}$ of subsets of S has a convergent subsequence. (*Hint.* See G. T. Whyburn, Analytic Topology, *American Mathematical Society Colloquium Publications*, vol. 28, p. 11, New York, 1948. It will repay you to study carefully the applications that Whyburn makes of his theorem.)

Bibliographical Comment

The works mentioned in our bibliographical comments on the previous chapter are also excellent references here. In addition, we should mention that the Polish journal *Fundamenta Mathematicae* lists problems, some of which might be suitable for term papers. The *Notices of the American Mathematical Society* also occasionally lists problems.

The proof we give on metrizability in Section 5 is an adaptation of the proof indicated in

André Weil, Sur les Espaces à structure uniforme et sur la topologie générale, *Actualités scientifiques et industrielles*, No. 551, Hermann et Cie, Paris, 1937.

For further information on the problem of metrizability, and, in particular, for the definition of a development, see

R. H. Bing, Metrization of Topological Spaces, *Canadian Journal of Mathematics*, vol. 3, pp. 175–186, 1951.

The literature concerning function spaces is voluminous. Perhaps the most appropriate references are:

S. Banach, *Théorie des opérations linéaires*, Warsaw, 1932.

M. Fréchet, *Les Espaces abstraits*, Gauthier-Villars, Paris, 1928.

At certain points in the text we have already referred to the reader to

Philip Franklin, *A Treatise on Advanced Calculus*, John Wiley & Sons, New York, 1940.

J. L. Kelley, Convergence in Topology, *Duke Mathematical Journal*, vol. 17, pp. 277–283, 1950.

Konrad Knopp, *Theory and Application of Infinite Series*, Blackie & Son, London, 1928.

Edmund Landau, *Grundlagen der Analysis*, Chelsea Publishing Co., New York, 1946.

E. J. McShane, *Integration*, Princeton University Press, Princeton, 1944.

G. T. Whyburn, Analytic Topology, *American Mathematical Society Colloquium Publications*, vol. 28, New York, 1948.

C H A P T E R   F I V E

ARCS AND CURVES

1. The Arc and Naturally Ordered Collections of Sets

An arc has already been defined as a homeomorph of the subspace J of the space R of real numbers, where J denotes the closed unit interval [0, 1]. From previous theorems and problems, we know that any homeomorph of J, that is, any arc, is a countably compact, non-degenerate, connected, locally connected, separable metric space. We know further that this list of properties is redundant in the sense that some of them imply some of the remaining ones. One fact about J not previously emphasized is that the deletion of either the point 0 or the point 1 or both these points from J leaves a connected set. Indeed these are the only points of J that have this property. It follows that any homeomorph of J must have two and only two points with these properties, and that any homeomorphism between J and the space must pair these points with 0 and 1, respectively. Such points are called non-cut points in accordance with the following definition.

DEFINITION 1.1. Let X be a subset of a space S. Then X is said to be a cutting of S if and only if $S - X$ is not connected. A single point of S is called a cut point of S if and only if it is a cutting of S; otherwise it is called a non-cut point of S.

Thus an arc has, among other properties, those of being a countably compact, non-degenerate, connected, separable metric space with at most two non-cut points. This section and the following one will be devoted to proving that the converse is also the case, and thus that this is a characterization of the arc.

To establish the result just mentioned, we must establish the existence of a homeomorphism between J and any space S which has these properties. A homeomorphism f is established by first showing that the space S has exactly two non-cut points. This is done by showing that every non-degenerate, countably compact, connected, separable T_1 space has at least two non-cut points. Since our space S has no more than two, it has exactly two non-cut points, and we denote these points by a and b. As remarked earlier, any homeomorphism must carry non-cut points into non-cut points; so, in defining our proposed homeomorphism $f: S \to J$, we define $f(a) = 0, f(b) = 1$. To define f for other points of S, we shall show that S can be ordered in a way that agrees with its topology. Since S can be ordered in this way, we say that S has a natural order. (We give a precise definition of a natural order presently.) The usual ordering of the real numbers R gives us what turns out to be a natural ordering of J. We then take a countable dense subset of S and a countable dense subset of J and show that, among the one-to-one correspondences that exist between any two countably infinite sets such as ours, we can, in this case, find one such correspondence that preserves the order. (The precise definition of what we mean by a correspondence or mapping preserving order appears shortly.) We use this one-to-one order preserving correspondence to define f on the countable dense subset of S. The fact that the orders on S and on J are both *natural* then comes into play in the proof that f, on the dense subset, is uniformly continuous, and thus possesses a unique continuous extension to all of S. This is the mapping f on S. The natural order again plays an important part in showing that f, as defined, is actually a homeomorphism.

We would like to suggest that the reader consult R. L. Wilder's article entitled The Origin and Growth of Mathematical Concepts, *Bulletin of the American Mathematical Society*, vol. 59, pp. 423–448, 1953, to get an idea of the history of the concept of the arc and some notion of the role it has played in the development of modern mathematics.

We now attend to the details of our proof. In obtaining the existence of non-cut points, and in our later work we need the concept of a separation of a set and related ideas.

DEFINITION 1.2. A subset X of a space S is said to separate a subset Y of S in S if and only if $S - X$ can be expressed as the union of two separated sets $M_1(X)$ and $M_2(X)$, such that $M_1(X) \cap Y \neq \emptyset$ and $M_2(X) \cap Y \neq \emptyset$. We call such a union a separation of Y. A subset X of a space S is said to separate two subsets A and B of S in S if and only if $S - X$ can be expressed as the union of two separated sets $M_1(X)$ and $M_2(X)$, such that A is contained in one of these sets and B is contained in the other. In this case, $M_1(X) \cup M_2(X)$ is called a separation of A and B in S.

It is important to realize that, given a cutting X of a connected space S, many different pairs of separated sets $M_1(X)$ and $M_2(X)$ may exist, so that $S - X = M_1(X) \cup M_2(X)$. Thus, in many of our proofs, we start by selecting one such pair $M_1(X)$ and $M_2(X)$ associated with a cutting X, and then, throughout the remainder of the proof, the sets $M_1(X)$, $M_2(X)$ are fixed by this selection. The reader should see, in this connection, Problem 1.

THEOREM 1.3. Let S be a countably compact, connected, non-degenerate, separable T_1 space. Then S has at least two non-cut points.

Proof. Since S is separable, there exists a sequence of points $\{p_n\}$ of S, such that the set $H = \bigcup_{n \in I} p_n$ is dense in S. We shall obtain a contradiction, under the assumption that S has at most one non-cut point. The contradiction appears in that we are able to construct, on the basis of this assumption, a connected set B which is a proper subset of S, contains H, and contains the one non-cut point of S if such a point exists. Since H is dense in S, and B contains H, B must be dense in S; i.e., $\bar{B} = S$. However, B is a proper subset of S, so that there must exist a point $p \in S - B$. This point p is a non-cut point of S, by Theorem III, 5.17, since $B \subset S - p \subset \bar{B} = S$. This is a contradiction, since the set B, as constructed, contains every non-cut point of S, and p is not a point of B.

Our construction of the set B depends on the following induction. Define n_1 to be the least integer such that the element p_{n_1} of H is a cut point of S. The set H must contain at least two points, since S is non-degenerate and T_1. But, by hypothesis, S has at most one non-cut point. Consequently, the point p_{n_1} must exist. Since p_{n_1} is a cut point of S, we have a separation of S given by

$$S - p_{n_1} = A_1 \cup B_1$$

where the notation is chosen in such a way that the non-cut point of S, if there is one, lies in B_1. Thus every point of the countably compact set $\bar{A}_1$ is a cut point of S. (Notice that $\bar{A}_1 = A_1 \cup p_{n_1}$, and $\bar{A}_1$ is a connected set by Corollary III, 5.20, since the space S is T_1.)

The set A_1 may be expressed in the form $A_1 = S - \bar{B}_1$. Therefore, A_1 is an open subset of S. There must thus exist a least integer n_2 such that the point p_{n_2} of H lies in A_1. Define A_2, B_2 to be the sets of a separation of S such that

$$S - p_{n_2} = A_2 \cup B_2$$

where the notation is chosen in such a way that the connected set $\bar{B}_1 = B_1 \cup p_{n_1}$ is contained in B_2. It is immediate that $\bar{A}_2 \subset A_1$.

In general, for $k \geq 2$, we define p_{n_k} to be the element of H with least index that lies in A_{k-1}, and define A_k, B_k to be the sets of a separation of S such that

$$S - p_{n_k} = A_k \cup B_k$$

where the notation is chosen in such a manner that the connected set $\bar{B}_{k-1} = B_{k-1} \cup p_{n_{k-1}}$ is contained in B_k, and, consequently, $\bar{A}_k \subset A_{k-1}$.

We define the set B by the equation $B = \bigcup_{k \in I} \bar{B}_k = \bigcup_{k \in I} [p_{n_k} \cup B_k]$, and note that this last set is the union of a collection of connected sets, each containing the point p_{n_1}. It follows that B is connected. But, for each index k, all elements of H having index less than n_{k+1} lie in B_{k+1}. Therefore, B contains H; hence $\bar{B} = S$. We also note that B contains the one non-cut point of S, if such a point exists.

The proof of the theorem will be complete if we show that B is a proper subset of S. To do this, define a set A by the equation $A = \bigcap_{k \in I} A_k$. Since $\bar{A}_{k+1} \subset A_k$, for every index k, it follows easily that $A = \bigcap_{k \in I} \bar{A}_k$, but, for each index k, the set $\bar{A}_{k+1}$ is a countably compact set contained in $\bar{A}_k$. Thus, by the Cantor product theorem, $A \neq \emptyset$. The set A is thus a non-empty set, and we see at once that A is a subset of $S - B$. Therefore, B is a proper subset of S, and the proof of the theorem is complete.

In line with the sketch of the manner in which we intend to characterize the arc, we now introduce the notion of a natural order.

DEFINITION 1.4. Let $\{A_\alpha\}$ be a collection of subsets of a space S, and let $<$ be an order relation defined on this collection. Then the order $<$ on $\{A_\alpha\}$ is said to be a natural order on $\{A_\alpha\}$ in S if and only if, for each index α, the sets L_α and R_α satisfy the conditions $\bar{L}_\alpha \cap R_\alpha = L_\alpha \cap \bar{R}_\alpha = \emptyset$, where L_α is the union of all elements A_β of $\{A_\alpha\}$ such that $A_\beta < A_\alpha$, and R_α is the union of all elements A_β of $\{A_\alpha\}$ such that $A_\alpha < A_\beta$. The symbol $<$ is usually read "precedes," and it is understood that $A_\beta > A_\alpha$ means $A_\alpha < A_\beta$ and that the symbol $>$ is to be read "follows." A collection of subsets of a space S is said to be naturally ordered in S if and only if there exists an order on the collection that is a natural order in S. Similarly, a space S is said to be naturally ordered if and only if the collection of all subsets of S, each consisting of a single point of S, is naturally ordered in S. A first element of an ordered collection is an element that precedes every other element of the collection in the given ordering. A last element of an ordered collection is an element that follows every other element of the collection in the given ordering. An end element is either a first element or a last element of an ordered collection.

As indicated previously, we call an order natural if it agrees with the topology. Suppose that S is a naturally ordered connected space. Then the set L_p of elements, which precede a given element p, and the set R_p

of elements, which follow p, are disjoint open subsets of the subspace $L_p \cup R_p$ of S. Furthermore, if p is a closed subset of S, then each of these sets is open in S. We also see that, if L_p and R_p are non-empty, then they are separated subsets of S; hence p is a cut point of S. Similar remarks hold for a naturally ordered collection of disjoint subsets of a space. Our next theorem shows that, for a certain type of collection of cuttings of a connected space, we can reverse the above remarks and obtain a natural order for the collection.

DEFINITION 1.5. Let S be a space, and G a collection of subsets of S. The collection G is said to be non-separated if and only if the elements of G are disjoint, and no element of G separates, in S, any two points of any other element of G.

We see at once that every collection of subsets of a space S, each member of which consists of a single point of S, is a non-separated collection of subsets of S.

THEOREM 1.6. Let A and B be disjoint non-empty subsets of a connected space S, and $K(A, B)$ a non-empty, non-separated collection of cuttings of S, each of which separates A and B in S. Then there is a natural order for $K(A, B)$ such that, if $X \in K(A, B)$ and

$$S - X = N_A(X) \cup N_B(X)$$

is any separation of A and B in S, $A \subset N_A(X)$, $B \subset N_B(X)$, then $L_X \subset N_A(X)$, $R_X \subset N_B(X)$.

Proof. Suppose that the conditions hold. We shall define an order on $K(A, B)$ in terms of a particular separation

$$S - X = M_A(X) \cup M_B(X), \qquad A \subset M_A(X), \qquad B \subset M_B(X) \quad (1)$$

of A and B in S, associated with each element X of $K(A, B)$. We shall prove that this order is a natural order, and that, if the separation

$$S - X = N_A(X) \cup N_B(X), \qquad A \subset N_A(X), \qquad B \subset N_B(X) \quad (2)$$

of A and B in S is due to an element X of $K(A, B)$, then $L_X \subset N_A(X)$, $R_X \subset N_B(X)$.

Consider the two particular separations (1) and

$$S - Y = M_A(Y) \cup M_B(Y), \qquad A \subset M_A(Y), \qquad B \subset M_B(Y) \quad (3)$$

chosen as associated with the distinct elements X and Y of $K(A, B)$. Since the collection $K(A, B)$ is non-separated, the set X must be contained in one of the sets $M_A(Y)$ and $M_B(Y)$. A similar statement holds for the set Y and the sets $M_A(X)$ and $M_B(X)$. It follows that exactly one of the statements

$$X \cup M_A(X) \subset M_A(Y) \tag{4}$$

and

$$Y \cup M_A(Y) \subset M_A(X) \tag{5}$$

is true. We define our order relation $<$ as follows:

$$X < Y \quad \text{if and only if} \quad M_A(X) \subset M_A(Y) \quad \text{and} \quad M_A(X) \neq M_A(Y)$$

$$Y < X \quad \text{if and only if} \quad M_A(Y) \subset M_A(X) \quad \text{and} \quad M_A(Y) \neq M_A(X)$$

It is clear from this definition and the preceding discussion that, given two elements X and Y of $K(A, B)$, then either $X < Y$ or $Y < X$. Also, if $X < Y$, then X and Y are distinct. To show that we have actually defined an order relation, the transitivity property remains to be established. Let X, Y, Z be elements of $K(A, B)$ such that $X < Y$ and $Y < Z$; i.e., $M_A(X) \subset M_A(Y)$, and $M_A(Y) \subset M_A(Z)$. Thus $M_A(X) \subset M_A(Z)$; hence $X < Z$. We have thus defined an order relation for the collection $K(A, B)$.

To show that this is a natural order, let X be an arbitrary element of $K(A, B)$. Clearly the union L_X of all elements Y of $K(A, B)$ satisfying $Y < X$ has the property that $L_X \subset M_A(X)$. Similarly, the union R_X of all elements Y of $K(A, B)$ such that $X < Y$ has the property that $R_X \subset M_B(X)$. Thus L_X and R_X are separated sets, and we have proved that the order we have defined is a natural order.

To complete the proof let $X \in K(A, B)$, and suppose that the separation (2) holds. Let Y be any element of $K(A, B)$ that precedes X. Then exactly one of the following statements must be true:

$$X \cup N_A(X) \subset M_A(Y) \tag{6}$$

$$Y \cup M_A(Y) \subset N_A(X) \tag{7}$$

From the definition of the order, it follows that (5) must hold; hence X is not contained in $M_A(Y)$. Therefore, (6) is false so that (7) is true. Since (7) holds for every element Y of $K(A, B)$ such that $Y < X$, it follows that $L_X \subset N_A(X)$. The fact that $R_X \subset N_B(X)$ can be established in a similar manner.

It is, of course, possible that other natural orders for the set $K(A, B)$ may exist. One such as we have defined will be called a standard natural order.

DEFINITION 1.7. Let A and B be disjoint non-empty subsets of a connected space S, and $K(A, B)$ a non-empty non-separated collection of cuttings of S, each of which separates A and B in S. A natural order for $K(A, B)$ is said to be a standard natural order from A to B if and only if, given any $X \in K(A, B)$ and any separation $S - X = N_A(X) \cup N_B(X)$ of S such that $A \subset N_A(X)$, $B \subset N_B(X)$, then $L_X \subset N_A(X)$, $R_X \subset N_B(X)$, where L_X and R_X are taken in the sense of this order. An order on the collection consisting of A, B, and the elements of $K(A, B)$ is called a standard natural order if and only if it is a natural order such that A is the first element, B is the last element, and this order on the collection $K(A, B)$ is a standard natural order from A to B.

COROLLARY 1.8. Let A and B be disjoint non-empty subsets of a connected space S, and $K(A, B)$ a non-empty non-separated collection of cuttings of S, each of which separates A and B in S. Then the collection consisting of A, B, and the elements of $K(A, B)$ has a standard natural order.

Proof. By Theorem 1.6, there is a standard natural order on $K(A, B)$ from A to B. Define the order on the collection $K(A, B)$ augmented by the elements A and B to be this standard natural order for elements of $K(A, B)$, and define $A < X$, for all $X \in K(A, B)$, $X < B$ for all $X \in K(A, B)$, and $A < B$. It follows easily that this is a natural order, and hence a standard natural order.

In the case that A and B are each single points, and $K(A, B)$ consists of sets each composed of a single point, we use a slightly different notation.

DEFINITION 1.9. Given two distinct points a and b of a connected space S, $E(a, b)$ denotes the set of all points of S that separate a and b in S.

As a special case of Theorem 1.6 and Corollary 1.8, we have the following theorem.

THEOREM 1.10. If a, b are distinct points of a connected space S, then the set $E(a, b) \cup a \cup b$ has a standard natural order.

The following theorem constitutes the first step in the characterization of the arc.

THEOREM 1.11. Let S be a non-degenerate, countably compact, separable, connected T_1 space which has at most two non-cut points. Then

(i) S has exactly two non-cut points, a and b.

(ii) $\overline{E(a, b)} = E(a, b) \cup a \cup b = S$.

(iii) If $x \in E(a, b)$, then the separation $S - x$ of S consists of exactly two connected sets, one containing a, the other containing b.

(iv) $S = E(a, b) \cup a \cup b$ has a standard natural order.

Proof. Suppose the conditions hold. Then (i) follows from Theorem 1.3. Consider (ii), and suppose $x \in S - (a \cup b)$, where a and b are the two non-cut points of S. We then have a separation

$$S - x = A \cup B$$

and we note that each of the sets $A \cup x$ and $B \cup x$ is a countably compact, connected, non-degenerate, separable subspace of S. Thus, in particular, $A \cup x$ has at least two non-cut points. Let c be a non-cut point of $A \cup x$ distinct from x. We may then write $S - c = [(A \cup x) - c] \cup [B \cup x]$, which is the union of two connected sets with the common point x. Thus $S - c$ is a connected set; hence c is a non-cut point of S. Thus c must be either the point a or the point b, and we may choose our notation in such a way that $c = a$. Thus $a \in A$. In exactly the same way, we can show that $b \in B$. Consequently, x separates a and b in S; i.e., $x \in E(a, b)$. Thus (ii) holds, since, in addition, we know that every point of S is a limit point of S, and thus every neighborhood of a point of S (S is a T_1 space) contains infinitely many points of S and hence of $E(a, b)$.

To establish (iii) we need only repeat the above argument about non-cut points and observe, for instance, that, since $A \cup x$ contains only one of the non-cut points of S, then x must be a non-cut point of $A \cup x$; hence A is connected set. Statement (iv) now follows directly from Theorem 1.10.

Our next step in the characterization of the arc is to find out more precisely how a natural order on a space is related to the topology of that space. First, however, we need to define some terms related to ordered sets.

DEFINITION 1.12. Let F be a set ordered by $<$. Then we say that a point z of F lies between two points x and y of F if and only if either $x < z < y$ or $y < z < x$. We say that y next follows x in F if and only if $x < y$ in F and no element of F lies between x and y in F. Similarly, we say that x next precedes y in F if and only if y next follows x in F. We say that x and y are adjacent elements in F or, equivalently, that x is next to y in F, if and only if no element of F lies between x and y in F.

DEFINITION 1.13. Let K be an ordered set. Given $x, y \in K$, $x < y$, then (x, y) denotes the set of all elements of K that lie between x and y and is called an open-interval in K. Given $x, y \in K$, y not preceding

x in K, then $[x, y]$ denotes the set of elements of K consisting of x, y and all elements of K that lie between x and y. The set $[x, y]$ is called a closed-interval of K. In either case, the elements x and y of K are called the end points of the interval. A subset of K consisting of an open-interval of K, together with exactly one of its end points is called a half-open-interval. If (x, y) is an open-interval of K, then $[x, y)$ denotes the half-open-interval that contains the point x, and $(x, y]$ denotes the half-open-interval that contains the point y.

The following elementary facts about ordered sets are assumed to be known. We state them as theorems, which are left to the reader to prove.

THEOREM 1.14. If F is a finite ordered set and x is an element of F which is not the last element of F, then there exists an element of F next following x. Similarly, if x is an element of F which is not the first element of F, then there exists an element of F next preceding x.

THEOREM 1.15. Given an element x of an ordered set F, then there is at most one element of F that next follows x and there is at most one element of F that next precedes x.

THEOREM 1.16. Let F be a finite ordered set containing exactly n elements. Then there is at least one way in which a subset of a set L of $m \geq n$ integers can be used to index F, such that x_{k_1} of F precedes x_{k_2} of F if and only if the elements k_1, k_2 of L are such that $k_1 < k_2$ in the usual order of the integers. Furthermore, if $m = n$, then there is one and only one way in which this indexing can be done, and it is such that, given any $k \in L$, there is an element of F that is indexed by k.

THEOREM 1.17. If F is a finite non-empty ordered set, then F has a first element and a last element.

With regard to these theorems, notice that the set of real numbers, as usually ordered, has neither a first element nor a last element, and that no element has a next element.

Since any non-degenerate countably compact connected metric space S with at most two non-cut points has a natural order, and we intend to prove it to be an arc, it is to be expected that we can show that the open-intervals in the natural order are not only open sets of S but also that these open sets form a basis for the topology of S. If we augment this collection by those half-open-intervals that contain the first and last points of S in this order, then this is indeed the case as a consequence of the following theorem.

THEOREM 1.18. Let S be a non-degenerate naturally ordered T_1 space, and denote by σ the collection of subsets of S consisting of all open-intervals of S, and all those half-open-intervals of S that contain an end point of S, where these intervals are taken in the sense of the natural order on S. In the event that S is either compact or both countably compact and separable, the collection σ is a basis for the topology of S.

Proof. The use we make of the compactness, or countable compactness with separability, is to establish that any non-empty closed subset of S has a first point and a last point. The proof will be written, employing compactness as an hypothesis, but is also valid if countable compactness with separability is employed. We show first that every element of σ is an open set of S, and then show that, given any point p of S and any neighborhood U of p in S, there exists an element A of σ such that $p \in A \subset U$. Thus σ is shown to be a basis for the topology of S.

Every non-empty element of σ is of one of the forms $[a, x)$, (x, y), $(y, b]$, where a and b denote the first and last elements of S (they exist since S is compact and T_1), and x, y are points of S for which these intervals are defined. For any point $x \in S - a$, the set $R_x \cup x$ is closed; hence its complement $[a, x)$ is open. Similarly, if $y \in S - b$, then $L_y \cup y$ is closed; hence its complement $(y, b]$ is open. If $x < y$, then (x, y) is the intersection of the two open sets $[a, y)$ and $(x, b]$, and is thus an open set.

To complete the proof, let p be a point of S and U an arbitrary neighborhood of p. Suppose first that $p = a$. Then there exists a neighborhood V of p, defined as the intersection of U with $[p, b)$, which is contained in U, but does not contain b. Thus $S - V$ is a non-empty closed subset of S, and hence compact. Let q be the first point of $S - V$. It is clear that $p \neq q$ and that $[p, q)$ is an element of σ containing p and contained in U. The proof is thus complete if $p = a$, and a similar proof holds if $p = b$.

In the case that p is distinct from both a and b, we define the neighborhood V of p to be the intersection of the three open sets $[a, b)$, U, and $(a, b]$. This neighborhood V is contained in U and contains neither the point a nor the point b. Thus each of the sets $K_1 = (L_p \cup p) \cap (S - V)$ and $K_2 = (R_p \cup p) \cap (S - V)$ is closed, non-empty, and hence compact. Let q be the last point of K_1, and r the first point of K_2. It is clear that $q < p < r$. Thus the element (q, r) of σ contains p and is contained in U. This completes the proof of Theorem 1.18.

Given a non-degenerate ordered set S, the topology for S which has the collection σ described in Theorem 1.18 as a basis is called the interval topology for S. It is, of course, necessary to prove that the collection does generate a topology, in general. This is a straightforward proof, which is left to the student.

THEOREM 1.19. Let S be a non-degenerate ordered set. Then the collection σ of subsets of S described in Theorem 1.18 is a basis for a topology for S. In this topology, the given order for S is a natural order for S.

DEFINITION 1.20. Let S be a non-degenerate ordered set. Then the interval topology for S is the topology for S that has the collection of subsets of S described in Theorem 1.18 as a basis. The intervals are taken in the sense of the given order on S.

It will be observed that Theorem 1.18 states that the interval topology of a non-degenerate naturally ordered compact (or countably compact and separable) T_1 space is the same as the given topology. We know that the interval topology of J is the same as the usual topology of J. For a non-degenerate, countably compact, connected, separable metric space with at most two non-cut points, we have shown the existence of a natural order, and that the interval topology in this order is the given topology. There are thus many similarities with J. A further similarity is the result, as a consequence of the next theorem, that every interval is a connected set.

THEOREM 1.21. Let S be a non-degenerate, naturally ordered, compact (or countably compact and separable) T_1 space. If S is connected, then S is locally connected, and every interval in S is a connected set. Furthermore, every non-degenerate interval in S contains infinitely many points, and every open interval in S is non-degenerate.

Proof. The interval topology (in the sense of the given natural order) for S is the same as the given topology of S. Let $[x, y]$ be any closed interval of S. Then the mapping $g: S \to [x, y]$ defined, for any $s \in S$, by $g(s) = s$ if $s \in [x, y]$, $g(s) = x$ if $s \in L_x$, $g(s) = y$ if $s \in R_y$, is easily seen to be continuous. Thus $[x, y]$ is a connected set. If $x < y$, then every point of (x, y) is a cut point of $[x, y]$. Since $[x, y]$ must have two non-cut points, each of the points x and y is a non-cut point of $[x, y]$. Consequently, each of the sets $[x, y)$ and $(x, y]$ is connected.

Since the non-degenerate closed interval $[x, y]$ is connected, it must contain infinitely many points (from which we see that every non-degenerate interval in S contains infinitely many points). Let p be a point of S between x and y. Then the open interval (x, y) is the union of the two half-open-intervals $(x, p]$ and $[p, y)$, each of which has been seen to be a connected set. Thus (x, y) is connected, and we have proved that every interval in S is a connected set.

Finally, S is locally connected, since it has a basis (namely, its open intervals), every element of which is a connected set.

We have thus seen how very similar a non-degenerate, compact, connected metric space having at most two non-cut points is to J in its properties. To establish the homeomorphism by the scheme indicated earlier, we shall make use of the concept of an ε subdivision of such a space to find a countable dense subset, and then complete our proof of the characterization in the next section.

DEFINITION 1.22. Let S be an ordered metric space with a first point a and a last point b, distinct from a. Given $\varepsilon > 0$, a finite subset F of S is said to be an ε subdivision of S if and only if $a \in F$, $b \in F$, and, given any two adjacent points $x, y \in F$, $x < y$, the diameter of each of the half-open-intervals $[x, y)$ and $(x, y]$ is less than ε.

THEOREM 1.23. Let S be a non-degenerate, compact, naturally ordered metric space having a first point a and a last point b. Given $\varepsilon > 0$, there exists an ε subdivision F of S. Furthermore, if S is infinite and F contains exactly N points, then, for any positive integer k, there exists an ε subdivision of S containing exactly $N + k$ points.

Proof. Suppose that we are given a real number $\varepsilon > 0$ and a space S satisfying the hypotheses of the theorem. Define a set K_a to consist of the point a and all points $x \in S$ for which there exists a finite set of points $x_0, x_1, \cdots, x_n$ of S satisfying $a = x_0 < x_1 < \cdots < x_n = x$, and, for $i = 1, 2, \cdots, n$, each of the half-open-intervals $[x_{i-1}, x_i)$ and $(x_{i-1}, x_i]$ has diameter less than ε. The first part of the theorem will be proved if we show that $b \in K_a$. This will be done by establishing the following three facts about the set K_a:

(i) If $x \in K_a$, then $L_x \subset K_a$.

(ii) If K_a has a last point, then b is the last point of K_a.

(iii) K_a is closed, hence compact; thus K_a has a last point; thus $b \in K_a$.

The proof of (i) is left as an exercise for the student. To prove (ii) suppose that c is a point of K_a distinct from b. We must show that c is not the last point of K_a; i.e., that there exists a point of K_a that follows c. Since the interval topology is the topology of S, there exists a point f of S, such that $c < f$, and the diameter of $[c, f)$ is less than the smaller of the two numbers $\rho(c, b)$ and ε. (We phrase the proof this way to include the case $c = a$.) Clearly, if (c, f) is empty, then $f > c$, and $f \in K_a$. On the other hand, if (c, f) is non-empty, then any point of this open interval is a point of K_a following c. Thus (ii) is established.

To prove (iii), assume that there exists a limit point d of K_a which is not in K_a. From (i), we see at once that $K_a \subset L_d$. We may therefore find a point c of K_a such that $c < d$, and each of the half-open-intervals

$[c, d)$ and $(c, d]$ has diameter less than ε. It follows at once, from the definition of K_a, that $d \in K_a$; this is a contradiction; hence K_a is closed. Thus K_a is compact. Therefore, by Problem 6, K_a has a last point. Hence, by (ii), $b \in K_a$. Thus (iii) is established. Hence S has an ε subdivision.

To complete the proof of the theorem, suppose that S is infinite, and that an ε subdivision F containing exactly N points is given. Then, between some two adjacent points x, y of F, $x < y$, there are infinitely many points. If we adjoin one of these points to F, the result is an ε subdivision of S containing exactly $N + 1$ points. Using this process, the final statement of the theorem is easily established by mathematical induction.

PROBLEMS

1. (*a*) Let S denote the subspace of the Cartesian plane E^2 consisting of the so-called x and y axes. Let p denote the point $(0, 0)$ of E^2. Show that S is connected, and that p is a cut point of S; so there exist separated sets A and B such that $S - p = A \cup B$. Show that, even with the requirement that $(-1, 0) \in A$ and $(1, 0) \in B$, the sets A and B are not unique.

(*b*) Let S be a space, and a, b two points of S. Prove that, if there exists a cut point p of S such that $S - p = A \cup B$, where $a \in A$, $b \in B$, with A and B separated, then, if both A and B are connected, the separation $S - p = A \cup B$ is unique. By this separation being unique we mean that, if A_1, B_1 is any pair of separated sets such that

$$S - p = A_1 \cup B_1, \qquad a \in A_1, \qquad b \in B_1$$

then $A_1 = A$, $B_1 = B$.

2. In connection with the proof of Theorem 1.3, show that all elements of H of index less than n_{k+1} lie in B_{k+1}, and thus that $H \subset B$.

3. In connection with the proof of Theorem 1.6, show that, if $X, Y \in K(A, B)$ so that, as in the proof,

$$S - X = M_A(X) \cup M_B(X)$$

$$S - Y = M_A(Y) \cup M_B(Y)$$

then exactly one of the statements

$$X \cup M_A(X) \subset M_A(Y)$$

$$Y \cup M_A(Y) \subset M_A(X)$$

holds. (*Hint*. Establish the validity of equations such as

$$S = M_A(X) \cup M_B(Y) \cup [M_A(Y) \cap M_B(X)] \cup X \cup Y$$

with varying arrangements of indices A and B, and use the hypotheses.)

4. Consider the proof of Theorem 1.11. How do we know that, given a separation $S - X = A \cup B$ of S, each of the sets $A \cup x$ and $B \cup x$ is a countably compact, connected, non-degenerate, separable T_1 space?

5. Give in detail the argument outlined in the proof of Theorem 1.11 for establishing the validity of (iii).

6. Let S be a non-degenerate naturally ordered T_1 space. Prove that, if S is compact, then S has a first point and a last point. (*Hint.* Show that, if S has no first point, then the collection $\{R_p\}_{p \in S}$ is an open covering of S.)

7. Let S be a non-degenerate naturally ordered T_1 space. Prove that, if S is countably compact and separable, then S has a first point and a last point.

8. Prove that J is naturally ordered under the usual ordering of the real numbers.

9. Let S be a naturally ordered non-degenerate connected space. What can you deduce concerning the topology of S? For example, is S a T_1 space?

10. Prove that every collection of disjoint connected subsets of a space S is non-separated.

2. The Arc

To complete the characterization of the arc we must establish the existence of a homeomorphism from any non-degenerate countably compact connected separable metric space S with at most two non-cut points onto the closed unit interval J. In the previous section, we observed that, if such a homeomorphism exists, then S must have exactly two non-cut points and must be naturally ordered. Our development in this section is based on the further observation that, if a homeomorphism exists, then a countable dense subset K of S corresponds to the rationals in J, and that this subset K of S can be indexed by the rationals in J in such a manner that, if x_r and x_s belongs to K, then x_r precedes x_s in S if and only if the rational numbers r and s of J satisfy the condition $r < s$. This indexing, of course, is such that, under the homeomorphism, the point x_r of K corresponds to the rational number r of J. So, to establish a homeomorphism, we attempt to reverse this process. Namely, we attempt to find a countable dense subset K of S and to index it by rational numbers of J in agreement with the order. Then, from the indexing, we define a mapping whose continuous extension to S is a homeomorphism of S onto J.

The basic theorem on indexing of a countable infinite ordered set K appears shortly. We will show later that we can find subsets of S that satisfy its hypotheses, and then use the indexing to establish the homeomorphism with J. In our proof of this indexing theorem, we suppose that the set K is written as the union of a countable collection of finite subsets F_i of K. This is done because, in a finite ordered set, after assigning an index to one element (not the last element), the next following element in the set is uniquely determined, and we can then assign an index to it. Thus an inductive definition of indices for all elements of $K = \bigcup_{i \in I} F_i$ is possible by first indexing F_1, then F_2, and, after indexing F_i, then indexing F_{i+1}. Notice in connection with this remark that, in a countable infinite ordered set such as the rational numbers in R, no element has a next element. We now state and prove the indexing theorem.

THEOREM 2.1 (*The Indexing Theorem*). Let $\{F_i\}$ be a collection of finite subsets of a set S. Define $K = \bigcup_{i \in I} F_i$, and suppose that the following conditions hold:

1. $F_i \subset F_{i+1}$, for $i \in I$, and associated with each F_i is a positive integer n_i.
2. F_1 contains exactly $2^{n_1} + 1$ points.
3. K is ordered by $<$ and has a first and a last element. Both these elements belong to F_1.
4. If x and y are adjacent elements in F_i with respect to the order $<$ in K, then exactly one of the following conditions holds:
   (i) No element of K lies between x and y.
   (ii) There are exactly $2^{n_{i+1}} - 1$ elements of K between x and y in F_{i+1}.

Then there is one and only one way in which each element of K can be denoted by x_k^i, i and k integers, $i \geq 1$, $k \geq 0$, such that

(*a*) An element of K is denoted by x_k^i if and only if that element lies in F_i.

(*b*) If x_k^i denotes an element of K, then $0 \leq k \leq 2^{n_1 + n_2 + \cdots + n_i}$.

(*c*) If $x_{k_1}^i$ and $x_{k_2}^i$ denote elements of K, then $x_{k_1}^i < x_{k_2}^i$ in K if and only if $k_1 < k_2$.

(*d*) If x_k^i and x_h^j denote the same element of K, where $i > j$, then $k = (2^{n_{j+1} + \cdots + n_i})h$.

(*e*) If x_k^i and x_h^i denote adjacent elements of F_i and there exists an element of K between these elements, then $|h - k| = 1$.

(*f*) x_0^1 and $x_{2^{n_1}}^1$ denote the first and last points of K, respectively.

Proof. Suppose that a collection of finite sets $\{F_i\}_{i \in I}$ and a set $K = \bigcup_{i \in I} F_i$ are given, satisfying the conditions of the hypothesis. Our proof is by mathematical induction. We show first that there is one and only one way in which indices can be assigned to F_1, satisfying conditions (*a*) through (*f*). Then we show that, if there is one and only one way in which indices can be assigned to F_i, for $i = 1, 2, \cdots, m$, satisfying conditions (*a*) through (*f*), then there is one and only one way in which indices can be assigned to F_i, $i = 1, 2, \cdots, m + 1$, satisfying these same conditions. When this has been done, the proof of the theorem is complete.

Since F_1 contains exactly $2^{n_1}+1$ elements, there is, by Theorem 1.16 exactly one way that subscripts k, $0 \leq k \leq 2^{n_1}$, can be assigned to its elements so that $x^1_{k_1} < x^1_{k_2}$ if and only if $0 \leq k_1 < k_2 \leq 2^{n_1}$. It then follows from (1), (2), and (3), that (*f*), as well as (*a*) through (*e*), is satisfied. Suppose then, that there is one and only one way in which indices can be assigned to F_i, $i = 1, 2, \cdots, m$, satisfying conditions (*a*) through (*f*). We then assign indices to the elements of F_{m+1} in the following way. Each element will be denoted by x^{m+1}_k, where the superscript is chosen as $m+1$ to satisfy condition (*a*). The subscripts k are determined as follows. Each element of F_{m+1} that occurs in F_m has already been denoted as an element of the form x^m_h. In F_{m+1}, this element is indexed as $x^{m+1}_{(2^{n_{m+1}})h}$. Not only can this be done, but also it is the only choice that can be made so that condition (*d*) is satisfied. Moreover, since $0 \leq h \leq 2^{n_1+\cdots+n_m}$, we have that $0 \leq (2^{n_{m+1}})h \leq 2^{n_1+\cdots+n_{m+1}}$, and condition (*b*) is satisfied. Also, the elements of F_{m+1} which lie in F_m satisfy (*c*) when indexed in this manner, since (*c*) holds when these elements are indexed in the form x^m_h. Now, by (1) and (3), every element of F_{m+1} either is an element of F_m or lies between two adjacent elements of this set. If it lies between two adjacent elements of F_m, then their subscripts differ by unity when the elements are written in the form x^m_h. Consequently, these elements may be denoted in F_{m+1} by $x^{m+1}_{(2^{n_{m+1}})h}$ and $x^{m+1}_{(2^{n_{m+1}})(h+1)}$. Furthermore, by (4), there are, in this instance, exactly $2^{n_{m+1}} - 1$ points between these two points, and thus, by Theorem 1.16, there is exactly one way in which the integers k, satisfying $(2^{n_{m+1}})h < k < (2^{n_{m+1}})(h+1)$, may be assigned as subscripts so that (*c*) is satisfied, not only for these points, but also for these points together with the two points $x^{m+1}_{(2^{n_{m+1}})h}$ and $x^{m+1}_{(2^{n_{m+1}})(h+1)}$. In this manner indices of the form x^{m+1}_k are assigned to every element of F_{m+1}. It is clear that this choice of indices is the only one for which conditions (*a*), (*c*), and (*d*) are satisfied. Furthermore, it follows that not only are (*a*), (*c*), and (*d*) satisfied for F_{m+1}, but also all of the conditions (*a*) through (*f*) are satisfied for this indexing of F_i, $i = 1, 2, \cdots, m+1$. Thus the theorem is established by mathematical induction.

Corollary 2.2. Let $\{F_i\}_{i\in I}$ be a collection of finite subsets of a set S. If $K = \bigcup_{i\in I} F_i$, and the members of the collection $\{F_i\}$ satisfy the hypotheses of Theorem 2.1, then there exists a subset D of the rational numbers in J containing 0 and 1 and a one-to-one mapping $f: K \to D$, which is onto D and satisfies the condition that, for $x, y \in K$, $f(x) < f(y)$ in D if and only if $x < y$ in K.

Proof. Suppose that the set K is given, satisfying the hypotheses of

Theorem 2.1. Then there is a unique indexing for K which satisfies conditions (a) through (f) of that theorem. Suppose that K has been indexed in this manner. Define a mapping $g: K \to J$ by $g(x_k^i) = k/2^{n_1+n_2+\cdots+n_i}$, for each element x_k^i of K. From conditions (c) and (d), we have that, if x_k^i and x_h^j denote the same element of K, then $g(x_k^i) = g(x_h^j)$. From (a), we see that g is defined for every point of K. From (b), we have $0 \leq g(x) \leq 1$ for every $x \in K$, whereas, from (c) and (d), we see that $g(x) < g(y)$ in J if and only if $x < y$ in K, and thus g is one-to-one. It is also clear that the set $D = g(K)$ is a set of rational numbers in J containing the point 0 and the point 1, and that the mapping $f: K \to D$, defined by $f(x) = g(x)$ for every $x \in K$, and the set D itself satisfy the requirements of the corollary.

We shall make such repeated use of the indexing theorem and its corollary that it is convenient to introduce the following terminology.

DEFINITION 2.3. Let S and T be ordered sets and $f: S \to T$ a mapping. Then f is said to be order-preserving if and only if, given $x, y \in S$, then $f(x)$ precedes $f(y)$ in the order of T if and only if x precedes y in the order of S.

DEFINITION 2.4. We say that an ordered set $K = \bigcup_{i \in I} F_i$ has the index property if and only if, in the order for K, the set K and the elements F_i of the union satisfy the hypotheses of the indexing theorem. If an ordered set $K = \bigcup_{i \in I} F_i$ has the index property and x_k^i denotes an element of K, then the indexing of this element is in accord with the particular indexing of K satisfying the conclusions of the indexing theorem.

DEFINITION 2.5. If an ordered set $K = \bigcup_{i \in I} F_i$ has the index property, then we refer to the set D and the mapping f of Corollary 2.2 as the associated set of rationals and the associated mapping onto this set, respectively.

Before giving the proof of the characterization of the arc, we study the continuity of certain maps related to the associated mapping onto the associated set of rationals of an ordered subset $K = \bigcup_{i \in I} F_i$ of a metric space S, where the set K is assumed to have the index property.

THEOREM 2.6. Let $K = \bigcup_{i \in I} F_i$ be an ordered subset of a metric space S, where K has the index property, D is the associated set of rationals, $g: D \to S$ is a mapping defined by $g(x) = f^{-1}(x)$ for $x \in D$, where f is the associated mapping of K onto D. If, for every $i \in I$, and every half-open-interval G_i of K whose end points are adjacent elements of F_i, the diameter of G_i satisfies $\delta(G_i) < 1/2^i$, then g is uniformly continuous.

Proof. Suppose the hypotheses are satisfied. Given $\varepsilon > 0$, there exists a positive integer k such that $1/2^k < \varepsilon$. Define $\delta = (1/2)(1/2^{n_1+n_2+\cdots+n_{k+1}})$. We shall show that, given $t, t_1 \in D$ satisfying $|t - t_1| < \delta$, then $\rho[g(t), g(t_1)] < \varepsilon$, and thus g is uniformly continuous. Suppose that these conditions hold, i.e., that $t, t_1 \in D$ and $|t - t_1| < \delta$. Then there exists an integer h satisfying $0 < h < 2^{n_1+\cdots+n_{k+1}}$ such that t and t_1 both lie in the closed interval $[(h-1)/2^{n_1+\cdots+n_{k+1}}, (h+1)/2^{n_1+\cdots+n_{k+1}}]$. Since

$$\frac{h+1}{2^{n_1+\cdots+n_{k+1}}} - \frac{h-1}{2^{n+\cdots+n_{k+1}}} = \frac{2}{2^{n_1+\cdots+n_{k+1}}} \leq \frac{1}{2^{n_1+\cdots+n_k}}$$

(because $n_{k+1} > 0$), it follows that $g(t)$ and $g(t_1)$ must lie in a set G_k. Accordingly, $\rho[g(t), g(t_1)] < \varepsilon$.

If, with the notation of Theorem 2.6, we know that S is a naturally ordered compact metric space and that the order on K is that of K as a subset of S, then the mapping $h\colon K \to J$ defined by $h(x) = f(x)$, for every $x \in K$, is also uniformly continuous.

THEOREM 2.7. Let $K = \bigcup_{i\in I} F_i$ be a subset of a naturally ordered compact metric space S where, with K ordered as a subset of S, K has the index property. Let D be the associated set of rationals to K, and $h\colon K \to J$ a mapping defined by $h(x) = f(x)$, for every $x \in K$, where f is the associated mapping of K onto D. Then the mapping h is uniformly continuous.

Proof. Suppose that $K = \bigcup_{i\in I} F_i$, a subset of a naturally ordered compact metric space S, is given so that the hypotheses are satisfied. Given $\varepsilon > 0$, we must show that there exists a $\delta > 0$, such that, for any pair of points t_1 and t_2 of K satisfying $t_2 \in S_\delta(t_1)$, we must have $|h(t_2) - h(t_1)| < \varepsilon$. We get at the existence of such a δ indirectly.

We first show that, for each $i \in I$, there exists a $\delta_i > 0$ such that, for $t \in K$, $S_{\delta_i}(t) \cap F_i$ is at most two adjacent elements of F_i if t is not in F_i, and $S_{\delta_i}(t) \cap F_i = t$ if $t \in F_i$. We then show that, for some integer i, δ_i is the δ we seek. Let t be a point of F_i which is neither the first point nor the last point of F_i, and suppose that x and y are the next preceding and next following point of t in F_i. Then $L_x \cup R_y \cup x \cup y$ is a compact subset of S disjoint with t. The distance from t to this set is not zero, and we define δ_t to be one-half the distance from t to that set. If t is the first point or the last point of F_i, then only one of the compact sets $L_x \cup x$, and $R_y \cup y$ is defined, and we define δ_t to be one-half the distance from t to that set. If t is a point of K not in F_i, then t lies between two adjacent elements x_h^i and x_{h+1}^i of F_i. Let x_j^i and x_k^i denote the elements next preceding and next following x_h^i and x_{h+1}^i, respectively, if such exist, in F_i. Then the closed interval $[x_h^i, x_{h+1}^i]$ of S is a compact subset of S disjoint with the compact subset $(L_{x_j^i} \cup x_j^i) \cup (R_{x_k^i} \cup x_k^i)$ of S. (Take $L_{x_j^i} \cup x_j^i = \emptyset$ if x_j^i does not exist, and take $R_{x_k^i} \cup x_k^i = \emptyset$ if x_k^i does not exist. Certainly at least one of these points exists, since F_1 contains at least three points.) The distance between these sets is not zero, and we define δ_t to be one-half this distance. Notice that, for each t, the number δ_t which we have defined depends only on the two adjacent points of F_i between which t lies. Since F_i is a finite set, it follows that the set of positive numbers $\bigcup_{t \in K} \delta_t$ is finite. There must thus be a least element of this set, and we denote this least element by δ_i. Clearly $\delta_i > 0$.

Now, let t_1, t_2 be any two points of K satisfying $t_1 \in F_i$, $t_2 \in S_{\delta_i}(t_1)$. Then one of the following conditions must hold:

(*a*) $t_2 = t_1$.

(*b*) $t_2 \in (t_3, t_1)$, where t_3 next precedes t_1 in F_i.

(*c*) $t_2 \in (t_1, t_4)$, where t_4 next follows t_1 in F_i.

In case (*a*), $|h(t_2) - h(t_1)| = 0$. In case (*b*), since $t_2 \in (t_3, t_1)$, it is not an end point of this interval, and thus the subscripts of the points t_3, t_1, when these points are expressed in the form x_k^i differ by exactly unity. Thus, from the definition of h in terms of f, it follows in case (*b*), and similarly in case (*c*), that

$$|h(t_2) - h(t_1)| < 1/2^{n_1 + \cdots + n_i}. \tag{1}$$

If neither t_1 nor t_2 lies in F_i, it follows by similar reasoning that $t_2 \in S_{\delta_i}(t_1)$ implies

$$|h(t_2) - h(t_1)| < 2/2^{n_1 + \cdots + n_i}. \tag{2}$$

Thus, in any event, $t_1, t_2 \in K$, $t_2 \in S_{\delta_i}(t_1)$ implies that (2) holds.

It is now a simple matter to show that h is uniformly continuous on K. Given $\varepsilon > 0$, we know, since $n_i \geq 1$ for $i \in I$, that there exists an $i \in I$ such that

$$2/2^{n_1 + \cdots + n_i} < \varepsilon \tag{3}$$

Thus, for this i, if we take $\delta = \delta_i$, then, for any two points t_1, t_2 of K satisfying $\rho(t_1, t_2) < \delta$, we must have $|h(t_2) - h(t_1)| < \varepsilon$, by (2) and (3). This completes the proof of Theorem 2.7.

We have now assembled the machinery needed for our characterization of the arc.

THEOREM 2.8. A topological space S is an arc if and only if it is a non-degenerate, countably compact, connected, separable metric space having at most two non-cut points.

Proof. If S is an arc, then the conditions hold as we have already seen. Suppose the conditions hold. Then, since S is non-degenerate and connected, it must have exactly two non-cut points, and we denote these points by a and b. Moreover, the space S has a natural order in which a and b are its first and last points, respectively. By Theorem 1.23 there must exist an $\varepsilon = 1/2$ subdivision of S containing exactly $2^{n_1} + 1$ points for some positive integer n_1. We denote by F_1 the set of points of this subdivision. Each closed interval of S which has adjacent points of F_1 as end points is easily seen to be a non-degenerate connected set, and thus contains infinitely many points. The hypotheses of Theorem 1.23 again apply, and there must thus exist an $\varepsilon = 1/2^2$ subdivision of each such interval. Since there are only a finite number of these intervals to be subdivided, there must exist a positive integer n_2 such that each subdivision of an interval contains exactly $2^{n_2} + 1$ points. The set of all the points of these subdivisions is defined to be the set F_2. In a similar manner, by induction, if F_i has been constructed to be an $\varepsilon = 1/2^i$ subdivision of S, then F_{i+1} is defined to be an $\varepsilon = 1/2^{i+1}$ subdivision of S containing F_i, and having the property that each interval of F_i which was subdivided had precisely $2^{n_{i+1}} + 1$ points in its subdivision, for some positive integer n_{i+1}. The set of points $K = \bigcup_{i \in I} F_i$ thus defined has the index property, and is clearly dense in S. Also the hypotheses of both Theorems 2.6 and 2.7 are satisfied so that, in the notation of those theorems, each of the mappings

$$g: D \to S$$

and

$$h: K \to J$$

is uniformly continuous. Clearly, $g^{-1}(g(x)) = h(g(x))$, and each of these mappings is one-to-one. Thus, by Problem IV, 79, the extensions of these mappings yield a homeomorphism $t: S \to D^\star$, where $D \subset D^\star \subset \bar{D}$. Since $t(a) = 0$, $t(b) = 1$, and S is connected, we must have $D^\star = J$. Thus t is a homeomorphism of S onto J, and S is therefore an arc.

PROBLEM

11. Let S be a non-degenerate ordered set. Prove that S, in its interval topology, is homeomorphic to the space R of all real numbers if and only if the following conditions hold:

(i) S has no first point and no last point.

(ii) S is connected.

(iii) There exists a sequence $\{p_n\}$ of points of S, such that, given any two points $x, y \in S$ such that $x < y$, there exists an integer N such $x < p_N < y$.

3. Simple Closed Curves

Just as the prototype of the arc is the closed unit interval J, so the prototype of the simple closed curve is the unit circle $x^2 + y^2 = 1$.

DEFINITION 3.1. The unit circle is the subspace of E^2 consisting of all points (x, y) that satisfy the equation $x^2 + y^2 = 1$. (Throughout this and the next section, we will use K to denote the unit circle.)

DEFINITION 3.2. A space C is said to be a simple closed curve if and only if it is homeomorphic with the unit circle K.

The following two theorems are easily established. The first deals with properties of the unit circle K itself; the second with properties of K as a subset of E^2.

THEOREM 3.3. The unit circle K is a compact, connected, non-degenerate metric space satisfying the following conditions:

(i) Every point of K is a non-cut point of K.

(ii) For every pair of distinct points a, b of K, the set $K - (a \cup b)$ is not connected.

THEOREM 3.4. The unit circle K as a subset of E^2 has the property that $E^2 - K$ is the union of two disjoint open sets A and B. Furthermore, each of the sets A, B is connected, and $K = \bar{A} - A = \bar{B} - B$.

It is natural to ask, in connection with the first theorem, whether the given conditions characterize simple closed curves in general. Similarly, does the result of the second theorem hold for any subset of E^2 that is a simple closed curve? The answer to both these questions is in the affirmative, as we shall prove in this and the next section. The indicated generalization of Theorem 3.4 is known as the Jordan curve theorem.

In our treatment of the characterization of simple closed curves, and the Jordan curve theorem, the concepts of components and regions are useful.

DEFINITION 3.5. Let S be a topological space. A subset X of S is said to be a component of S if and only if it satisfies the following conditions:

(i) X is non-empty.

(ii) X is connected.

(iii) If Y is any connected subset of S satisfying $Y \cap X \neq \emptyset$, then $Y \subset X$.

DEFINITION 3.6. Let S be a topological space. A subset G of S is said to be a region in S if and only if G is both connected and open in S.

The following basic theorems concerning these concepts are easily established.

THEOREM 3.7. Let M be a subset of a space S, and p a point of M. Then the component of M containing p is the union of all connected subsets of M that contain the point p.

THEOREM 3.8. Let M be a subset of a space S. Then every component of M is closed in M.

THEOREM 3.9. Let H_1 and H_2 be components of a subset M of a space S, and suppose that $H_1 \neq H_2$. Then H_1 and H_2 are disjoint.

THEOREM 3.10. A subset M of a space S is locally connected if and only if, given any subset G of M which is open in M, then every component of G is a region in M. (See Problem IV, 16.)

THEOREM 3.11. Every countably compact locally connected set M in a T_1 space S has a finite number of components.

The proof of the characterization theorem for simple closed curves is based on our characterization of arcs. For our first theorem in this direction, the concept of independent arcs is needed.

DEFINITION 3.12. Let a and b be distinct points of a space S, and suppose that F_1 and F_2 are arcs in S, each having a and b as end points. Then F_1 and F_2 are said to be independent arcs from a to b if and only if $F_1 \cap F_2 = a \cup b$.

THEOREM 3.13. A non-degenerate space C is a simple closed curve if and only if, for any pair of distinct points a and b of C, the space C may be expressed as the union of two independent arcs from a to b.

Proof. First of all suppose that C is a simple closed curve, i.e., that there exists a homeomorphism $f: K \to C$ of K onto C. Let a, b be any pair of distinct points of C, and define points a', b' of K by $a' = f^{-1}(a)$, $b' = f^{-1}(b)$. The points a', b' are distinct points of K since f is one-to-one. It is a simple exercise to show that K is the union of two independent arcs H_1 and H_2 joining a' and b', and that $f(H_1)$, $f(H_2)$ are independent arcs of C joining a and b. Since K is the union of H_1 and H_2, C is the union of $f(H_1)$ and $f(H_2)$.

To prove the converse, suppose that C satisfies the given condition, and let a, b be a pair of distinct points of C. Express C as the union of two independent arcs F_1 and F_2 from a to b. Since F_1 and F_2 are arcs, there exist homeomorphisms

$$f_1: F_1 \to J, \qquad f_2: F_2 \to J$$

satisfying $f_1(a) = f_2(a) = 0$, $f_1(b) = f_2(b) = 1$ of F_1 and F_2 with the unit interval J. Now K, the unit circle, can clearly be expressed as the union of two independent arcs H_1 and H_2, joining the points $(-1, 0)$, $(1, 0)$. Thus there exist homeomorphisms of J with each of H_1 and H_2, namely:

$$g_1: J \to H_1, \qquad g_2: J \to H_2$$

satisfying $$g_1[(1, 0)] = (1, 0), \qquad g_2[(1, 0)] = (1, 0)$$

$$g_2[(0, 0)] = g_2[(0, 0)] = (-1, 0)$$

We now define a mapping f: $C \to K$, of C into K, in the following way:

$$\text{for} \quad x \in F_1, \quad f(x) = g_1[f_1(x)]; \qquad \text{for} \quad x \in F_2, \quad f(x) = g_2[f_2(x)]$$

It is easily seen that the mapping f thus defined is a homeomorphism of C onto K. Therefore, C is a simple closed curve.

COROLLARY 3.14. If a space C contains two distinct points a and b, and two independent arcs H_1 and H_2 from a to b such that $C = H_1 \cup H_2$, then C is a simple closed curve.

THEOREM 3.15. A space C is a simple closed curve if and only if it is a non-degenerate, countably compact, connected, separable metric space satisfying the following condition:

(i) For every pair of distinct points a, b of C, the set $C - (a \cup b)$ is not connected.

Moreover, every simple closed curve C has the property that every point of C is a non-cut point of C.

Proof. It is easily seen that every simple closed curve satisfies all the requirements of the theorem. We assume, therefore, that C is a space satisfying all the requirements of the theorem, and shall prove that C is a simple closed curve. We first show that C satisfies the following condition:

(ii) Every point of C is a non-cut point of C.

Suppose that (ii) is false; i.e. that there exists a point $z \in C$ such that z is a cut point of C. There must thus exist a separation $C - z = H \cup K$. It is easily seen that $\bar{H} = H \cup z$, $\bar{K} = K \cup z$, and it follows at once, from Corollary III, 5.20, that each of the sets $\bar{H}$ and $\bar{K}$ is connected. From Theorem 1.3, we see that each of the sets $\bar{H}$ and $\bar{K}$ must have at least two non-cut points. There must thus exist points a, b of $\bar{H} - z$ and $\bar{K} - z$, respectively, such that each of the sets $\bar{H} - a$ and $\bar{K} - b$ is connected. It follows at once that the set $C - (a \cup b) = [\bar{H} - a] \cup [\bar{K} - b]$ is connected, which contradicts (i). This completes the proof of (ii).

Now let a, b be any pair of distinct points of C. By (i), there exists a separation $C - (a \cup b) = A \cup B$. Using (ii), it is easily seen that $\bar{A} = A \cup a \cup b$, $\bar{B} = B \cup a \cup b$. Since each of the sets $\bar{A}$ and $\bar{B}$ is closed, while $\bar{A} \cup \bar{B} = S$, a connected set, and $\bar{A} \cap \bar{B}$ consists of exactly two points, it follows at once, from Problem III, 50, that $\bar{A}$ or $\bar{B}$ is a connected set. We assume that the notation has been so chosen that $\bar{A}$ is connected.

Suppose that $\bar{B}$ is not connected; i.e., that there exists a separation

$$\bar{B} = B_1 \cup B_2$$

Now $\bar{A}$, B_1, and B_2 are each closed subsets of C and $C = (\bar{A} \cup B_1) \cup B_2$. Hence $(\bar{A} \cup B_1) \cap B_2$ must contain at least two points by Problem 22. Since B_1 and B_2 are disjoint, it follows that $(\bar{A} \cup B_1) \cap B_2 = a \cup b$. Accordingly, $(a \cup b) \subset B_2$. Repeating the argument for B_1, it follows that $(a \cup b) \subset B_1$; hence $B_1 \cap B_2 \neq \emptyset$. This contradiction to the assumption that B_1 and B_2 were separated sets proves that the set $\bar{B}$ is connected.

Now each of the connected sets $\bar{B}$ and $\bar{A}$ is countably compact, nondegenerate, separable, and metric, and contains the two points a and b. We shall show that each of these sets is an arc from a to b. Suppose first that neither of the sets $\bar{A}$, $\bar{B}$ is an arc from a to b. Then, by our characterization of the arc, there must exist points c and d such that $c \in \bar{A} - (a \cup b)$, $d \in \bar{B} - (a \cup b)$ such that each of the sets $\bar{A} - c$, $\bar{B} - d$ is connected. Since these sets each contain the point a, their union, namely $C - (c \cup d)$ must be connected. This is a contradiction to (i), since the pair of distinct points c, d of C must separate C.

We may thus assume that $\bar{A}$ is an arc from a to b. Let c be any point of $\bar{A} - (a \cup b)$. Then $\bar{A} - c$ consists of exactly two components A_1 and A_2, containing a and b, respectively. Assuming that $\bar{B}$ is not an arc from a to b and choosing the point d as before, we see at once that $\bar{B} - d$ is a connected set, containing both a and b. The union of this connected set with the set $\bar{A} - c$ again gives us the contradiction in that the pair of points c, d does not separate the space C. Thus both $\bar{A}$ and $\bar{B}$ must be arcs from a to b, and the theorem follows from Corollary 3.14.

Our characterization of simple closed curves is thus established. We now turn our attention to the proof of the Jordan curve theorem. The proof is accomplished by first establishing certain properties of line segments and polygons in E^2. On these properties is based a proof that no subset of E^2 that is an arc separates E^2. This important fact leads fairly directly to the final steps of the proof of the Jordan curve theorem itself.

DEFINITION 3.16. (*a*) Let $p_1 = (x_1, y_1)$ and $p_2 = (x_2, y_2)$ be two distinct points of E^2. Then $L(p_1, p_2)$ denotes the subset of E^2 consisting of all points (x, y) such that $x = \lambda x_1 + (1 - \lambda)x_2$ and $y = \lambda y_1 + (1 - \lambda)y_2$ for some λ satisfying $0 \leq \lambda \leq 1$. The set $L(p_1, p_2)$ is called the line segment joining p_1 and p_2. The line segment $L(p_1, p_2)$ is said to be parallel to the y axis if and only if the abscissa of p_1 equals the abscissa of p_2; that is, if and only if $x_1 = x_2$. The line segment $L(p_1, p_2)$ is said to be parallel to the x axis if and only if the ordinate of p_1 equals the ordinate of p_2; that is, if and only if $y_1 = y_2$.

(*b*) Let $p_0, p_1, \cdots, p_n$ be a finite collection of distinct points of E^2. Define the set $H = L(p_0, p_1) \cup L(p_1, p_2) \cup \cdots \cup L(p_{n-1}, p_n)$. Then H is said to be a polygonal arc from p_0 to p_n if and only if the following conditions hold:

(i) $L(p_i, p_{i+1}) \cap L(p_{i+1}, p_{i+2}) = p_{i+1}$, for $i = 0, 1, \cdots, n-2$.

(ii) $[L(p_i, p_{i+1}) \cup L(p_{i+1}, p_{i+2})] \not\subset L(p_i, p_{i+2})$, for $i = 0, 1, \cdots, n-2$.

(iii) $L(p_i, p_{i+1}) \cap L(p_j, p_{j+1}) = \emptyset$ if $|i - j| > 1$, $i = 0, 1, \cdots, n-1$; $j = 0, 1, \cdots, n-1$.

If H is a polygonal arc from p_0 to p_n, then each of the line segments $L(p_i, p_{i+1})$, $i = 0, 1, \cdots, n-1$ is called a side of H, and each of the points $p_0, p_1, \cdots, p_n$ is called a vertex of H. The two points p_0 and p_n are called the end points of the polygonal arc. Given a vertex p_i, $i = 0, 1, \cdots, n$, a vertex p_j, $j = 0, 1, \cdots, n$, is said to be adjacent to p_i if and only if $|i - j| = 1$.

(*c*) Given two points $a, b \in E^2$. Then a polygonal arc H from a to b is said to be a special polygonal arc from a to b if and only if every side of H is parallel either to the x axis or to the y axis.

Definition 3.17. (*a*) A non-degenerate subset P of E^2 is said to be a polygon if and only if, given any two distinct points a and b of P, P is the union of two independent polygonal arcs from a to b. A point p of a polygon P of E^2 is said to be a vertex of P if and only if every line segment contained in P that contains p has p as an end point. A line segment contained in P is said to be a side of P if and only if its end points are vertices of P. Two distinct vertices of P are said to be adjacent if and only if there is a side of P that contains them.

(*b*) A non-degenerate subset P of E^2 is said to be a special polygon if and only if, given any two distinct points a, b of P, P is the union of two independent special polygonal arcs from a to b.

It is evident that a special polygon is a polygon, and the following theorem is immediate.

Theorem 3.18. (*a*) Every polygonal arc from a to b is an arc from a to b.

(*b*) Every polygon is a simple closed curve.

Given two points of a set, it is frequently important to know conditions under which there exists an arc in this set joining the two points.

THEOREM 3.19. Let G be any region in E^2, and $a, b \in G$. Then there exists a special polygonal arc joining a and b in G.

Proof. Define M as the subset of G, consisting of the point a and all points of G that can be joined to a by a special polygonal arc in G. The proof that M is both open and closed in G is an almost immediate consequence of Problem 23, and is left to the student. Since G is connected, no proper subset of G can be both open and closed in G. Hence $M = G$, which proves the theorem.

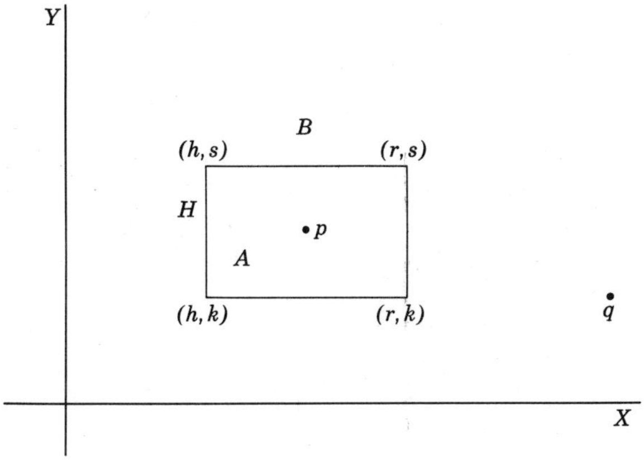

Figure 10

Among the simplest polygons is the special rectangle.

DEFINITION 3.20. A subset H of E^2 is said to be a special rectangle if and only if H is a special polygon having exactly four sides.

We now show that the Jordan curve theorem holds for all simple closed curves that are special rectangles.

THEOREM 3.21. Let H be a special rectangle in E^2. Then $E^2 - H$ has exactly two components A and B. Moreover, $H = \bar{A} - A = \bar{B} - B$. If M is any special polygonal arc joining two points a, b of $E^2 - H$ and having exactly one point in H, and if this point is not a vertex of M, then a and b do not lie in the same component of $E^2 - H$.

Proof. The four vertices of H may be taken as the points (h, k), (h, s), (r, k), (r, s), where $k < s$, $h < r$. Let p, q be points of $E^2 - H$ defined as follows: $p = [(h + r)/2, (k + s)/2]$, $q = [h + 2(r - h), k]$. (See Figure 10.) Let A be the set of all points of $E^2 - H$ that can be joined to p by a special polygonal arc contained in $E^2 - H$, and B the set of all points of $E^2 - H$ that can be joined to q by a special polygonal arc contained in $E^2 - H$. It follows at once that each of the sets A, B is connected. With the hints given, the student should have no difficulty in establishing the following results:

(i) A point (x, y) of $E^2 - H$ lies in A if and only if both the conditions $h < x < r$ and $k < y < s$ hold. A point (x, y) of $E^2 - H$ lies in B if and only if at least one of these conditions does not hold.

(ii) The sets A and B are disjoint.

(iii) $E^2 - H = A \cup B$ is a separation, and each of the sets A, B is a component of $E^2 - H$.

(iv) $H = \bar{A} - A = \bar{B} - B$.

(*Hints.* To establish (i) for A, show that, if $h < x < r$ and $k < y < s$, then the union of the two line segments $L[(x, y), ((h + r)/2, y)]$ and $L[((h + r)/2, y), ((h + r)/2, (k + s)/2)]$ contains a special polygonal arc from (x, y) to p which is contained in $E^2 - H$. On the other hand, if N, with vertices $p_1 = p, p_2, \cdots, p_n = (x, y)$, is a special polygonal arc joining p to (x, y) and contained in $E^2 - H$, show that p_1 satisfies the inequalities of (i), and, if p_i satisfies these conditions, then p_{i+1} does also. Similar techniques prove (i) for B. Statement (ii) follows directly from the inequalities of (i), as also does (iii), since we already know that each of the sets A and B is connected. To establish (iv), consider a point (h, y), $k < y < s$, of H. This point is the limit of each of the sequences $\{a_n\}$ and $\{b_n\}$, where $a_n = (h + (h + r)/2n, y)$, $b_n = (h - 1/n, y)$, for $n \in I$. It is clear that $a_n \in A$, $b_n \in B$, for $n \in I$. Similarly for other points of H.)

To complete the proof of the theorem, let a, b be points of $E^2 - H$ such that there exists a special polygonal arc M joining a and b which has exactly one point v in common with H, where v is not a vertex of M. Denote by D the component of $E^2 - H$ that contains the point a; thus D is the one of the sets A and B of the preceding paragraph in which the point a lies. In the polygonal arc M, let $L(p_i, p_{i+1})$ be the side containing the point v, where we have chosen the notation in such a way that $p_0 = a$, $p_n = b$. In M there is a polygonal arc joining a to p_i and lying in $E^2 - H$. Since the points a and p_i thus lie in a connected subset of $E^2 - H$, namely, an arc, and $a \in D$, it follows that $p_i \in D$. By use of the inequalities of (i), it follows that $p_{i+1} \in F$, where F denotes the component of $E^2 - H$ distinct from D. But p_{i+1} and b must lie in the same component of $E^2 - H$, since M contains an arc joining p_{i+1} and b which lies in $E^2 - H$. Consequently, $b \in F$, and the proof of the theorem is complete.

Looking at statement (i) in the proof of Theorem 3.21, we see that one of the components A, B of $E^2 - H$, namely, the component that was denoted by B in that proof, is unbounded, whereas the other component A is bounded. Thus every special rectangle has the property that exactly one component of its complement is unbounded. This result is true for simple closed curves in general.

THEOREM 3.22. Let C be any simple closed curve in E^2. Then exactly one component of $E^2 - C$ is unbounded.

Proof. The set C is compact; hence, being a subset of E^2, it must be closed and bounded. It follows at once from this fact that there exists a special rectangle H in E such that C is contained in the bounded component of $E^2 - H$. By Theorem 3.21, we easily see that the union of H and the unbounded component of H is a connected set, which must lie in a single unbounded component of $E^2 - C$. Every other component of $E^2 - C$ (if there are any!) must lie in the bounded component of $E^2 - H$, and hence be a bounded set. This proves Theorem 3.22.

DEFINITION 3.23. If C is any simple closed curve in E^2, then the unbounded component of $E^2 - C$ is called the exterior of C. The set of all points of E^2 (if there are any) which lie neither in C nor in the exterior of C is called the interior of C.

A still closer look at statement (i) in the proof of Theorem 3.21, reveals the following theorem, which is conveniently stated in terms of the concept of the center of a special rectangle.

DEFINITION 3.24. Let H be a special rectangle with vertices (h, k), (h, s), (r, k), (r, s), where $h < r$, $k < s$. Then the point $((h + r)/2, (k + s)/2)$ is called the center of H. If H is a special rectangle with center q, then, by $H(q)$, we denote the component of $E^2 - H$ that contains q.

THEOREM 3.25. Let H be a special rectangle with center q. Then

(i) $H(q)$ is the interior of H.

(ii) $H(q)$ is a connected neighborhood (i.e., a region) containing the point q.

(iii) If $r \in H(q)$, then $L(r, q) \subset H(q)$.

(iv) The collection of all interiors of special rectangles with center at q is a basis at the point q.

The properties of special rectangles established in the next theorem are fundamental to our use of special rectangles in establishing the Jordan curve theorem. The statement of this theorem involves the concept of the interior of an arc.

DEFINITION 3.26. Let a, b be the end points of an arc A. Then the set $A - (a \cup b)$ is called the interior of A. We shall denote the interior of an arc A by the symbol A'.

THEOREM 3.27. Let W be a special rectangle with center at q, and s, t two distinct points of $E^2 - W(q)$, where $W(q)$ is the interior of W. If $H = L(s, q) \cup L(q, t)$ is a special polygonal arc from s to t, then the following conditions hold:

(i) H has as its vertices either the two points s and t or the three points s, q, t.

(ii) $H \cap W$ consists of exactly two points u and v.

(iii) H contains a special polygonal arc M from u to v and $M = L(u, q) \cup L(q, v)$.

(iv) $M' \subset H'$.

(v) If $H = L(s, t)$, then $W \cup L(u, v)$ contains exactly three special rectangles W, W_1, W_2 and $W(q) = W_1(q_1) \cup W_2(q_2) \cup L'(u, v)$, where q_i is the center of W_i, for $i = 1, 2$. Furthermore,

$$W_1(q_1) \cap W_2(q_2) = \emptyset, \quad W_1 \cap W_2 = \overline{W_1(q_1)} \cap \overline{W_2(q_2)} = M = L(u, v)$$

If $x, y \in E^2 - W(q)$, $L(x, y) \cap L(s, t) = q$, where $L(x, y)$ is a special polygonal arc, then, if the indices 1, 2 are chosen properly, $L(x, y)$ may be written as the union of the four line segments $L(x, q_1)$, $L(q_1, q)$, $L(q, q_2)$, $L(q_2, y)$, where the interiors of no two of these line segments have a point in common.

(vi) $W(q)$ is the union of three disjoint non-empty sets. One of these sets is M', while each of the other two is a region in E^2. Furthermore, every point of M' is a limit point of each of these regions.

The proof, which is left to the student, is probably most easily made by first showing that, if $H = L(s, t)$, then $W \cup M$ is the union of two special rectangles W_1 and W_2 with the line segment M as a common side; thus the theorem holds. (See Figure 11.) The other case, $H \neq L(s, t)$, is then treated by splitting W, as in the first instance, by line segments parallel to the axes, thus forming four special rectangles such that the sides of M are among the sides of these rectangles. The result then follows by use of the first result. For the details, the inequalities and technique of the proof of Theorem 3.21 are vital.

One of our principal problems in the proof of the Jordan curve theorem consists in showing that every simple closed curve C in E^2 has a non-empty connected interior. This fact has already been established for special rectangles in Theorem 3.21. Our next step will be to prove it for special polygons in general. In fact, we shall prove a bit more; namely that every special polygon has property J.

DEFINITION 3.28. A special polygon H in E^2 is said to have property J if and only if the following conditions hold:

(i) There exists a separation $E^2 - H = A \cup B$, where each of the sets A, B is a region.

(ii) $H = \bar{A} - A = \bar{B} - B$.

(iii) If N is any special polygonal arc joining two points a, b of $E^2 - H$ and having exactly one point in H, and if this point is not a vertex of N, then a and b do not lie in the same component of $E^2 - H$.

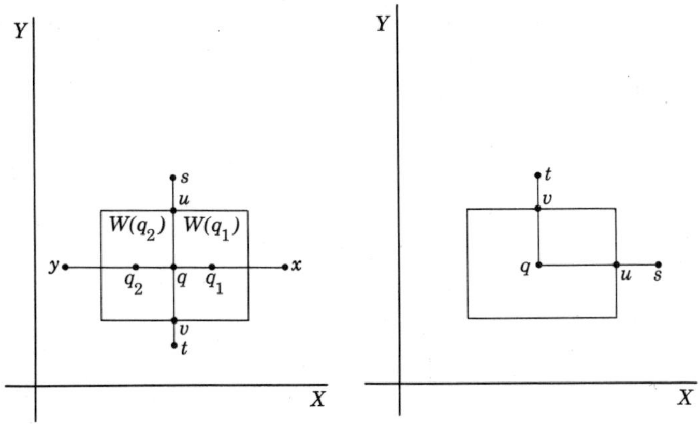

Figure 11

In our proof that every special polygon has property J, we shall make use of the concept of the amalgamation of two special polygons across a common line segment. This is a special case of the amalgamation of two simple closed curves across a common line segment.

DEFINITION 3.29. Let H_1 and H_2 be simple closed curves, having as their intersection a line segment $L(a, b)$. The amalgamation of H_1 and H_2 across the common line segment $L(a, b)$ is the set $H = (H_1 \cup H_2) - L'(a, b)$.

In connection with this definition as applied to special polygons see Figure 12.

Our proof is by contradiction. We argue that, if there is a special polygon that does not have property J, then there is an integer n, and a special polygon P with n vertices that does not have property J, but every special polygon with fewer than n vertices does have property J. We will show that:

(a) There exist two special polygons P_1 and P_2, each with fewer than n vertices, such that P is the amalgamation of P_1 and P_2 across a common side.

(b) The amalgamation of two special polygons with property J across a common line segment is a special polygon having property J.

This contradicts the assumption, thus every special polygon has property J. The argument for (b) is the content of the next theorem.

THEOREM 3.30. Let H_1 and H_2 be two special polygons having as their intersection a line segment $L(a, b)$, and suppose that each of the polygons H_1, H_2 has property J. Then the amalgamation of H_1 and H_2 across the line segment $L(a, b)$ is a special polygon having property J.

Proof. The proof that H is a special polygon is left as an exercise for the student.

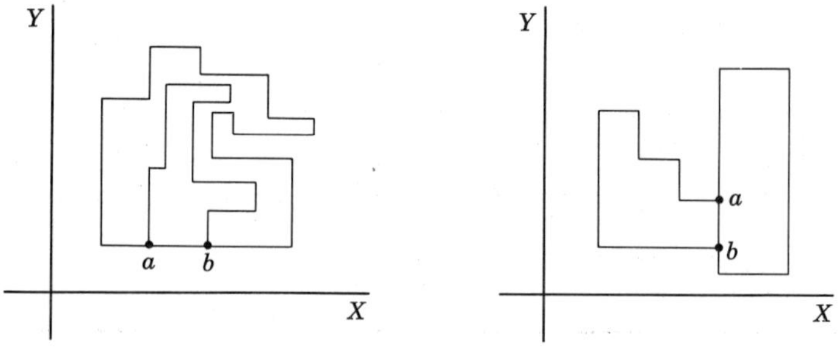

Figure 12

The remainder of the proof divides naturally into the demonstration of the following four facts:

1. $E^2 - H$ is not connected.
2. If A is any component of $E^2 - H$, then $\bar{A} - A = H$.
3. $E^2 - H$ has exactly two components.
4. Condition (iii) of the definition of property J holds.

From the first three facts it is immediate that (i) and (ii) of the definition of property J hold. We first prove (1).

For $i = 1, 2$ the special polygon H_i is the union of $L(a, b)$ and a special polygonal arc K_i joining a and b, where the special polygon H is the union of the special polygonal arcs K_1 and K_2. First of all, it should be noted that each of the sets K_1' and K_2' is connected, and that $K_1' \subset E^2 - H_2$, $K_2' \subset E^2 - H_1$. The hypothesis that each of the sets H_1 and H_2 has property J gives us the separations

$$E^2 - H_1 = A_1 \cup B_1$$

$$E^2 - H_2 = A_2 \cup B_2$$

where A_1, B_1 are the components of $E^2 - H_1$, A_2, B_2 are the components of $E^2 - H_2$, and the notation is chosen in such a way that $K_1' \subset A_2$,

$K_2' \subset A_1$. Having agreed on this notation, we now establish the following results:

(i) $A_1 \cap A_2 \neq \emptyset$.
(ii) $B_1 \cup B_2 \cup L'(a, b) \neq \emptyset$.
(iii) $A_1 \cap A_2$ is an open set.
(iv) B_1 and B_2 are disjoint open sets.
(v) $B_1 \cup B_2 \cup L'(a, b)$ is an open set.
(vi) The sets $B_1 \cup B_2 \cup L'(a, b)$ and $A_1 \cap A_2$ are disjoint.
(vii) $E^2 - H = [B_1 \cup B_2 \cup L'(a, b)] \cup [A_1 \cap A_2]$.

Statement (i) follows, since, by property J, every point of K_2' is a limit point of A_2, and A_1 is a neighborhood of each point of K_2'.

Statement (ii) follows, since, by property J, the set B_1 is non-empty. Again, by property J, each of the sets A_1, A_2, B_1, B_2 is open; hence (iii) holds, and, to establish (iv), it remains to be shown that $B_1 \cap B_2 = \emptyset$. Suppose $B_1 \cap B_2 \neq \emptyset$. Then, since B_2 is a connected subset of $E^2 - H_1$, it follows that $B_2 \subset B_1$. A similar argument shows that $B_1 \subset B_2$. Consequently, $B_1 = B_2$, and, by property J, $H_1 = \bar{B}_1 - B_1 = \bar{B}_2 - B_2 = H_2$, contrary to the hypotheses of the theorem. Thus (iv) holds.

We establish (v) by showing that, if $q \in B_1 \cup B_2 \cup L'(a, b)$, then there is a neighborhood of q that is contained in this set. If $q \in B_1 \cup B_2$, then $B_1 \cup B_2$ is such a neighborhood of q. If $q \in L'(a, b)$, then there exists, by Theorem 3.25, a special rectangle W, having its center at q, and with the property that the special polygon H is contained in the exterior of W. In particular, both a and b lie in the set $E^2 - W(q)$. Thus, by Theorem 3.27, $L(a, b) \cap W$ consists of exactly two points u and v. Furthermore, by the same theorem, $W(q) = W_1(q_1) \cup W_2(q_2) \cup L'(u, v)$, where $W_1(q_1)$ and $W_2(q_2)$ are disjoint regions. By property J, $W(q)$ must intersect B_1; hence there is a point of B_1 in one of the sets $W_1(q_1)$, $W_2(q_2)$. We suppose the notation is such that this particular set is $W_1(q_1)$. Since $W_1(q_1) \subset E^2 - H_1$, it follows that $W_1(q_1) \subset B_1$. By property J, $W(q) \cap B_2 \neq \emptyset$; hence there is a point of B_2 in one of the sets $W_1(q_1)$, $W_2(q_2)$. This set must be $W_2(q_2)$ since the sets B_1 and B_2 are disjoint. Since $W_2(q_2) \subset E^2 - H_2$, we must have $W_2(q_2) \subset B_2$. Finally, $L'(u, v) \subset L'(a, b)$. Hence $W(p) = W_1(q_1) \cup W_2(q_2) \cup L'(u, v)$ is contained in $B_1 \cup B_2 \cup L'(a, b)$, and (v) holds.

Statement (vi) follows, since no point of A_1 is in B_1 or $L'(a, b)$, and no point of A_2 is in B_2. Statement (vii) follows immediately from the definition of H as the amalgamation of H_1 and H_2 across the line segment $L(a, b)$.

From the results (i) through (vii), we conclude that the set $E^2 - H$ is not connected. Thus the proof of (1) is complete.

We next show that (2) holds, and use this fact to prove (3). Suppose that A is a component of $E^2 - H$. Then A is open since it is a component of an open subset of the locally connected space E^2. Since A is connected, $\bar{A}$ is connected. Thus, if $\bar{A} \subset E^2 - H$; then $\bar{A} = A$, and A is a proper subset of the connected space E^2, and A is both open and closed. This is impossible; therefore, $\bar{A} \cap H$ is non-empty. Consequently, $(\bar{A} - A) \cap H$ is non-empty. The set $(\bar{A} - A) \cap H$ is closed, since it is the intersection of two closed sets. We will now show that $(\bar{A} - A) \cap H$ is open in H. It then follows that $(\bar{A} - A) \cap H = H$, since H is connected.

Let $q \in (\bar{A} - A) \cap H$; then q is either a vertex of H or a point of H which is interior to a side of H. In the case that q is a vertex, let s, t denote the adjacent vertices of H; in the case that q is interior to a side of H, let s, t denote the end points of that side. There exists a special rectangle W, with center q, such that the arc of H from s to t which does not contain q lies in the exterior of W. By Theorem 3.27, the set $W \cap H$ consists of exactly two distinct points u, v, and there exists a special polygonal arc M from u to v such that $q \in M'$, the interior of the arc M. Furthermore, $W(q)$ is the union of the sets D, E, and M', where D and E are disjoint regions contained in $W(q)$, each disjoint with M', and where every point of M' is a limit point of each of these regions. Since q is a limit point of A, $W(q)$ contains a point x of A. Thus x must lie either in D or in E. If $x \in D$, then $D \subset A$, since D is a connected subset of $E^2 - H$. Every point of M', however, is a limit point of D; hence every point of M' is a limit point of A. Similarly, if $x \in E$, then every point of M' is a limit point of A. It follows that $q \in M' \subset (\bar{A} - A) \cap H$, so that $(\bar{A} - A) \cap H$ is both open and closed in H. Therefore, $(\bar{A} - A) \cap H = H$. Since every point of $\bar{A}$ not in A must lie in H, this implies that $\bar{A} - A = H$. The proof of (2) is thus complete.

To prove (3), we note that, by (1), there are at least two components A and B of $E^2 - H$. We shall show that there can be no others. Let q be a point of H which is not a vertex of H, and s, t the end points of the side of H which contains q. There exists a special rectangle W with center q such that the arc of H joining s and t, which does not contain q, lies in the exterior of W. By Theorem 3.27, the set $W \cap H$ consists of exactly two points u and v, and the interior of W is the union of $L'(u, v)$ and two disjoint non-empty regions D and E. Since q is a limit point of A, one of the sets D, E must lie in A. The other must then lie in B. Thus $[W(q) \cap (E^2 - H)] \subset A \cup B$. If F is a component of $E^2 - H$ distinct from A and B, then q is also a limit point of F. But this is impossible, since $W(q)$ is a neighborhood of q containing no point of F. This completes the proof that $E^2 - H$ has exactly two components; i.e., that (3) holds.

To complete the proof that H has property J, we need only show that condition (iii) of property J holds. Let N be a special polygonal arc joining two points a, b of $E^2 - H$ such that $N \cap H$ is exactly one point q, and q is not a vertex of N. We shall show that a and b do not lie in the same component of $E^2 - H$. Let x, y be the vertices of the side of N containing q, where the notation is chosen in such a manner that N contains an arc from a to x and an arc from b to y, such that neither of these arcs contains the point q. Thus a and b lie in different components of $E^2 - H$ if and only if x and y lie in different components of $E^2 - H$. The fact that q is not a vertex of N implies that q is not a vertex of H. Thus let s and t denote the end points of the side of H containing q. There exists a special rectangle W with center q such that x, y, s, t lie in the set $E^2 - W(q)$, and the arc of H joining s to t which does not contain q is exterior to W. Thus, using the notation of Theorem 3.27, $W(q) = W_1(q_1) \cup W_2(q_2) \cup L'(u, v)$, and each of the sets $L(x, q_1)$ and $L(y, q_2)$ lies in $E^2 - H$. Since $W_1(q_1)$ and $W_2(q_2)$ lie in different components of $E^2 - H$, it follows that q_1 and q_2, and, consequently, x and y lie in different components of $E^2 - H$. Thus a and b lie in different components of $E^2 - H$. This completes the proof of Theorem 3.30.

THEOREM 3.31. Every special polygon has property J.

Proof. Let H be a special polygon. Then H has at least four vertices. If H has exactly four vertices, then H is a special rectangle, and thus has property J, by Theorem 3.21. The proof we give that an arbitrary special polygon H has property J is by contradiction. We assume that there is a special polygon that does not have property J. Then there is an integer $n > 4$, such that every special polygon having fewer than n vertices has property J, and, by our assumption, a special polygon H with exactly n vertices that does not have property J. Let x_0 be the smallest abscissa of any point of H. Then among the points of H which have abscissa x_0, there is one whose ordinate is smallest, and we denote this ordinate by y_0. It is clear that the point $p_0 = (x_0, y_0)$ is a vertex of H. There exists a largest real number y_2 such that every point of the line segment joining p_0 to the point $p_1 = (x_0, y_2)$ is a point of H. Also there exist largest real numbers x_1, x_2 such that every point on the line segment joining p_0 to the point $p_3 = (x_1, y_0)$ and every point on the line segment joining p_1 to the point $p_2 = (x_2, y_2)$ is a point of H. Clearly, each of the points p_1, p_2, p_3 is a vertex of H. We shall show that H is the amalgamation across a common line segment of two special polygons, each having fewer than n vertices. From Theorem 3.30, it then follows that H has property J, which is a contradiction. This contradiction shows that every special polygon has property J. In establishing the fact that H is the amalgamation of two such polygons, there are two cases to be considered.

Case 1. $x_2 \leq x_1$. (See Figure 13.)

If the rectangle having as vertices p_0, p_1, p_2 and the point (x_2, y_0) has no point of H in its interior, then we consider the line segment L joining the point (x_2, y_0) and the point p_2. The set L', defined to be the interior of L, is not contained in H, since H has more than four vertices. Thus there exists a point $p_4 \in L' - H$. There is a point v_1 of $L \cap H$ having least ordinate greater than the ordinate of p_4, and a point v_2 of $L \cap H$ having greatest ordinate less than the ordinate of p_4. Thus the intersection of $L(v_1, v_2)$ and H consists of precisely the two points v_1 and v_2.

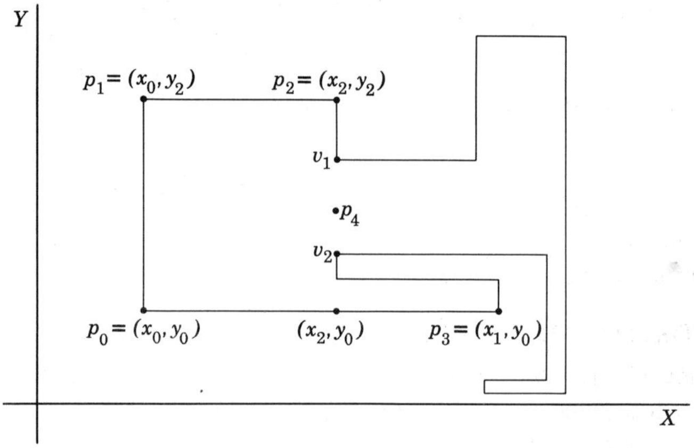

Figure 13

Moreover, H is the union of two independent special polygonal arcs A_1 and A_2 joining v_1 and v_2. Thus $H_1 = A_1 \cup L(v_1, v_2)$ and $H_2 = A_2 \cup L(v_1, v_2)$ are special polygons, whose amalgamation across the common line segment $L(v_1, v_2)$ is precisely the special polygon H. Every vertex of H_1 and H_2 is a vertex of H, with the possible exception of v_1 and v_2. The point v_1 is actually a vertex of H and is a vertex of exactly one of H_1 and H_2. The point v_2 may or may not be a vertex of H; it is, however, a vertex of at least one of H_1 and H_2. The polygons H_1 and H_2 have, therefore, at most one vertex in common, and, if they have a common vertex, this vertex is v_2. On the other hand, every vertex of H is a vertex either of H_1 or of H_2. Since H_1 and H_2 have at most one vertex in common, and since the set consisting of all vertices of H_1 and all vertices of H_2 contains at most one point (the point v_2) not a vertex of H, it follows that

$$k_1 + k_2 \leq n + 2$$

where k_1 is the number of vertices of H_1, and k_2 is the number of vertices of H_2. Thus $k_1 \leq n + 2 - k_2$ and $k_2 \leq n + 2 - k_1$. Since each of the integers k_1 and k_2 is at least 4, this means that $k_1 \leq n - 2$, $k_2 \leq n - 2$. Thus H_1 and H_2 are special polygons with property J, and, consequently, by Theorem 3.30, H has property J, the desired contradiction.

If the rectangle having as vertices p_0, p_1, p_2 and the point (x_2, y_0) does have a point of H in its interior, then we may find a point $p_5 = (x_5, y_5)$ interior to this rectangle which has the smallest possible abscissa x_5 of any point of H interior to this rectangle, and which has the smallest ordinate y_5 of any point of H interior to this rectangle that has abscissa x_5. Taking v_1 as (x_5, y_5) and v_2 as (x_0, y_5), the argument given in the previous paragraph, using the segment $L(v_1, v_2)$ yields the desired result.

Case 2. $x_1 < x_2$.

The discussion of this case is similar to that of case 1, and is left to the reader.

The indicated contradiction being obtained in either case, the theorem is established.

PROBLEMS

12. Prove Theorem 3.3.

13. Prove Theorem 3.4.

14. Prove Theorem 3.7.

15. Prove Theorem 3.8.

16. Prove Theorem 3.9.

17. Prove Theorem 3.10.

18. Prove Theorem 3.11.

19. In the proof of Theorem 3.13, verify that, if a' and b' are distinct points of K, then K is the union of two independent arcs H_1 and H_2 joining a' and b'.

20. In the last part of the proof of Theorem 3.13, verify that the mapping f is a homeomorphism as asserted.

21. In the first part of the proof of Theorem 3.15, show that $\bar{A} = A \cup a \cup b$ and $\bar{B} = B \cup a \cup b$.

22. Let C be a non-degenerate connected space having no cut point, and suppose that A and B are closed non-empty subsets of C such that $C = A \cup B$. Prove that $A \cap B$ contains at least two points.

23. Given a spherical neighborhood $S_\varepsilon(p)$ of a point $p \in E^2$, and a point $q \in S_\varepsilon(p)$. Show that there exists a special polygonal arc joining p and q which lies entirely in $S_\varepsilon(p)$.

24. Let a, b, c be distinct points of E^2, M a polygonal arc joining a and b, N a polygonal arc joining b and c. Show that there exists a polygonal arc H joining a and c which lies in the union of M and N. Furthermore, show that, if M and N are special polygonal arcs, then H is a special polygonal arc.

25. Prove that, if M is any compact subset of E^2, then $E^2 - M$ has exactly one unbounded component.

26. Prove Theorem 3.25.

27. Prove Theorem 3.27.

28. State and prove the analogue of Corollary 3.14 both for polygons and for special polygons.

29. In the first part of the proof of Theorem 3.30, it is stated that conditions (i) and (ii) of the definition of property J follow immediately from (1), (2), and (3). Why is this the case?

30. Suppose that H is a special polygon, and M a special polygonal arc such that $M' \cap H = q$, and q is not a vertex of M. Prove that q is not a vertex of H.

31. Suppose that H_1 and H_2 are special polygons having a line segment L as their intersection, each of which separates E^2 between two points a and b. Prove that the amalgamation of H_1 and H_2 across their common line segment L does not separate E^2 between a and b.

32. Let H be a polygon and L a line segment having its end points, and only its end points, in common with H. Show that H is the amalgamation of two polygons H_1 and H_2 across their common line segment L. Denote by k_1 and k_2 the number of vertices of H_1 and H_2, respectively, and determine a least upper bound for $k_1 + k_2$ in terms of n, the number of vertices of H.

33. Let H be a polygonal arc. Prove that every arc contained in H is a polygonal arc. Show also that every arc contained in a special polygonal arc is a special polygonal arc.

34. In the proof of Theorem 3.31, the inequality $k_1 + k_2 \leq n + 2$ appears. Sketch a special polygon for which the construction given in the proof leads to a case for which $k_1 + k_2 = n + 2$. Also sketch a special polygon for which this is not the case.

4. The Jordan Curve Theorem

In this section we complete the proof of the Jordan curve theorem. The main steps of the argument consist in proving the following statements:

1. No arc contained in E^2 cuts E^2.
2. If C is a simple closed curve contained in E^2 and A is a component of $E^2 - C$, the $C = \bar{A} - A$.
3. If C is a simple closed curve contained in E^2 and if C contains a line segment $L(c, d)$ which is parallel to an axis, then $E^2 - C$ consists of exactly two components.

From these statements we deduce the Jordan curve theorem. The proofs make use of the notion of subarc which we proceed to define.

DEFINITION 4.1. Let A and B be arcs. Then B is said to be a subarc of A if and only if $B \subset A$.

To aid in distinguishing between subarcs of an arc A, we introduce the following notational device.

DEFINITION 4.2. Let A be an arc with end points a and b. Then we denote A by ab or by $ap_1p_2 \cdots p_nb$, where $p_i \in A$, for $i = 1, 2, \cdots, n$, and, in the natural order of A from a to b, we have $a < p_1 < p_2 < \cdots < p_n < b$. When we write A as ab or as $ap_1p_2 \cdots p_nb$ and refer to an order in A, we refer to the natural order in A from a to b.

Thus, in using this notation, if $axyb$ denotes an arc A it follows that the subarc xb of A contains y in its interior, and has end points x and b. It also follows that a does not belong to the subarc xb of A and that each of the points x and y is an interior point of A.

In our proof of statement 1, we make use of the following theorem concerning locally connected spaces. It represents, as it appears in the proof of statement 1, a more adroit use of the fact that E^2 is locally connected than we have made previously.

THEOREM 4.3. Let S be a locally connected space, H a closed subset of S, and A, B two non-empty connected subsets of $S - H$. If H does not separate A and B in S, then $A \cup B$ is contained in a region in $S - H$.

Proof. Since S is locally connected and H is closed, it follows, from Theorems 3.8 and 3.10, that each component of $S - H$ is both open and closed in $S - H$. Let M denote the component of $S - H$ which contains A. Then

$$S - H = M \cup [S - (M \cup H)]$$

is a separation. The connected set B cannot lie in $[S - (M \cup H)]$ since H does not separate A and B. Therefore, $A \cup B \subset M$, which is a region in $S - H$, and the proof of the theorem is complete.

Our proof of statement 1 is by contradiction. We suppose that there is an arc F contained in E^2 which cuts E^2. Therefore, F cuts E^2 between some two components A and D of $E^2 - F$. We show that there must then exist a proper subset S of F, which is both open and closed in F, and which does not cut E^2 between A and D. But, since the arc F is connected, it cannot contain a proper subset that is both open and closed in F. This contradiction establishes statement 1. It turns out that the proof that S as defined is a non-empty, proper, open subset of F is short. The presentation of the technical details of the proof that S is also closed in F is long.

THEOREM 4.4. If F is an arc contained in E^2, then $E^2 - F$ is connected.

Proof. Suppose the theorem false, i.e., that there exists an arc $F = ab$ which cuts E^2. Since F is closed and E^2 is locally connected, it follows that each component of $E^2 - F$ is both open and closed in $E^2 - F$, and that each such component is open in E^2. Let A denote the unbounded component of $E^2 - F$, and let D denote some other component of $E^2 - F$. Then

$$E^2 - F = A \cup B$$

is a separation, where $B = E^2 - (A \cup F)$. It is clear that $D \subset B$, that B is a bounded set, and that F separates E^2 between A and D.

Define S to be the subset of F consisting of the point a and all points x of F, such that the subarc ax of F does not separate E^2 between A and D. We will show that S is a proper subset of F which is both open and closed in F.

Since $a \in S$, the set S is non-empty. But F separates E^2 between A and D; hence b is not in S. Therefore, S is a proper subset of F.

The point a of F does not cut E^2. Thus there exists, by Theorem 3.19, a special polygonal arc joining a point of A to a point of D in the region $E^2 - a$. Since the point a and this special polygonal arc are disjoint compact sets, there exists a neighborhood of a in E^2, and, consequently, a neighborhood of a in F, which does not meet this special polygonal arc. Since F is an arc, there is then a point x, interior to F, such that the subarc ax of F does not meet this special polygonal arc. Hence, ax does not separate E^2 between A and D. Thus $ax - x$ is a neighborhood of a in F contained in S. In the case that $x \in S$ and $x \neq a$, the subarc ax of F does not separate A and D in E^2. Hence, by using Theorem 4.3, we deduce that there is a special polygonal arc, disjoint with ax, which joins a point of A and a point of D. Thus, not only does every point of ax lie in S, but also, as in the case $x = a$, there is a neighborhood of x in F which is contained in S. This completes the proof that S is open in F.

We now show that S is closed in F. In our proof, we shall make use of the fact, established above, that $x \in S$ and $x \neq a$ implies $ax \subset S$. For the proof, let z be any limit point of S. Since S is a subset of the closed set F, $z \in F$. We consider the following possibilities:

(i) $z = a$.

(ii) $z \neq a$, and there exists a point $x \in S$ such that, in the natural order in F from a to b, z precedes x.

(iii) $z \neq a$, and there does not exist a point $x \in S$ such that z precedes x.

In case (i), $z \in S$. In case (ii), $z \in ax \subset S$. In case (iii), since every neighborhood of z contains points of S, it follows that every point of $az - z$ is in S. On the other hand, there is no point of $F - az$ in S. It suffices to show, therefore, to complete the proof that S is closed in F, that, if az is a subarc of $F = ab$ such that a and all interior points of az lie in S, then $z \in S$.

Suppose that az is such a subarc of $F = ab$. The component D of $E^2 - F$, being an open subset of E^2, must contain a point $q = (x_1, y_1)$ whose abscissa is not that of a or that of z. Since B is bounded, the point $p = (x_1, y_1 + \delta(B) + 1)$ is well defined and lies in A. The line segment $L(p, q)$ is thus parallel to the y axis, and z does not lie on $L(p, q)$. Furthermore, $z \in S$ if and only if az does not separate E^2 between p and q.

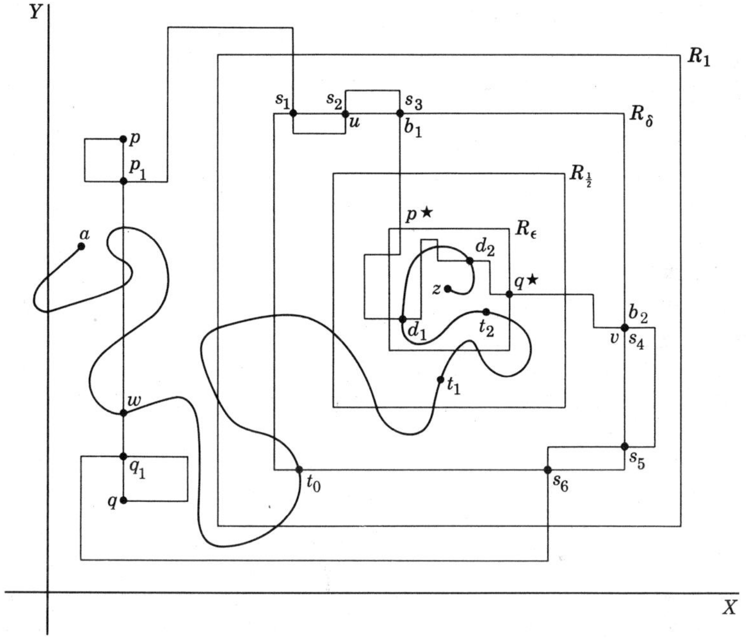

Figure 14

There are thus two cases to consider:

(iv) $az \cap L(p, q) = \emptyset$.

(v) $az \cap L(p, q) \neq \emptyset$.

In case (iv), it is immediate that $z \in S$. In case (v), we show that $z \in S$ by proving the existence of a special polygonal arc joining p and q, which does not intersect az. The student may find Figure 14 helpful as he reads the remainder of the proof.

Considering case (v), we know that the set $az \cap L(p, q)$ is a compact non-empty subset of az, and thus there is a last point w of this set on the arc az. Furthermore, $w \neq z$. Therefore, there exists a special rectangle R_1 with center $z = (x_0, y_0)$ such that $\overline{R_1(z)} \cap [L(p, q) \cup aw] = \emptyset$, where $R_1(z)$ denotes the interior of R_1. There also exist two positive real numbers h and k such that $(x_0 - h, y_0 - k)$, $(x_0 - h, y_0 + k)$, $(x_0 + h, y_0 + k)$, $(x_0 + h, y_0 - k)$ are the vertices of R_1. For each real number λ, $0 < \lambda \leq 1$, we denote by R_λ the special rectangle with vertices $(x_0 - \lambda h,\ y_0 - \lambda k)$, $(x_0 - \lambda h,\ y_0 + \lambda k)$, $(x_0 + \lambda h,\ y_0 + \lambda k)$, $(x_0 + \lambda h,\ y_0 - \lambda k)$. It is immediate that the center of R_λ, for $0 < \lambda \leq 1$, is z, and it follows, from the results and techniques of the previous section, that

$$R_{\lambda_1} \subset R_{\lambda_2}(z) \quad \text{if and only if} \quad 0 < \lambda_1 < \lambda_2 \leq 1$$

where $R_{\lambda_2}(z)$ denotes the interior of R_{λ_2}. Furthermore, $\{R_\lambda(z)\}, 0 < \lambda \leq 1$, is a basis at the point z.

Since $R_{1/2}(z)$ is a neighborhood of z, there exists a point $t_1 \in R_{1/2}(z)$ which is interior to the arc az and such that the subarc t_1z of awt_1z is contained in $R_{1/2}(z)$. There also exists a real number ε, $0 < \varepsilon < \frac{1}{2}$, such that the arc awt_1 lies in the exterior of R_ε, and a point t_2 interior to az such that the subarc t_2z of az is contained in $R_\varepsilon(z)$. We see, therefore, that the following conditions must hold:

(vi) $az = awt_1t_2z$.

(vii) $[aw \cup L(p, q)] \cap \overline{R_\lambda(z)} = \emptyset$, $0 < \lambda \leq 1$.

(viii) The collection $\{R_\lambda\}$, $\frac{1}{2} < \lambda < 1$, is infinite.

(ix) $t_1z \subset R_\lambda(z)$, $\frac{1}{2} \leq \lambda \leq 1$.

(x) $awt_1 \cap \overline{R_\varepsilon(z)} = \emptyset$.

(xi) $t_2z \subset R_\varepsilon(z)$.

(xii) $t_2 \in S$.

From (xii), we see that there exists a special polygonal arc P from p to q which is disjoint with the subarc at_2 of at_2z. If P is disjoint with t_2z, then P is disjoint with at_2z, and, consequently, $z \in S$. If P is not disjoint with t_2z, then we make use of the collection $\{R_\lambda\}$ and the special rectangle R_ε to find another special polygonal arc from p to q which does have the property of being disjoint with at_2z.

We suppose, therefore, that P intersects t_2z. This implies that the arc $P = pq$ has a first point d_1 and a last point d_2 in t_2z, in the natural order from p to q on P. It follows that in the subarc pd_1 of P there is a last point p_1 on $L(p, q)$. In the subarc p_1d_1 of P there is a first point $p^\star$ on R_ε. In the subarc d_2q of P there is a first point q_1 on $L(p, q)$. In the subarc d_2q_1 there is a last point $q^\star$ on R_ε. The following three conditions must thus hold:

(xiii) $z \in S$ if and only if az does not separate p_1 and q_1 in E^2.

(xiv) The subarcs $p_1p^\star$ and $q^\star q_1$ of P are disjoint.

(xv) $p_1p^\star \cap L(p, q) = p_1$, $p_1p^\star \cap R_\varepsilon = p^\star$, $q^\star q_1 \cap L(p, q) = q_1$, $q^\star q_1 \cap R_\varepsilon = q^\star$.

The points $p^\star$ and $q^\star$ are distinct points of R_ε, by (xiv) and (xv).

Let $p^\star q^\star$ denote one of the two independent arcs of the simple closed curve R_ε joining $p^\star$ and $q^\star$. It is clear that $p^\star q^\star$ is a special polygonal arc joining $p^\star$ and $q^\star$ whose interior is disjoint with $p_1p^\star$ and with $q^\star q_1$. Thus, using (xiv), the set $Q = p_1p^\star \cup p^\star q^\star \cup q^\star q_1$ is a special polygonal arc joining p_1 and q_1.

We shall show next that the collection $M = \{R_\lambda\}$, $\frac{1}{2} < \lambda < 1$, contains a special rectangle R_δ satisfying the following conditions:

(xvi) No vertex of Q is on R_δ.

(xvii) No vertex of R_δ is on Q.

To do this we notice that Q has a finite number of vertices (x_1, y_1), (x_2, y_2), $\cdots$, (x_n, y_n). If a vertex (x_j, y_j) of Q is to lie on a side of R_λ parallel to the x axis, then we must have

$$\lambda = (y_0 \pm y_j)/k$$

Similarly, if (x_j, y_j) is to lie on a side of R_λ parallel to the y axis, then we must have

$$\lambda = (x_0 \pm x_j)/h$$

Let α be the set of all real numbers of the form $(x_0 \pm x_j)/h$ or $(y_0 \pm y_j)/k$, $j = 1, 2, \cdots, n$. The set α is finite; hence there exists a real number δ such that $\frac{1}{2} < \delta < 1$, and δ is not in α. Clearly no vertex of Q is on R_δ. On the other hand, suppose R_δ has a vertex on Q. Then this vertex is interior to a side of Q since it cannot be a vertex of Q. Since this vertex of R_δ must be of the form $(x_0 \pm \delta h, y_0 \pm \delta k)$, for some choice of signs, it then follows that δ lies in α, which is a contradiction. Thus no vertex of R_δ is on Q.

We shall complete the proof by modifying the arc Q, using a subarc of R_δ, in such a way that we obtain a special polygonal arc joining p_1 to q_1 which is disjoint with az. To do this, it is convenient to list a number of important properties of the arc Q. Some of these are listed here in the nature of a summary of what we have done.

(xviii) Q is a special polygonal arc joining p_1 and q_1.

(xix) Q is disjoint with the arc at_1.

(xx) $Q = Q_1 \cup p^\star q^\star \cup Q_2$, where Q_1 is an arc joining p_1 and $p^\star$, Q_2 is an arc joining $q^\star$ and q_1, and the following conditions hold:

(*a*) The arcs Q_1 and Q_2 are disjoint.

(*b*) $Q_1 \cap L(p, q) = p_1$, $Q_2 \cap L(p, q) = q_2$.

(*c*) $Q_1 \cap R_\varepsilon = p^\star$, $Q_2 \cap R_\varepsilon = q^\star$.

(*d*) $(Q_1 \cup Q_2) \cap az = \emptyset$.

(xxi) No vertex of Q lies on R_δ, and no vertex of R_δ lies on Q.

(xxii) Each of the sets $Q_1 \cap R_\delta$ and $Q_2 \cap R_\delta$ is a non-empty finite set consisting of an odd number of points. Thus $(Q_1 \cup Q_2) \cap R_\delta$ is non-empty and finite and consists of an even number of points.

Using the definition of Q, we see that only (xxii) requires proof. Let K be a side of Q_1 parallel to the y axis, and suppose that K has a point in common with R_δ. From (xxi), we see that K can only intersect sides of R_δ that are parallel to the x axis. There are at most two points of these sides of R_δ which have the same abscissa as an arbitrary point of K; hence $K \cap R_\delta$ consists of at most two points. A similar argument shows that, if K is parallel to the x axis, then $K \cap R_\delta$ consists of at most two points. Since Q_1 has a finite number of sides, it follows that $Q_1 \cap R_\delta$ is a finite set. By a similar argument, we see that $Q_2 \cap R_\delta$ is a finite set. It remains to be shown that each of these sets is non-empty and consists of an odd number of points.

Since $p_1 \cup q_1$ lies in the exterior of R_δ, while $p^\star \cup q^\star$ lies in the interior of this rectangle, it follows that each of the sets $Q_1 \cap R_\delta$ and $Q_2 \cap R_\delta$ is non-empty. Denote by $u_1, u_2, \cdots, u_k$ the finite number of intersections of Q_1 with R_δ, where the indices are chosen in such a way that $Q_1 = p_1p^\star = p_1u_1u_2 \cdots u_kp^\star$. Choose points $\alpha_1, \alpha_2, \cdots, \alpha_k, \beta_1, \beta_2, \cdots, \beta_k$ of Q_1, such that $Q_1 = p_1\alpha_1u_1\beta_1\alpha_2u_2\beta_2\alpha_3u_3\beta_3 \cdots \beta_{k-1}\alpha_ku_k\beta_kp^\star$. Denote by V_E the exterior of R_δ, and by V_I the interior of this rectangle. Thus V_E and V_I are the two components of $E^2 - R_\delta$. Clearly, $p_1 \in V_E$, $p^\star \in V_I$. Moreover, since R_δ has property J, it follows that, for every i satisfying $1 \leq i \leq k$, the points α_i, β_i lie in different components of $E^2 - R_\delta$. From our definition of the points α_i, β_i, it is clear that the subarc $\beta_i\alpha_{i+1}$ lies in a single component of $E^2 - R_\delta$. Since $\alpha_1 \in V_E$, we must thus have $\beta_1\alpha_2 \subset V_I$, $\beta_2\alpha_3 \subset V_E$, $\beta_3\alpha_4 \subset V_I$, and so on. From this fact, it is clear that every β_i, with i even, lies in V_E, while every β_i, with i odd, lies in V_I. However, β_k is known to lie in V_I; hence k must be odd. This completes the proof that $Q_1 \cap R_\delta$ consists of an odd number of points. The proof that $Q_2 \cap R_\delta$ consists of an odd number of points is similar. Thus the proof of (xxii) is complete.

We now observe that Q and $L(p_1, q_1)$ are independent special polygonal arcs joining p_1 and q_1. Thus the union $Y = Q \cup L(p_1, q_1)$ is a special polygon. Using (xix) and the definition of the point w, we have at once $Y \cap [wt_1 - w] = \emptyset$. Thus, since w and t_1 lie in different components of $E^2 - R_\delta$, it follows that $wt_1 \cap R_\delta \neq \emptyset$, and that no point of wt_1 that is on R_δ lies in Y. Let t_0 be the first point of wt_1 that lies on R_δ. Clearly $wt_1 = wt_0t_1$. From Theorem 3.31, we see that the set $E^2 - Y$ has exactly two components. Denote by Z_0 the component of $E^2 - Y$ that contains the point t_0, and by Z_1 the other component of this set.

Denote by u, v points of the non-empty sets $Q_1 \cap R_\delta$, $Q_2 \cap R_\delta$, respectively. From (xxii) we know that $(Q_1 \cup Q_2) \cap R_\delta$ is a non-empty finite set consisting of an even number (which we denote by $2n$) of points. The arc uvt_0 of R_δ has the point v in common with $(Q_1 \cup Q_2)$. We denote by s_{2n} the last point of uvt_0 in $Q_1 \cup Q_2$. Denote the points of $(Q_1 \cup Q_2) \cap R_\delta$ by $s_1, s_2, \cdots, s_{2n}$, where the notation is chosen in such a way that the arc t_0us_{2n} of R_δ may be written $t_0s_1s_2 \cdots s_{2n}$. Denote by s_1s_{2n} the subarc of t_0us_{2n} joining these points. It is clear that s_1s_{2n} contains $(Q_1 \cup Q_2) \cap R_\delta$. Thus, from (xxii), we have at once

(xxiii) Each of the sets Q_1, Q_2 intersects s_1s_{2n} in an odd number of points.

We shall prove the existence of an even integer i_0 such that $s_{i_0-1} \cup s_{i_0}$ is contained neither in the set Q_1 nor in the set Q_2. For the particular case illustrated in Figure 14, we have $i_0 = 4$. Suppose that no such even integer i_0 exists. Then each of the subarcs $s_1s_2, s_3s_4, \cdots, s_{2n-1}s_{2n}$ of s_1s_{2n} must intersect the set Q_1, either in the empty set or in a set consisting of an even number of points. Thus the union of these subarcs, and thus the arc s_1s_{2n} must intersect Q_1 in an even number of points. This contradiction to (xxiii) establishes the existence of the required even integer i_0.

Denote by F_i the subarc $s_{i-1}s_i$ of s_1s_{2n}, for $i = 2, 3, \cdots, 2n$, and by F'_i the interior of this subarc. It is clear, using Theorem 3.31, that $F'_i \subset Z_0$, for $i = 3, 5, \cdots, 2n - 1$, while $F'_i \subset Z_1$, for $i = 2, 4, \cdots, 2n$. Thus, since i_0 is even, $F'_{i_0} \subset Z_1$. Define b_1 as the first point of the arc $Q_1 = p_1p^\star$ which lies in the set $s_{i_0-1} \cup s_{i_0}$, $b_2 = (s_{i_0-1} \cup s_{i_0}) - b_1$. Thus $b_2 \in Q_2$. Denote by p_1b_1, b_2q_1 subarcs of Q_1 and Q_2, respectively, and define $N = p_1b_1 \cup F_{i_0} \cup b_2q_1$. It is immediate that N is a special polygonal arc joining p_1 and q_1. The proof will be complete if we show that N is disjoint with az. We have at once

$$\begin{aligned} N \cap az &= (p_1b_1 \cup F_{i_0} \cup b_2q_1) \cap az \subset (Q_1 \cup F'_{i_0} \cup Q_2) \cap az \\ &= F'_{i_0} \cap az, \text{ (by xx}d\text{)} \\ &= F'_{i_0} \cap [aw \cup (wt_1 - w) \cup t_1z] \\ &= \emptyset \end{aligned}$$

since F'_{i_0}, being a subset of R_δ, cannot intersect either aw or t_1z, and $(wt_1 - w) \subset Z_0$, the component of $E^2 - Y$ not intersecting F'_{i_0}. This completes the proof of Theorem 4.4.

Theorem 4.4 is probably the biggest single step in the proof of the Jordan curve theorem. One of its many useful consequences is given as our next theorem.

THEOREM 4.5. Let C be any simple closed curve in E^2, and A any component of $E^2 - C$. Then $\bar{A} - A = C$.

Proof. The set A, being a component of $E^2 - C$, is open in E^2 and closed in $E^2 - C$. From these facts, and Theorem 4.4, it follows that

$\bar{A} - A = C$. To see this, notice first that $\bar{A} - A = \bar{A} \cap C$ and thus $\bar{A} - A$ is a closed subset of C. Suppose there exists a point $p \in C - [\bar{A} - A]$. Then this set $C - [\bar{A} - A]$ is a neighborhood of p in C, and, consequently, there exist two distinct points a, b of $C - p$, such that C is the union of two independent arcs apb, aqb and $[\bar{A} - A] \subset aqb$. Thus A is closed in $E^2 - aqb$. But A is open in $E^2 - aqb$, since A is contained in this set and A is open in E^2. We have thus found a non-empty subset A of $E^2 - aqb$ which is both open and closed in this set. The set A is a proper subset of $E^2 - aqb$, since it does not contain the point p. This is impossible, since, by Theorem 4.4, $E^2 - aqb$ is connected. We conclude that $\bar{A} - A = C$.

THEOREM 4.6. Let C be a simple closed curve in E^2 containing a line segment $L(c, d)$ which is parallel either to the x axis or to the y axis. Then $E^2 - C$ consists of exactly two components A and B. Moreover, $C = \bar{A} - A = \bar{B} - B$.

Proof. Using Theorem 4.5, it follows that the proof will be complete if we show that $E^2 - C$ has exactly two components.

Denote by q a point of $C - L(c, d)$, and by $p = (x_0, y_0)$ a point of $L'(c, d)$. The point p and the arc cqd of C are disjoint compact sets. Hence there exists a special rectangle W with center p such that $\overline{W(p)} \cap cqd = \emptyset$. If $L(c, d)$ is parallel to the x axis, let r, s denote the exactly two points of W with abscissa x_0; otherwise, let them denote the exactly two points of W with ordinate y_0. It then follows that $L(r, s) \cap L(c, d) = p$, and that $L(r, s) \cap cqd = \emptyset$.

To prove that $E^2 - C$ is not connected, we assume that it is connected, and hence a region, and show that this assumption leads to a contradiction to Theorem 3.31. Since, under this assumption, $E^2 - C$ is a region, there exists a special polygonal arc M in $E^2 - C$ joining r and s. Let r_1, s_1 denote the last point of $M = rs$ in $L(r, p)$, and the first point of $M = rs$ in $L(p, s)$, respectively. The subarc $M_1 = r_1s_1$ of M and the line segment $L(r_1, s_1)$ are two independent special polygonal arcs joining r_1 and s_1; hence their union, P, is a special polygon. We note that $L(c, d)$ is a special polygonal arc joining c and d, which has exactly the point p in common with P. Moreover, p is not a vertex of $L(c, d)$. Thus, by Theorem 3.31, c and d lie in different components of $E^2 - P$. But the arc cqd joins c and d in $E^2 - P$, so that we have a contradiction. Therefore, $E^2 - C$ has at least two components. Denote by A, B any two components of $E^2 - C$.

We now show that A and B are the only components of $E^2 - C$. Using Theorem 3.27, and the notation of that theorem, we have $W(p) = W_1(p_1) \cup W_2(p_2) \cup L'(u, v)$, where $L'(u, v) \subset L(c, d)$. By Theorem

4.5, the set $W(p)$ must intersect both the component A and the component B of $E^2 - C$. Consequently, $W_1(p_1)$ must lie in one of these components, and $W_2(p_2)$ must lie in the other, so that we have $W_1(p_1) \cup W_2(p_2) \subset A \cup B$. No other component D of $E^2 - C$ can exist, for, otherwise, p would not be a limit point of D, which is a contradiction to Theorem 4.5. The proof of our theorem is thus complete.

We now prove the Jordan curve theorem.

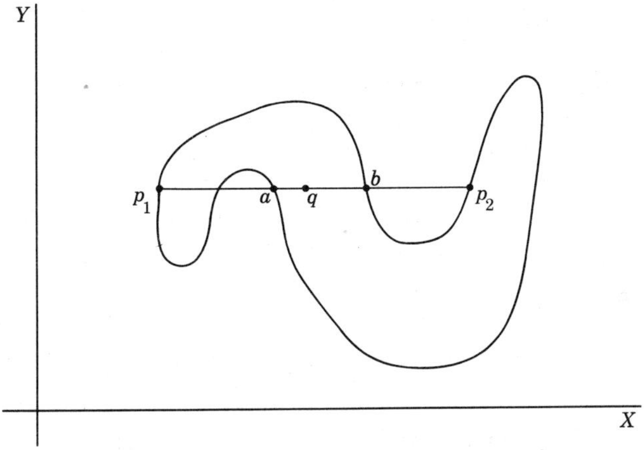

Figure 15

THEOREM 4.7 (*The Jordan Curve Theorem*). Let C be a simple closed curve in E^2. Then $E^2 - C$ consists of exactly two components A and B. Moreover, $C = \bar{A} - A = \bar{B} - B$.

Proof. If C contains a line segment parallel either to the x axis or to the y axis, then the theorem is immediate from Theorem 4.6. We thus assume that this is not the case.

We next show that there exist two simple closed curves C_1 and C_2 having as their intersection a line segment $L(a, b)$ which is parallel either to the x axis or to the y axis, such that C is the amalgamation of C_1 and C_2 across the common line segment $L(a, b)$.

Since C is a non-degenerate connected set in E^2, it contains three distinct points $p_1 = (x_1, y_1)$, $p_2 = (x_2, y_2)$, $p_3 = (x_3, y_3)$. If any pair of these points has the same ordinate, then the line segment joining this pair is parallel to the x axis. Suppose, for example, that p_1 and p_2 are such a pair; i.e., that $y_1 = y_2$. Then $L(p_1, p_2)$ is parallel to the x axis. By our assumption that C contains no line segment parallel to either axis, it follows that there must be a point $q \in L'(p_1, p_2) - C$. In the natural order on $L(p_1, p_2)$ from p_1 to p_2, let a be the last point of C preceding q, and b the first point of C following q. (See Figure 15.) Then C is the union of two independent arcs L_1 and L_2 joining a and b. Define C_1 and C_2, respectively, as the union of L_1 and L_2 with $L(a, b)$. Then C_1 and C_2 are simple closed curves having the line segment $L(a, b)$ in common, and having as their amalgamation across this line segment the given simple closed curve C. Moreover, $L(a, b)$ is parallel to the x axis.

If no pair of the points p_1, p_2, p_3 have the same ordinate, then we arrange the notation in such a way that $y_1 < y_2 < y_3$. Consider the two independent arcs $p_1p_2p_3$ and p_1qp_3 of C. The mapping $g: p_1qp_3 \to R$, defined by $g[(x, y)] = y$, for each point (x, y) of p_1qp_3, is continuous. Since p_1qp_3 is a connected set, $g(p_1qp_3)$ is a connected subset of R containing the point $g(p_1) = y_1$ and the point $g(p_3) = y_3$. Since $y_1 < y_2 < y_3$, there must therefore exist a point $p_4 \in p_1qp_3$ such that $g(p_4) = y_2$; i.e., the point p_4 has ordinate y_2. Applying the argument of the previous paragraph to the pair of distinct points p_2, p_4 of C again shows that C is the amalgamation of two simple closed curves C_1 and C_2 across their intersection which is a line segment parallel to an axis.

By Theorem 4.6, we have the separations

$$E^2 - C_1 = A_1 \cup B_1$$

$$E^2 - C_2 = A_2 \cup B_2$$

where A_1, B_1 are components of $E^2 - C_1$, A_2, B_2 are components of $E^2 - C_2$, and $C_1 = \bar{A}_1 - A_1 = \bar{B}_1 - B_1$, $C_2 = \bar{A}_2 - A_2 = \bar{B}_2 - B_2$. Define $L_1 = C_1 - L(a, b)$, $L_2 = C_2 - L(a, b)$ and choose the notation in such a way that $L_1 \subset A_2$, $L_2 \subset A_1$. Then, precisely as in the proof of Theorem 3.30, it follows that

$$E^2 - C = [B_1 \cup B_2 \cup L'(a, b)] \cup [A_1 \cap A_2]$$

and that these sets are separated. Thus $E^2 - C$ is not connected.

It follows from Theorem 4.5 that, for every component A of $E^2 - C$, we must have $\bar{A} - A = C$. We know that $E^2 - C$ has at least two components. The proof of the Jordan curve theorem will thus be complete if we show that $E^2 - C$ cannot have three components. The student may find Figure 16 helpful as he reads the remainder of the proof.

Let A and B be distinct components of $E^2 - C$. Pick two arcs cpd and xqy on C which are disjoint. Denote the interior of cpd by F_1' and the interior of xqy by F_2'. Let $r \in F_1'$. There exists a special rectangle W having r as center and with the property that the closed set $[W \cup W(r)]$ is disjoint with $C - F_1'$. The set $W(r)$ is a neighborhood of r; hence there exists a point $r' \in W(r) \cap A$. The line segment $L(r', r)$ lies in $W(r)$ and has a first point, a_1, in the order from r' to r, on the arc cpd. Then the line segment $L(r', a_1)$ lies in A except for its end point a_1. In exactly the same manner we may find a line segment $L(s', a_2)$ which lies in A except for its end point a_2 which is a point of F_2'. Moreover, the line segments $L(r', a_1)$ and $L(s', a_2)$ may be so chosen that they do not intersect. Now the points r' and s' lie in the region A of E^2. They may thus be joined by a special polygonal arc in A. In the union of this special polygonal arc and the two sets $L(r', a_1)$ and $L(s', a_2)$, we may find a polygonal arc M joining a_1 and a_2 with the property that $M' \subset A$ and M contains at least one side parallel to the x axis or the y axis. We note that $a_1 \in F_1'$, $a_2 \in F_2'$.

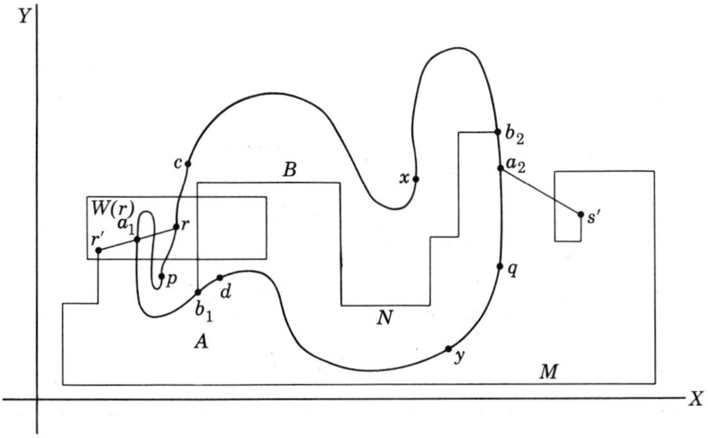

Figure 16

By the same process we can find two points b_1, b_2 such that $b_1 \in F_1'$, $b_2 \in F_2'$, and there exists a polygonal arc N joining b_1 and b_2 with the property that $N' \subset B$. Also N contains a side parallel to the x axis or the y axis. Define C_1 as the union of M, N, the subarc a_1b_1 of cpd, and the subarc a_2b_2 of xqy. Clearly, C_1 is a simple closed curve. Moreover, C_1 contains a line segment parallel either to the x axis or to the y axis. Thus, by Theorem 4.6, $E^2 - C_1$ consists of exactly two components D_1 and D_2, and we know that $C_1 = \bar{D}_1 - D_1 = \bar{D}_2 - D_2$.

Now $\bar{D}_1 - D_1 = C_1$. Hence, since A is a neighborhood of a point $m \in M \subset C_1$, it follows that $A \cap D_1 \neq \emptyset$. Likewise, since B is a neighborhood of a point $n \in N \subset C_1$, it follows that $B \cap D_1 \neq \emptyset$. Thus, using a similar argument for D_2, we conclude that each of the sets D_1 and D_2 contains points of both A and B. If D_1 lies in $E^2 - C$, then, since D_1 is connected and contains a point of A, $D_1 \subset A$. This is impossible, since A and B are disjoint subsets of $E^2 - C$ and $D_1 \cap B \neq \emptyset$. Thus $D_1 \cap C \neq \emptyset$. Similarly, $D_2 \cap C \neq \emptyset$. Now $C - C_1$ is composed of exactly two sets G_1 and G_2, each of which is the interior of an arc of C

and hence a connected set. The set D_1 must intersect the union of G_1 and G_2; hence we may choose the notation in such a way that D_1 intersects G_1. Since G_1 is a connected subset of $E^2 - C_1$, it follows that G_1 is contained in D_1; hence G_1 and D_2 are disjoint. From the fact that D_2 intersects C, we now see that G_2 is contained in D_2.

Suppose now that $E^2 - C$ contains a component S distinct from both A and B. It follows at once that S cannot intersect either M or N. Consequently, S is a connected subset of $E^2 - C_1$; hence S must be contained in either D_1 or D_2. Suppose S contained in D_1. Then

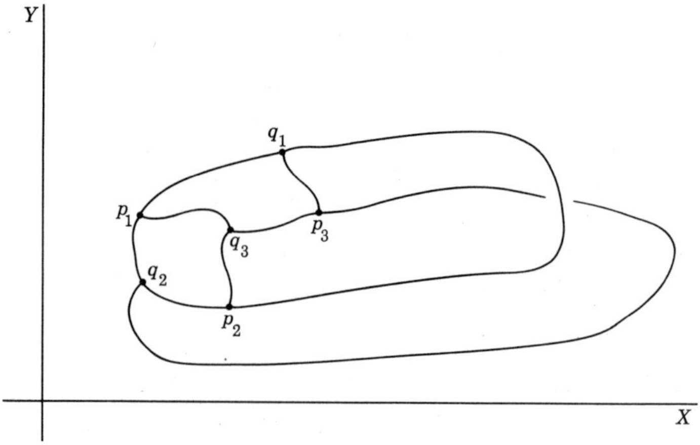

Figure 17

$\bar{S} - S \neq C$, since no limit point of S can lie in D_2, and D_2 contains G_2 which is a non-empty subset of C. This contradiction (or a similar one if S lies in D_2) proves the non-existence of S, and completes the proof of the Jordan curve theorem.

It is difficult to comment adequately on the role of the Jordan curve theorem, since it is used in almost every proof of theorems in analysis involving regions in E^2. Moreover, since the Argand plane (i.e., the plane of complex numbers) is homeomorphic to E^2, the theorem plays a vital role in the theory of functions of a complex variable.

PROBLEMS

35. Let ab be an arc in E^2, and V a neighborhood of a in E^2. Show that there exists a point $x \in ab$ such that the subarc ax of ab is contained in V.

36. Prove that, for any point $a \in E^2$, the set $E^2 - a$ is a region in E^2.

37. Outline the argument used in the proof of Theorem 4.4.

38. Let S be a subset of E^2 which contains two distinct points p and q such that $S = C_1 \cup C_2 \cup C_3$, where each C_i, $i = 1, 2, 3$, is an arc joining p and q, and each pair of these arcs is a pair of independent arcs from p to q. Prove that S contains exactly three simple closed curves T_1, T_2, T_3, and that $E^2 - S$ has exactly three components A_1, A_2, A_3. Show also that, if the notation is properly chosen, then $\bar{A}_i - A_i = T_i$, for $i = 1, 2, 3$.

39. Let S be a space which is the union of nine arcs p_1q_1, p_1q_2, p_1q_3, p_2q_1, p_2q_2, p_2q_3, p_3q_1, p_3q_2, p_3q_3 such that the interiors of any pair of these arcs are disjoint and the points $p_1, p_2, p_3, q_1, q_2, q_3$ are distinct. Such a space is called a primitive skew curve of type I. Prove that a primitive skew curve of type I is not homeomorphic to any subset of E^2. (See Figure 17.)

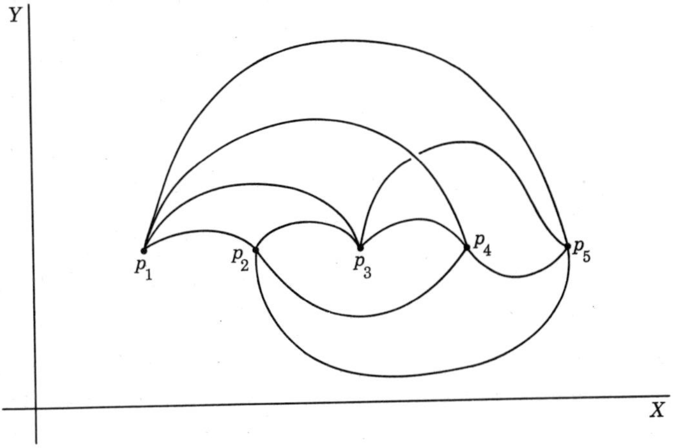

Figure 18

40. Let S be a space which is the union of ten arcs p_1p_2, p_1p_3, p_1p_4, p_1p_5, p_2p_3, p_2p_4, p_2p_5, p_3p_4, p_3p_5, p_4p_5, such that the interiors of any pair of these arcs are disjoint and the points p_1, p_2, p_3, p_4, p_5 are distinct. Such a space is called a primitive skew curve of type II. Prove that a primitive skew curve of type II is not homeomorphic to any subset of E^2. (See Figure 18.)

5. Peano Spaces

In Section 2, we obtained a characterization of all spaces that are images of J, the closed unit interval, under *homeomorphisms*. Such spaces are known as arcs. In this section, we obtain a characterization of all Hausdorff spaces which are images of J under *continuous mappings*. Such spaces are known as Peano spaces. Throughout this section the letter J will denote the closed unit interval $[0, 1]$ of the space R of real numbers.

Definition 5.1. Let S be a Hausdorff space. Then S is said to be a Peano space if and only if there exists a continuous mapping of J onto S. Peano spaces are also called continuous curves.

From the definition and earlier results on continuous mappings, we deduce immediately that a Peano space is compact and connected. It is not so obvious that every Peano space is also locally connected and metrizable, but these facts will follow from the next two theorems.

THEOREM 5.2. Let S and T be spaces and $f: S \to T$ a mapping which is both continuous and closed. Then, if S is locally connected, $f(S)$ is locally connected.

Proof. Suppose that S is locally connected. To prove that $f(S)$ is locally connected, it is sufficient, by Problem IV, 16, to show that, if U is any subset of $f(S)$ that is open in $f(S)$, and if C is any component of U, then C is open in $f(S)$. From the continuity of f, and the fact that U is open in $f(S)$, it follows at once that $f^{-1}(U)$ is open in S. Let A be any component of $f^{-1}(U)$. Then, since A is connected, and C is a component of U, we must have either $f(A) \subset C$ or $f(A) \cap C = \emptyset$. From this, it follows that $f^{-1}(C)$ is the union of a collection of components of $f^{-1}(U)$. But every component of $f^{-1}(U)$ is open in S, since S is locally connected; hence $f^{-1}(C)$ is open in S. Consequently, $S - f^{-1}(C)$ is closed. Therefore, since f is a closed mapping, $f[S - f^{-1}(C)] = f(S) - C$ is closed. This proves that C is open; hence $f(S)$ is locally connected.

THEOREM 5.3. Let S and T be compact Hausdorff spaces, and f a continuous mapping of S onto T. If S is second countable, then T is second countable.

Proof. Since S and T are compact Hausdorff spaces and f is continuous, it follows that f is a closed mapping. Let $\{G_n\}_{n\in I}$ be a countable basis for S. Then the collection M of all sets V_n, each of which is either a G_n, for some $n \in I$, or the union of a finite number of such sets, is not only a basis for S, but also a countable basis for S. Let $M = \{V_n\}_{n\in I}$. It is evident that the union of any finite number of elements of M is an element of M. Since f is a closed mapping, each of the sets $f(S - V_n)$ is closed. The proof will be complete if we show that the collection $W = \{T - f(S - V_n)\}_{n\in I}$ is a countable basis for T.

It is clear that the collection W is countable, and that each of its elements is open in T. It remains to be shown that W is a basis for T. Let U be any non-empty open subset of T and $x \in U$. Then $f^{-1}(x)$ is a closed, and hence compact, subset of the open set $f^{-1}(U)$ in S. For each $y \in f^{-1}(x)$, there exists an integer n_y such that $y \in V_{n_y} \subset f^{-1}(U)$. A finite number of sets of the open covering $\{V_{n_y}\}_{y\in f^{-1}(x)}$ of the compact set $f^{-1}(x)$ must cover $f^{-1}(x)$, and, by our construction of the collection M, the union of this finite collection of elements of M is an element V_k of M. Thus $f^{-1}(x) \subset V_k \subset f^{-1}(U)$; hence $S - f^{-1}(x) \supset S - V_k \supset$

$S-f^{-1}(U)$. However, $S-f^{-1}(U)=f^{-1}(T-U)$ and $S-f^{-1}(x)= f^{-1}(T-x)$. Consequently,

$$f^{-1}(T-x)\supset S-V_k\supset f^{-1}(T-U)$$

so that
$$T-x\supset f(S-V_k)\supset T-U$$

i.e.,
$$x\in T-f(S-V_k)\subset U$$

Since $T-f(S-V_k)$ is an element of W, the proof that W is a countable basis for T is complete.

From these two theorems, our previous results on continuous mappings, Problem IV, 55, and the fact that a second countable, regular T_1 space is metrizable, we obtain the following theorem.

THEOREM 5.4. Every Peano space S is a non-empty, compact, connected, locally connected, metrizable space.

We shall now show that the converse of Theorem 5.4 is also true: i.e., that every non-empty, compact, connected, and locally connected metrizable space S is a Peano space. This result is the celebrated Hahn-Mazurkiewicz characterization of these spaces. The method of proof has much in common with the methods used to establish the characterization of the arc. We suppose that S is a non-empty, compact, connected, locally connected, metrizable space. In the case that S is a single point, it is obvious that S is a Peano space. In the case that S is non-degenerate, we use the compactness and metrizability of S to obtain a countable dense subset K of S. We show that we can order K and define, for each $n\in I$, a subset F_n of K such that $K=\bigcup_{n\in I} F_n$, and the hypotheses of the index theorem as well as those of Theorem 2.6 are satisfied. Thus we show that there exists a uniformly continuous mapping g of a dense subset D of J onto K. The continuous extension f of g from D to J is then a mapping of J onto S, and hence S is a Peano space.

Actually, we determine K, the sets F_n, and an order for K simultaneously. The results concerning order, ε nets, and ε chains which we shall need for our proof are contained in several lemmas. It is particularly convenient to use an M metric in our constructions. The existence of such a metric for our space S is guaranteed by the hypotheses that S is connected, locally connected, and metrizable.

LEMMA 5.5. Let S be a non-degenerate compact, connected metric space, and a and b distinct points of S. Then there exists a countable collection $\{P_n\}_{n \in I}$ of subsets of S such that

(i) $P_i \cap P_j = \emptyset$ if $i \neq j$.

(ii) For each $n \in I$, P_n is a $1/2^{n+1}$ net in S.

(iii) $\bigcup_{n \in I} P_n$ is a countable subset of S, dense in S, contained in $S - (a \cup b)$.

Proof. For each $n \in I$, define $\varepsilon_n = 1/2^{n+1}$. By Theorem IV, 4.14, there exists an ε net in S, where ε is the minimum of $\frac{1}{32}$ and $\rho(a, b)/4$. Define P_1 to be the points of this net in $S - (a \cup b)$. Since $\varepsilon \leq \rho(a, b)/4$, $P_1 \neq \emptyset$. Since $\varepsilon \leq \frac{1}{32}$, the set P_1 is an ε_1 net in S. Thus $P_1 \subset S - (a \cup b)$, and P_1 satisfies (i) and (ii). Suppose that we can find sets P_j contained in $S - (a \cup b)$ and satisfying (i) and (ii) for $j = 1, 2, \cdots, k$. The set P_{k+1} is defined in the following way. Since S is compact, we can find a subset $q_1, q_2, \cdots, q_m$ of S, such that the spherical neighborhoods $S_{\varepsilon_{k+2}}(q_1)$, $S_{\varepsilon_{k+2}}(q_2)$, $\cdots$, $S_{\varepsilon_{k+2}}(q_m)$ covers S. Since S is a non-degenerate connected metric space, every neighborhood in S contains infinitely many points of S. For each integer $j = 1, 2, \cdots, m$, we may thus find a point $r_j \in S_{\varepsilon_{k+2}}(q_j) - \{\bigcup_{i=1}^{k} P_i \cup a \cup b\}$. We define P_{k+1} to be the set consisting of the m points $r_1, r_2, \cdots, r_m$. Clearly, the sets $P_1, P_2, \cdots, P_k$, P_{k+1} satisfy (i). To show that (ii) is satisfied, we must show that P_{k+1} is a $1/2^{k+2}$ net in S. Let $x \in S$. Then there exists an integer j_0, satisfying $1 \leq j_0 \leq m$, such that $\rho(x, q_{j_0}) < \varepsilon_{k+2} = 1/2^{k+3}$. Thus $\rho(x, r_{j_0}) \leq \rho(x, q_{j_0}) + \rho(q_{j_0}, r_{j_0}) < 1/2^{k+3} + 1/2^{k+3} = 1/2^{k+2}$. This proves that P_{k+1} is a $1/2^{k+2}$ net in S. The theorem thus follows by mathematical induction, since (iii) is obvious.

LEMMA 5.6. Let a and b be distinct points of the connected metric space S, and P a finite subset of S, containing neither a nor b. Then, given $\varepsilon > 0$, there exists an ε chain of distinct points from a to b in $S - P$. Moreover, this chain can be so chosen that it contains at least three points.

Proof. Denote by k the number of points of the finite set P, and define $\varepsilon_1 = \varepsilon/(k + 2)$. Note that $k + 2 \neq 0$. The proof will be based on the result of Problem IV, 34: namely, that every connected subset of a metric space is well-chained. Denote by L_1 an ε_1 chain of distinct points $a = x_0, x_1, \cdots, x_n = b$. Since L_1 is an ε_1 chain, it follows at once from the triangle property of the metric, that, if x_i and x_{i+k+1} are points of L_1, then $\rho(x_i, x_{i+k+1}) \leq (k + 1)\varepsilon_1 < \varepsilon$. Thus, if we remove from L_1 all points of P that it contains, and note that there are at most k such points, it is immediate that the remaining points, taken in the order in which they occurred in L_1, form an ε chain of distinct points from a to b in $S - P$. This can be done for every $\varepsilon > 0$. In particular, if we take $\varepsilon < \rho(a, b)$, the chain must contain at least three points.

LEMMA 5.7. Let a and b be distinct points of the connected metric space S, and P, Q disjoint finite subsets of $S - (a \cup b)$. Then, given $\varepsilon > 0$, there exists an ε chain of distinct points from a to b in $S - P$, and this ε chain contains the set Q. Furthermore, there is a positive integer N such that, for any $n \geq N$, there is an ε chain of distinct points satisfying these conditions and containing exactly n points.

Proof. We show first the existence of an ε chain of points from a to b in $S - P$ which contains the set Q. If Q is empty, this is immediate from Lemma 5.6. We may thus assume that Q contains n distinct points $q_1, q_2, \cdots, q_n$ for some $n \in I$. Define $q_0 = a$, $q_{n+1} = b$. From Lemma 5.6, it is immediate that there exist ε chains $L_1, L_2, \cdots, L_{n+1}$ in $S - P$ which satisfy the following conditions:

(i) L_i is an ε chain from q_{i-1} to q_i, for $i = 1, 2, \cdots, n + 1$.

(ii) $L_1 \subset S - P$.

(iii) $L_i \subset q_i \cup [S - (P \cup \bigcup_{j=1}^{i-1} L_j)]$, for $i = 2, 3, \cdots, n + 1$.

From (i) and (iii), we have at once

(iv) $L_i \cap L_{i+1} = q_i$ for $i = 1, 2, \cdots, n$, and $L_i \cap L_j = \emptyset$ for $i = 1, 2, \cdots, n + 1$; $j = 1, 2, \cdots, n + 1$; $|i - j| > 1$.

We now index the points of each L_i in the order in which these points appear in this ε chain. The points of L_1 we denote by $a = x_0, x_1, \cdots, x_{n_1} = q_1$; the points of L_2 by $q_1 = x_{n_1}, x_{n_1+1} \cdots, x_{n_2} = q_2$; and, in general, the points of L_j by $q_{j-1} = x_{n_{j-1}}, x_{n_{j-1}+1}, \cdots, x_{n_j} = q_j$. Clearly the points $a = x_0, x_1, \cdots, x_{n_{n+1}} = b$ form an ε chain of distinct points from a to b in $S - P$ which contains the set Q.

To prove the final statement of the lemma, define $N = n_{n+1} + 1$, and note that we have found an ε chain satisfying the required conditions and containing exactly N points. To complete the proof, it suffices to show that, given an ε chain containing exactly M points, for some $M \geq N$, which satisfies the required conditions, we can find an ε chain satisfying these conditions which contains exactly $M + 1$ points. In the chain L containing exactly M points, the last two points are x_{M-2} and $x_{M-1} = b$. We know that $\rho(x_{M-2}, x_{M-1}) = \varepsilon - 2\delta \neq 0$ for some real number $\delta > 0$. The space S is connected, and thus $S_\delta(x_{M-1})$ contains infinitely many distinct points. In particular, it contains a point y which is not contained in the finite set consisting of the union of L and the set P. The chain containing exactly $M + 1$ points which we require is now defined as the points $a = x_0, x_1, \cdots, x_{M-2}, y, b$. Clearly, it is an ε chain satisfying all of our requirements. This completes the proof.

It is clear that, if $x_0, x_1, \cdots, x_n$ is an ε chain of distinct points in a metric space, then we can make this chain a simply ordered set by defining $x_i < x_j$ if and only if $i < j$. With this fact in mind, we have the following corollary to Lemma 5.7.

COROLLARY 5.8. Let a and b be distinct points of the connected metric space S, and P, Q disjoint finite subsets of $S - (a \cup b)$. Then, given $\varepsilon > 0$, there is a finite simply ordered subset L of S such that

(i) L is disjoint with P.

(ii) L contains Q.

(iii) a is the first point of L, and b is the last point of L.

(iv) For any pair of adjacent points x and y of L, $\rho(x, y) < \varepsilon$.

(v) $L \neq a \cup b$.

Furthermore, there is a positive integer N such that for any $n \geq N$ there is a finite simply ordered subset L of S containing exactly n points which satisfies (i)–(v).

We now prove the theorem of Hahn and Mazurkiewicz.

THEOREM 5.9. A space S is a Peano space if and only if S is non-empty, compact, connected, locally connected, and metrizable.

Proof. In view of Theorem 5.4, it is sufficient to prove that, if S is non-empty, compact, connected, locally connected, and metrizable, then S is a Peano space. This is immediate if S is a single point; hence we shall consider only the case in which S is non-degenerate. Since S is connected, locally connected, and metrizable, it follows, from Theorem IV, 2.4, that S has a metric ρ which is an M metric, and this will be the metric that we shall use for S throughout our proof.

Let a, b be two distinct points of S. Denote by $\{P_n\}_{n \in I}$ the countable collection of subsets of $S - (a \cup b)$ given by Lemma 5.5. Define $\varepsilon_n = 1/2^n$, for $n \in I$. Using Corollary 5.8, with $\varepsilon = \varepsilon_1/2$, we can find a finite simply ordered subset F_1 of S satisfying the following conditions:

(i) F_1 contains P_1.

(ii) a and b are, respectively, the first and last points of F_1.

(iii) There exists an integer n_1 such that F_1 contains exactly $2^{n_1} + 1$ points.

(iv) For any two adjacent points x and y of F_1, $\rho(x, y) < \varepsilon_1/2$.

Every pair of adjacent points x, y of F_1, $x < y$ has the property that $y \in S_{\varepsilon_1/2}(x)$. Furthermore, every point of S lies in $S_{\varepsilon_1/2}(x)$ for some

$x \in F_1$, since F_1 contains P_1. Let the points of F_1 be denoted by $x_0, x_1, \cdots, x_{2^{n_1}}$, where $x_i < x_j$ if and only if $i < j$, for $i = 0, 1, \cdots, 2^{n_1}$; $j = 0, 1, \cdots, 2^{n_1}$.

Since $x_1 \in S_{\varepsilon_1/2}(x_0)$, which is a connected set since we are using an M metric, there must exist, by Corollary 5.8, with $\varepsilon = \varepsilon_2/2$, a finite simply ordered subset F_1^1 of $S_{\varepsilon_1/2}(x_0) - [F_1 - (x_0 \cup x_1)]$, having x_0, x_1 as its first and last points, respectively, containing $S_{\varepsilon_1/2}(x_0) \cap [P_2 - (F_1 - [x_0 \cup x_1])]$, and such that the distance between any pair of adjacent elements of F_1^1 is less than $\varepsilon_2/2$.

Similarly there exists a finite simply ordered subset F_2^1 of $S_{\varepsilon_1/2}(x_1) - [(F_1 \cup F_1^1) - (x_1 \cup x_2)]$, having x_1, x_2 as its first and last points, respectively, containing $S_{\varepsilon_1/2}(x_1) \cap [P_2 - ([F_1 \cup F_1^1] - [x_1 \cup x_2])]$, and such that the distance between any pair of adjacent elements of F_2^1 is less than $\varepsilon_2/2$.

In precisely the same manner, the sets F_i^1 for $i = 1, 2, \cdots, 2^{n_1}$ are defined. Let n_2 be an integer having the property that $2^{n_2} + 1$ is larger than the number of points in F_i^1, for $i = 1, 2, \cdots, 2^{n_1}$. From Corollary 5.8, and the last part of the proof of Lemma 5.7 we see that, for each i satisfying $1 \leq i \leq 2^{n_1}$, we may now define a set G_i^1, which satisfies all the conditions satisfied by F_i^1, and the additional condition that it contains exactly $2^{n_2} + 1$ points. Define $F_2 = \bigcup_{i=1}^{2^{n_1}} G_i^1$. Define the order in F_2 to be that order whose restriction to G_i^1 is the previous order obtained for that G_i^1. Clearly, $F_1 \subset F_2$, and the order of F_2 restricted to F_1 is the previously defined order for F_1. Also $F_1 \supset P_1$, $F_2 \supset P_1 \cup P_2$, and the distance between adjacent elements of F_2 is less than $\varepsilon_2/2$, while the diameter of any closed interval of $F_1 \cup F_2$ having as its end points adjacent elements of F_1 is less than ε_1.

In like manner, we define F_3, and, in general, by induction, F_n for $n \in I$. Denote by K the set $\bigcup_{n \in I} F_n$, and define the order in K to be that order whose restriction to any F_n is the prescribed order in that F_n. The ordered set $K = \bigcup_{n \in I} F_n$ satisfies the hypotheses of the index theorem, and the diameter of any closed interval of K having end points which are adjacent elements of F_n is less than ε_n, so that the hypotheses of Theorem 2.6 are fulfilled. Furthermore, $K \supset \bigcup_{n \in I} P_n$; hence K is dense in S. Moreover, the set D of associated rationals is dense in J. Since the mapping $g: D \to S$ is uniformly continuous, and S is compact, and therefore complete, there is a unique continuous extension f of g from D to J. Finally, $f(J) \supset g(D) = K$, and $f(J)$ is compact. Consequently, $f(J) = S$, and we have proved that S is a Peano space. The proof of the Hahn-Mazurkiewicz theorem is thus complete.

COROLLARY 5.10. If a and b are distinct points of a Peano space S, then there exists a continuous mapping f of J onto S such that $f(0) = a$, $f(1) = b$.

The Hahn-Mazurkiewicz characterization makes it easy to recognize as Peano spaces certain spaces that do not satisfy the definition in terms of mappings of J in an obvious manner. A classic example of this variety is the subspace Z of E^2, where Z consists of all points (x, y) such that $0 \leq x \leq 1$, $0 \leq y \leq 1$. Since Z, the so-called unit square, is the same as $J \times J$, it is a connected, locally connected, compact metric space, and hence a Peano space. Explicit mappings of J onto Z are known, although their description is somewhat complicated. One such mapping is described in *A Treatise on Advanced Calculus* by Philip Franklin, p. 56, John Wiley & Sons, New York, 1940. It should be noted that a continuous mapping f of J onto Z *cannot* be one-to-one, since, otherwise, f would be a homeomorphism, and, consequently, Z would be an arc. (We know that every point of Z is a non-cut point, so that Z is not an arc.) This, and related examples, as pointed out in some detail by R. L. Wilder in The Origin and Growth of Mathematical Concepts, *Bulletin of the American Mathematical Society*, vol. 59, pp. 423–448, 1953, inspired much of the work we are presently concerned with.

We now turn to the proof that every Peano space is arcwise connected.

DEFINITION 5.11. Let S be a space. Then S is said to be arcwise connected if and only if, given any two distinct points a and b of S, there exists an arc axb in S.

Certainly every space that is arcwise connected is connected. On the other hand, we have exhibited connected spaces (e.g. Example III, 1.3) that are not arcwise connected. The property of being arcwise connected is not peculiar to Peano spaces. The space E^2 is arcwise connected (every two of its points can be joined by a special polygonal arc), but E^2 is not a Peano space since it is not compact. An example of a compact, connected, arcwise connected, metric space which is not locally connected, and hence not a Peano space, can be formed by a slight modification of Example IV, 3.22. Let A be an arc in E^2 with one end point in F, the other in G, and with no other points in F or G. Then the union of S and A is such a space.

Our proof of the arcwise connectivity of Peano spaces is based on the following idea. If a and b are two distinct points of a Peano space T, and there is an arc A in T joining a and b, then, since there is a mapping f of J onto T, we should be able to find and describe $f^{-1}(A)$, or at least a subset F of $f^{-1}(A)$ with the property that $f(F) = A$. We break our proof into two theorems. The first shows the existence of a particular subset F of J, while the second gives the conclusion that $f(F)$ is an arc joining a and b. Thus F plays the role indicated.

In our theorems and proof, we will make use of the fact that the components of an open subset of J are either open-intervals, half-open-intervals, or closed-intervals of J in the natural order of J from 0 to 1. It thus makes sense to speak of the end points of a component of an open subset of J.

THEOREM 5.12. Let a and b be distinct points of a Peano space M, and f a continuous mapping of J onto M. Then there exists a closed subset F of J such that

(i) $a \cup b \subset f(F)$.

(ii) If C is any component of $J - F$ and x, y are the end points of C, then $f(x) = f(y)$.

(iii) No proper closed subset of F satisfies both (i) and (ii).

Proof. The set J satisfies (i) and (ii). If J satisfies (iii), we define $F = J$. In the case in which J does not satisfy (iii), denote by $\{V_n\}_{n \in I}$ the collection of all non-empty elements of a countable basis for J. Since J does not satisfy (iii), there exists a proper closed subset K_1 of J satisfying (i) and (ii). Let $p_1 \in J - K_1$. Since $J - K_1$ is open, there exists an integer n, such that $V_n \subset J - K_1$; i.e., $K_1 \cap V_n = \emptyset$. We denote by n_1 the smallest positive integer such that there exists a proper closed subset of J satisfying (i) and (ii) and disjoint with V_{n_1}, and by F_1 such a proper closed subset of J. Thus F_1 satisfies (i) and (ii), and $F_1 \cap V_{n_1} = \emptyset$.

We know that there exist integers $n > n_1$, such that V_n is disjoint with F_1. In particular, any $V_n \subset V_{n_1}$ will satisfy this condition, and there are infinitely many V_n contained in V_{n_1}. We denote by n_2 the least integer greater than n_1 for which there exists a closed subset F_2 of F_1 satisfying (i) and (ii) and disjoint with V_{n_2}. Continuing by induction, suppose that F_k has been defined satisfying (i) and (ii) and disjoint with V_{n_k}. We then define n_{k+1} as the least integer greater than n_k such that there exists a closed subset $F_{n_{k+1}}$ of F_{n_k} disjoint with $V_{n_{k+1}}$ and satisfying (i) and (ii). By induction, we have thus defined a monotone decreasing sequence of sets $\{F_i\}$ and a corresponding subsequence $\{n_i\}$ of the sequence of all positive integers.

For each $i \in I$, F_i is non-empty. Moreover, each F_i is compact, since it is a closed subset of the compact set J. Thus, by the Cantor product theorem, the set $F = \bigcap_{i \in I} F_i$ is non-empty. The remainder of the proof will consist in showing that F satisfies (i), (ii), and (iii).

For each $i \in I$, define $K_i = f^{-1}(a) \cap F_i$, $L_i = f^{-1}(b) \cap F_i$. It is immediate that each of the sequences $\{K_i\}_{i \in I}$ and $\{L_i\}_{i \in I}$ is a monotone decreasing sequence of non-empty compact sets. Thus there exist points a', b' in $\bigcap_{i \in I} K_i$ and $\bigcap_{i \in I} L_i$, respectively. Clearly, $f(a') = a$, $f(b') = b$. Thus $a \cup b \subset f(F)$; hence F satisfies (i).

Let C be a component of $J - F$ having end points x and y, and p a point of C. Since $p \in J - F = J - \bigcap_{i \in I} F_i = \bigcup_{i \in I} (J - F_i)$, there exists an integer i_0 such that $p \in J - F_{i_0}$. For each $i \in I$, it is clear that $p \in J - F_{i_0+i}$, and we denote by C_i the component of $J - F_{i_0+i}$ which contains the point p. Let x_i, y_i denote the end points of C_i, where the notation is chosen in such a way that $x_i < y_i$ for $i \in I$. It is easily verified that the sequences $\{x_i\}_{i \in I}$, $\{y_i\}_{i \in I}$ converge to the points x and y, respectively, $x < y$. Since $f(x_i) = f(y_i)$, for $i \in I$, it is immediate from the continuity of f that $f(x) = f(y)$; i.e., F satisfies (ii).

To complete the proof, we must show that F satisfies (iii). Suppose that this is not the case. Then there must exist a proper closed subset Q of F which satisfies (i) and (ii). Let $q \in F - Q$. Then there exists an integer k, such that $q \in V_k$ and $V_k \cap Q = \emptyset$. Now $V_k \cap F \neq \emptyset$; hence $k \neq n_i$ for any $i \in I$. Now $k < n_1$ is impossible by the definition of n_1. Hence, since $n_{i+1} > n_i$ for every $i \in I$, there exists $i_1 \in I$ such that $n_{i_1} < k < n_{i_1+1}$. But this is likewise impossible by the definition of n_{i_1+1}. Thus no such proper closed subset Q of F can exist, and the proof of the theorem is complete.

THEOREM 5.13. Every Peano space is arcwise connected.

Proof. Let a and b be two distinct points of a Peano space M. By Corollary 5.10, there exists a continuous mapping f of J onto M with the property that $f(0) = a$, $f(1) = b$. Let F be a closed subset of J satisfying (i), (ii), and (iii) of Theorem 5.12. Define $A = f(F)$. We shall show that A is an arc in M with end points a and b.

From (i), and the fact that F is a closed (and hence compact) subset of J, we conclude that A is a non-degenerate, compact, metric space. It remains, in order to prove that A is an arc with end points a and b, to show that A is connected and every point of $A - (a \cup b)$ is a cut point of A.

We first show that A is connected. Suppose that this is not the case: i.e., that A is the union of two disjoint closed sets A_1 and A_2. Let $x \in f^{-1}(A_1)$, $y \in f^{-1}(A_2)$, and assume that the notation has been chosen in such a manner that $x < y$. The sets $f^{-1}(A_1)$ and $f^{-1}(A_2)$ are disjoint compact subsets of J. Let x_1 be the last point of the closed interval $[0, y]$ that lies in the set $f^{-1}(A_1)$, and y_1 the first point of the closed interval

$[x_1, 1]$ that lies in the set $f^{-1}(A_2)$. Thus the open interval (x_1, y_1) is a component of $J - F$ having end points x_1 and y_1. Consequently, by (ii), $f(x_1) = f(y_1)$. This is impossible, since A_1 and A_2 are disjoint. We conclude that A is connected.

It remains to be shown that, given any point $p \in A - (a \cup b)$, then p is a cut point of A. On the closed interval $J = [0, 1]$ denote by a_0, b_0 the first and last points, respectively, of the non-empty compact set F. Clearly, $F \subset [a_0, b_0]$, while, from (ii), we see that $f(a_0) = a$, $f(b_0) = b$. We also note that, for any point x of the open interval (a_0, b_0), $f(x) \neq a$ and $f(x) \neq b$. For, if $f(x) = a$, and $a_0 < x < b_0$, then $F \cap [x, b_0]$ is a proper closed subset of F satisfying (i) and (ii), which is impossible by (iii). We conclude that $f(x) \neq a$. Similarly, $f(x) \neq b$. Now for any point $p \in A - (a \cup b)$, let c_0 and c_1, respectively, denote the first and last points of $J = [0, 1]$ which lie in the non-empty compact set $F \cap f^{-1}(p)$. Since $F \cap ([a_0, c_0] \cup [c_1, b_0])$ satisfies (i) and (ii), we see, by (iii), that $F \subset ([a_0, c_0] \cup [c_1, b_0])$. Define $F_1 = F \cap [a_0, c_0)$, $F_2 = F \cap (c_1, b_0]$. Clearly, $A - p = f(F_1) \cup f(F_2)$, and each of these sets is non-empty. We will show that this is a separation of $A - p$. Suppose first that there exists a point $d \in f(F_1) \cap f(F_2)$. Then there must exist points $d_0 \in (a_0, c_0)$, $d_1 \in (c_1, b_0)$ such that $f(d_0) = f(d_1) = d$. Clearly $F \subset ([a_0, d_0] \cup [d_1, b_0])$, and thus $c_0 \in ([a_0, d_0] \cup [d_1, b_0])$, which is impossible. Thus $f(F_1)$ and $f(F_2)$ are disjoint. Now the set $f(F_1) \cup p = f(F \cap [0, c_0])$ is compact; hence $\overline{f(F_1)} \subset f(F_1) \cup p$. Thus $\overline{f(F_1)} \cap f(F_2) = \emptyset$. Similarly, $\overline{f(F_2)} \cap f(F_1) = \emptyset$. Therefore, $A - p$ is not connected for any $p \in A - (a \cup b)$. This completes the proof that A is an arc with end points a and b, and we have established the fact that M is arcwise connected.

We will obtain further results on Peano spaces in succeeding sections. There are several reasons for studying such spaces.

1. Compact, connected, locally connected, metric spaces abound in analysis.
2. The proofs of theorems concerning Peano spaces probably provide the richest single source of illustrations of the various mathematical techniques used in topology.
3. Some understanding of Peano spaces and their properties is helpful in understanding many current research interests.

The merit of each of these reasons undoubtedly depends on one's point of view. But, certainly, the most valuable asset a mathematician can have is knowledge of a wide variety of techniques and methods of proof. Thus, though we do not maintain that the statement of every theorem is important in itself, we do maintain that methods of proof are always worth examination.

PROBLEMS

41. Explain why the set D of associated rationals, as it occurs in the proof of Theorem 5.9, is dense in J.

42. Explain why the mapping f defined in the proof of Theorem 5.9 has exactly the property required in Corollary 5.10.

43. Prove Corollary 5.10 directly from the definition of a Peano space by establishing the result in the case $S = J$.

44. Let S and T be Peano spaces, S non-degenerate. Prove that there exists a continuous mapping of S onto T.

45. Let S and T be non-degenerate Peano spaces; a, b distinct points of S; c, d distinct points of T. Prove that there exists a continuous mapping f of S onto T with the property that $f(a) = c$, $f(b) = d$.

46. Prove that E^n is arcwise connected for every $n \in I$.

47. In the proof of Theorem 5.12, show that the sequences $\{x_i\}_{i\in I}$, $\{y_i\}_{i\in I}$ are monotone decreasing and monotone increasing respectively, and that they converge to the points x and y, respectively.

6. Property S

The closed unit interval J is a Peano space having a basis σ with the property that every element of σ is a connected set, and the closure of every non-empty element of σ is a Peano space. Now every Peano space is somewhat like J in that it is the continuous image of J. This suggests that every Peano space has a basis with like properties. We shall prove that this is the case. This result is not, however, so easily obtained as our remark concerning the Peano space J might imply. The difficulty is that, although a Peano space (since it is locally connected) has a basis whose elements are connected and locally connected, it is not necessarily true that the closure of each non-empty element of such a basis is a Peano space. To be sure, the closure will be a compact, connected, metric space. But is it locally connected? Consider the following example and Figure 19.

EXAMPLE 6.1. Let K be the subspace of E^2 consisting of all points (x, y) such that $x^2 + y^2 \leq 1$. Certainly K is a Peano space. For each $n \in I$, let W_n be the set of all points (x, y) of K satisfying

$$(1 - 1/(2n + 1))^2 < x^2 + y^2 < (1 - 1/(2n + 2))^2$$

and define T as the subset of K consisting of all points (x, y) satisfying

$$x^2 + y^2 < 1, \qquad y < 0$$

It is easily established that the set $U = T \cup \bigcup_{n \in I} W_n$ is an open connected subset of K. We observe that $\bar{U}$ is *not* a Peano space since it is not locally connected at the point $p = (0, 1)$. To see this, we need only notice that every open subset of $\bar{U}$ containing the point p and having diameter less than unity has infinitely many components.

We shall show, nonetheless, that a basis for any given Peano space does exist which is of the type described. The proof is based on the concept known as property S, which was introduced by the Polish mathematician Sierpinski in 1920. Our results concerning property S will also be useful in other connections.

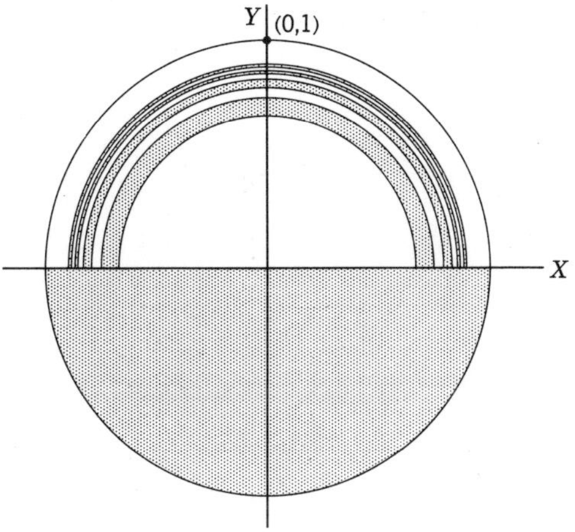

Figure 19

DEFINITION 6.2. Let K be a subset of a metric space T. Then K is said to have property S if and only if, for any $\epsilon > 0$, K can be expressed as the union of a finite number of connected sets, each of diameter less than ε.

We first observe that, if a subset K of a metric space has property S, then K is bounded. Thus no unbounded subset of T can have property S. Actually, the fact that K has property S implies that K is totally bounded.

DEFINITION 6.3. A subset K of a metric space T is said to be totally bounded if and only if, for any $\varepsilon > 0$, there exists a finite set of points $p_1, p_2, \cdots, p_n$ of T such that $K \subset \bigcup_{i=1}^{n} S_\varepsilon(p_i)$.

THEOREM 6.4. Let K be a subset of a metric space T. If K has property S, then K is totally bounded.

It is, of course, immediate that a totally bounded set is bounded. The converse is not necessarily true as the next example shows.

EXAMPLE 6.5. Let T denote the space of real numbers with metric $\rho^\star$ defined in terms of the standard metric ρ for T by

$$\rho^\star(x, y) = \rho(x, y)/(1 + \rho(x, y))$$

(See the proof of Theorem IV, 1.6). Since the mapping $f\colon R \to T$ (where R is the space of real numbers with the standard metric ρ), defined by $f(x) = x$, for $x \in R$, is a homeomorphism, and because of the relationship between $\rho^\star$ and ρ, it is easily deduced that T is a complete metric space. Furthermore, T is closed and bounded, but *not* compact. Thus T cannot be totally bounded, since every complete, totally bounded metric space is compact. To see this, suppose T is totally bounded. It follows that every sequence of distinct points has a Cauchy subsequence. Since T is complete, this sequence converges. Thus T is countably compact, and hence compact. This contradiction proves T is not totally bounded.

By the argument given in Example 6.5 the following theorem is established.

THEOREM 6.6. A metric space W is complete and totally bounded if and only if it is compact.

This suggests the following result which is also easily demonstrated.

THEOREM 6.7. If T is a totally bounded metric space, then the completion of T is totally bounded and compact.

Another immediate result, related to the finiteness condition in the definition of property S, is our next theorem.

THEOREM 6.8. Let T be a metric space. Then the union of a finite number of subsets of T, each having property S, is a set having property S.

To return to the definition of property S, we now consider its relationship to local connectedness. We first note the following theorem which is a special case of Theorem 3.10.

THEOREM 6.9. Let S be a space. Then S is locally connected if and only if, given any point x of S and any neighborhood U of x, then there is a neighborhood V of x which is contained in a single component of U.

THEOREM 6.10. Let K be a subset of a metric space T. If K has property S, then K is locally connected.

Proof. Let x be any point of K, and U any neighborhood of x in K. Then there exists $\varepsilon > 0$, such that $[K \cap S_\varepsilon(x)] \subset U$. Since K has property S, there exists a collection $G = \{G_1, G_2, \cdots, G_n\}$ of connected subsets of K, such that $K = \bigcup_{i=1}^{n} G_i$ and $\delta(G_i) < \varepsilon/3$, for $i = 1, 2, \cdots, n$. Define H as the union of all the G_i, $i = 1, 2, \cdots, n$, such that $x \in \bar{G}_i$, and L as the union of the remaining G_i (if there are none, define $L = \emptyset$). Certainly the following statements are true:

(i) $x \in H$; x is not in $\bar{L}$.

(ii) H is connected.

(iii) $\delta(H) < \varepsilon$, and thus $H \subset U$.

(iv) $x \in K - \bar{L} \subset H \subset U$.

Thus the neighborhood $K - \bar{L}$ of x in K is contained in a component of U. Hence, by the previous theorem, K is locally connected.

The next example shows that the converse is not true.

EXAMPLE 6.11. Let R be the space of real numbers and $q = 0$ the left end point of J. Consider the mapping $f: J - q \to R$ defined by $f(x) = \sin(\pi/x)$, for $0 < x \leq 1$ (see Example IV, 2.6). Then the graph F of f as a subset of the metric space E^2 is homeomorphic to $J - q$, and thus locally connected. But F does not have property S. For, take $\varepsilon = \frac{1}{4}$, and suppose $F = \bigcup_{i=1}^{n} G_i$, where G_i is a connected subset of F and $\delta(G_i) < \frac{1}{4}$ for $i = 1, 2, \cdots, n$. Now no G_i can contain more than one of the infinitely many points of F of the form $(1/m, 0)$, where $m \in I$. Thus $F = \bigcup_{i=1}^{n} G_i$ as described is impossible. We conclude that F does not have property S.

This example also serves to point out the important fact that property S is not a topological property, since $J - q$ is homeomorphic to F, and $J - q$ has property S in the standard metric for R restricted to $J - q$. Indeed, we see that whether or not a locally connected subset of a metric space has property S depends in general on the metric. For instance, in the case at hand, F has property S in the metric $\rho^\star$ defined by $\rho^\star(x, y) = |h(x) - h(y)|$, where $h: F \to J - q$ is a homeomorphism.

As a consequence of this example, it is necessary that we re-examine our theorems on B metrics and M metrics (Sections IV, 1 and IV, 2) to determine what happens to property S when we replace a given metric by a B metric or an M metric.

DEFINITION 6.12. Let ρ and $\rho^\star$ be two metrics for a metrizable space T.

Then ρ and $\rho^\star$ are said to be equivalent relative to property S in T if and only if a subset K of T having property S with respect to either one of these metrics has property S with respect to the other also.

THEOREM 6.13. Let T be a metric space with metric ρ. Then there is a B metric $\rho^\star$ for T which is equivalent to ρ relative to property S in T.

Proof. The metric $\rho^\star$ for the space T defined by $\rho^\star(x, y) =$ minimum $[\rho(x, y), 1]$ for any two points x, y of T is such a B metric.

THEOREM 6.14. Let T be a connected and locally connected metric space with metric ρ. Then there is an M metric ρ_M for the space T which is equivalent to ρ relative to property S in T. Furthermore, ρ_M may be so defined that it is also a B metric.

Proof. Given T with metric ρ, there is a B metric $\rho^\star$ which is equivalent to ρ relative to property S in T. Define $\rho_M: T \times T \to R$ as follows:

$\rho_M(x, y) =$ greatest lower bound of $\delta^\star(K)$ [the diameter of K using the metric $\rho^\star$] for all connected subsets K of T that contain both x and y

As in the proof of Theorem IV, 2.4, it follows that ρ_M is an M metric. Clearly ρ_M is also a B metric. Suppose that a subset K of T has property S with respect to $\rho^\star$. Then, given $\varepsilon > 0$, there exist connected sets G_i, $i = 1, 2, \cdots, n$, such that $K = \bigcup_{i=1}^{n} G_i$, and $\delta^\star(G_i) < \varepsilon$, $i = 1, 2, \cdots, n$. By the definition of ρ_M, it follows that $\delta_M(G_i) < \varepsilon$, for $i = 1, 2, \cdots, n$, where δ_M denotes the diameter using the metric ρ_M. Consequently, K has property S with respect to the metric ρ_M. On the other hand, if K has property S with respect to ρ_M, then it must have property S with respect to $\rho^\star$, since $\rho_M(x, y) \geq \rho^\star(x, y)$ for all $x, y \in T$. This completes the proof.

Thus far we have not examined how the ideas of property S and local connectivity are related in a compact set.

THEOREM 6.15. Let K be a compact subset of a metric space T. Then K is locally connected if and only if K has property S.

Proof. If K has property S, then, by Theorem 6.10, K is locally connected. Suppose K is locally connected. Given $\varepsilon > 0$, there is, for each point $p \in K$, a region $V(p)$ of K, containing the point p, and having diameter less than ε. Since K is compact, it is the union of a finite number of such regions, and hence has property S.

In particular, Theorem 6.15 implies that every Peano space has property S.

As a consequence of the fact that, in a metric space, every set containing a given connected set K and contained in $\bar{K}$ is also connected and has diameter equal to $\delta(K)$, we have the following result.

THEOREM 6.16. Let K be a subset of a metric space T and $K^\star$ a subset of T satisfying $K \subset K^\star \subset \bar{K}$. Then, if K has property S, $K^\star$ has property S.

A special case of this theorem will now be stated because of its importance in this section.

THEOREM 6.17. If K is a subset of a metric space T, and if K has property S, then $\bar{K}$ has property S. Thus, if K has property S, then $\bar{K}$ is locally connected.

We see, therefore, that, in order to find a basis of connected sets for a Peano space such that the closure of each non-empty element of this basis is a Peano space, it suffices to find a basis of connected sets, each of which has property S. We will be able to find such a basis, using the concept of ε growth.

DEFINITION 6.18. Let H, K be subsets of a metric space T, and $\varepsilon > 0$. Then H is said to ε-grow into K (or K is said to be an ε growth of H) if and only if

(*a*) $x \in K$ implies that there exists a connected subset M of T, such that $x \in M$, $M \cap H \neq \emptyset$, and $\delta(M) < \varepsilon$.

(*b*) There exists $\alpha > 0$, such that, if M is any connected subset of T satisfying $\delta(M) < \alpha$ and $M \cap H \neq \emptyset$, then $M \subset K$.

From this definition the next theorem is apparent.

THEOREM 6.19. If H and K are subsets of a metric space T such that, for some $\varepsilon > 0$, K is an ε growth of H, then

(i) $H \subset K$.

(ii) If H is connected, then K is connected.

(iii) $K \subset S_\varepsilon(H)$; hence $\delta(K) < \delta(H) + 2\varepsilon$.

(iv) If $H = \emptyset$, then $K = \emptyset$.

We next answer the question of the existence of ε growths for subsets of metric spaces.

THEOREM 6.20. Let H be a subset of a metric space T. Then, for any $\varepsilon > 0$, there exists a subset K of T such that K is an ε growth of H.

Proof. If $H = \emptyset$, then clearly $K = \emptyset$ is an ε growth of H. If $H \neq \emptyset$, then, for each point p of H, denote by $W(p)$ the union of all connected subsets of T of diameter less than ε, each of which contains the point p.

Define $K = \bigcup_{p \in H} W(p)$. It is easily verified that K is an ε growth of H, if we take $\alpha = \varepsilon$ in statement (*b*) of Definition 6.18.

In order to relate the concepts of ε growth and property S, we use a sequence of ε growths.

DEFINITION 6.21. Let H and K be subsets of a metric space T, and $\varepsilon > 0$ a real number. Then K is said to be an ε-sequential growth of H (or H is said to grow ε-sequentially into K) if and only if there is a sequence of positive numbers $\{\varepsilon_i\}$ and a sequence of subsets $\{H_i\}$ of T such that

(*a*) H_1 is an ε_1 growth of H, and H_{j+1} is an ε_{j+1} growth of H_j, for $j \in I$.

(*b*) $K = \bigcup_{i \in I} H_i$.

(*c*) $\sum_{i \in I} \varepsilon_i \leq \varepsilon$.

From Theorems 6.19 and 6.20 and the definition, the following result is almost immediate.

THEOREM 6.22. Let T be a metric space, H and K subsets of T, and $\varepsilon > 0$ a real number. Then

(i) There is an ε-sequential growth of H.

(ii) If K is an ε-sequential growth of H, then

(*a*) $H \subset K$.

(*b*) If H is connected, then K is connected.

(*c*) $K \subset S_\varepsilon(H)$; hence $\delta(K) < \delta(H) + 2\varepsilon$.

(*d*) If $H = \emptyset$, then $K = \emptyset$.

THEOREM 6.23. Let T be a metric space with property S, H and K subsets of T, and $\varepsilon > 0$ a real number. If K is an ε-sequential growth of H, then K has property S and is open in T.

Proof. Suppose that K is an ε-sequential growth of H, and let $\{\varepsilon_i\}$, $\{H_i\}$ be the sequences given in Definition 6.21. We first show that K is open in T. If $x \in K$, then there exists an integer i_0 such that $x \in H_{i_0}$. Since T is locally connected (Theorem 6.10), there is a region V in T which contains x and has diameter less than α_{i_0+1}, where α_{i_0+1} is the $\alpha > 0$ given by Definition 6.18*b* for the set H_{i_0+1} which is an ε_{i_0+1} growth of H_{i_0}. Thus $V \subset H_{i_0+1}$, and, accordingly, $x \in V \subset K$. Thus K is open in T.

To establish the fact that K has property S, we shall show that, given $\beta > 0$, K can be expressed as the union of a finite number of connected sets, each having diameter less than β. First of all, there is an integer k such that $\sum_{i \in I,\, i \geq k} \varepsilon_i < \beta/4$. By Definition 6.18*b*, there is a real number

$\alpha_{k+1} > 0$, such that any connected subset of T of diameter less than α_{k+1} containing points of H is contained in H_{k+1}. Since T has property S, it may be expressed as the union of a finite collection of connected sets, each having diameter less than the minimum of $\beta/4$ and α_{k+1}. Let $G_1, G_2, \cdots, G_n$ denote the sets of this finite collection which contain points of H_k. Certainly $H_k \subset \bigcup_{i=1}^{n} G_i \subset H_{k+1}$. For each i satisfying $1 \leq i \leq n$, denote by W_i the set of all points of K that lie in some connected subset of K, of diameter less than $\beta/4$, intersecting G_i. Clearly $G_i \subset W_i$, $\delta(W_i) < \beta$, and W_i is connected for $i = 1, 2, \cdots, n$.

We shall show that $K = \bigcup_{i=1}^{n} W_i$. It is immediate that $\bigcup_{i=1}^{n} W_i \subset K$. To show that $K \subset \bigcup_{i=1}^{n} W_i$, let $x \in K$. Then there exists an integer i_0 such that $x \in H_{i_0}$. If $i_0 \leq k$, then $x \in H_k \subset \bigcup_{i=1}^{n} G_i \subset \bigcup_{i=1}^{n} W_i$. If $i_0 \geq k$, then there exists, by the definition of ε growth, a collection of connected sets $L_{k+1}, L_{k+2}, \cdots, L_{i_0}$ such that $L_{k+i} \cap L_{k+i+1} \neq \emptyset$ for $1 \leq i \leq i_0 - 1$, $x \in L_{i_0}$, $L_{k+1} \cap H_k \neq \emptyset$, and $\delta(L_j) < \varepsilon_j$ for $k + 1 \leq j \leq i_0$. Thus $L = \bigcup_{j=k+1}^{i_0} L_j$ is a connected subset of K containing x and a point y of H_k, and having diameter less than $\beta/4$. Since y lies in some G_j, x must lie in the corresponding set W_j. Thus $K = \bigcup_{i=1}^{n} W_i$. Consequently, K has property S, and the proof of the theorem is complete.

By an easy application of this result, we obtain the key theorem on bases for metric spaces that have property S.

THEOREM 6.24. Let T be a metric space having property S. Then T has a basis, every element of which is an open connected set having property S.

Proof. Let $\{V_n(p)\}_{n \in I}$ be a countable basis at a point $p \in T$. For each $n \in I$, there exists a real number $\varepsilon_n > 0$, such that $S_{\varepsilon_n}(p) \subset V_n(p)$. Since T is locally connected, there exists, for each $n \in I$, a connected neighborhood $W_n(p)$ satisfying $W_n(p) \subset S_{\varepsilon_n/2}(p) \subset V_n(p)$. There is an ε-sequential growth $U_n(p)$ of $W_n(p)$ for $\varepsilon = \varepsilon_n/2$. Since T has property S and $W_n(p)$ is connected, $U_n(p)$ is an open connected subset of T having property S. Clearly

$$W_n(p) \subset U_n(p) \subset S_{\varepsilon_n/2}[W_n(p)] \subset S_{\varepsilon_n/2}[S_{\varepsilon_n/2}(p)] \subset S_{\varepsilon_n}(p) \subset V_n(p)$$

By our construction, the collection $\{U_n(p)\}_{n \in I}$ is a basis at p. Hence the collection $\{U_n(p)\}_{\substack{n \in I \\ p \in T}}$ is a basis for T consisting of open connected sets, each of which has property S.

Since every Peano space satisfies the hypotheses of this theorem, we have our desired result on bases for Peano spaces. Before stating this result formally, however, we will obtain some results on local arcwise connectivity.

Definition 6.25. Let T be a space. Then T is said to be locally arcwise connected if and only if, given any point $p \in T$ and any neighborhood U of p in T, there exists a neighborhood V of p in T such that $x, y \in V$, $x \neq y$, implies that there exists an arc in U joining x and y.

We shall also need the concept of local compactness.

Definition 6.26. Let T be a space. Then T is said to be locally compact if and only if, given any point $p \in T$ and any neighborhood U of p, there is a neighborhood V of p such that $V \subset U$ and $\bar{V}$ is compact.

Theorem 6.27. Every locally compact metric space T having property S is locally arcwise connected.

Proof. Suppose that T is a locally compact metric space having property S. Let p be a point of T and U any neighborhood of p. Since T is locally connected, there exists a connected neighborhood W of p contained in U. We shall show that W is arcwise connected. For each point $z \in W$, there is, by Theorem 6.24, the local compactness of T, and the fact that T is regular, a neighborhood $U(z)$ such that

(*a*) $z \in U(z) \subset \overline{U(z)} \subset W$.

(*b*) $U(z)$ is connected and has property S.

(*c*) $\overline{U(z)}$ is compact.

(*d*) The collection $\{U(z)\}_{z \in W}$ is an open covering of W.

Let x, y be distinct points of W. Then there is a simple chain $U(z_1)$, $U(z_2), \cdots, U(z_n)$ of sets of $\{U(z)\}_{z \in W}$ from x to y. Furthermore, the set $M = \bigcup_{i=1}^{n} U(z_i)$ is connected and has property S. Thus $\bar{M}$ is connected and has property S. Since $\bar{M} = \bigcup_{i=1}^{n} \overline{U(z_i)}$, $\bar{M}$ is compact. Therefore, $\bar{M}$ is a Peano space contained in W. By Theorem 5.13, there is an arc joining x and y in $\bar{M}$, and hence in W. This completes the proof that T is locally arcwise connected.

As a corollary to the proof of this theorem, we obtain the next result.

Corollary 6.28. If T is a locally compact metric space having property S, then every region in T is arcwise connected.

We know that every compact space is locally compact. Thus the results that we have obtained apply to Peano spaces.

THEOREM 6.29. Let T be a Peano space. Then T has a basis, each non-empty element of which is connected; has property S; and has a Peano space as its closure. Furthermore, T is arcwise connected and locally arcwise connected, and every region in T is arcwise connected.

In spite of the fact that a region M of a Peano space T is arcwise connected, it is not necessarily true that, if a is a point of $\bar{M} - M$, and b is a point of M, then there exists an arc ab in T such that $ab \subset M \cup a$. Our next example illustrates such a situation. See Figure 20.

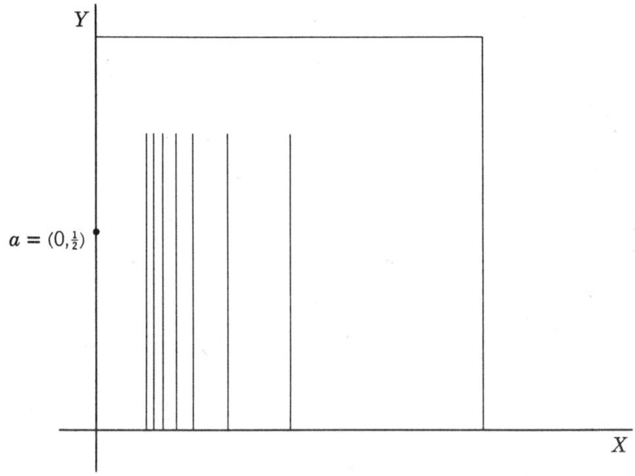

Figure 20

EXAMPLE 6.30. Let T be the subspace of E^2 consisting of all points (x, y) of E^2 such that $0 \leq x \leq 1$, $0 \leq y \leq 1$, and suppose that the metric for T is the standard metric for E^2 restricted to T. For each $n \in I$, define K_n to be the subset of T consisting of all points (x, y) of T, such that $x = 1/(n + 1)$, $0 \leq y \leq \frac{3}{4}$. Denote by W the set of all points (x, y) of T that do not satisfy both the conditions $0 < x < 1$, $0 < y < 1$, and define $K = W \cup \left(\bigcup_{n \in I} K_n\right)$, $M = T - K$. It is easily established that K is a compact subset of T, and that M is a non-empty region in T. We observe, in passing, that neither K nor M has property S. Denote by a the point $(0, \frac{1}{2})$ of K, and by b some point of M. Suppose that there exists an arc ab in $M \cup a$. Then, for $\varepsilon = \frac{1}{16}$, ab can be expressed as the union of a finite number of connected sets $G_1, G_2, \cdots, G_n$, each of diameter less than $\frac{1}{16}$. There is a point c on ab, $c \neq a$, such that the subarc ac of ab is contained in $S_{1/16}(a)$. There is an $n_0 \in I$, such that $1/(n_0 + 1)$ is less than the abscissa of c. It follows that, for each $i \in I$, there is a point (x_i, y_i) of ac such that $1/(n_0 + i + 1) < x_i < 1/(n_0 + i)$, $\frac{7}{16} < y_i < \frac{9}{16}$.

The points (x_i, y_i) are all distinct, and no G_j can contain more than one point of the sequence (x_i, y_i). This is a contradiction since $\bigcup_{i \in I} (x_i, y_i) \subset \bigcup_{j=1}^{n} G_j$. Thus no such arc *ab* exists. On the other hand, in any neighborhood of *a* in *T*, there is a point *d* of *K*, such that an arc *db* exists in $M \cup d$. This is the content of our next theorem.

THEOREM 6.31. Let M be a region in a Peano space T, a a point of $\bar{M} - M$, b a point of M, and U any neighborhood of a in T. Then there is a point d of U in $\bar{M} - M$ and an arc bd in $M \cup d$.

Proof. Suppose a neighborhood U of a in T is given. There is a region V of T containing a and contained in U. Since a is a limit point of M, $M \cap V \neq \emptyset$; thus $M \cup V$ is a region. By Corollary 6.28, there exists an arc ab in this region. The arc ab and the set $\bar{M} - M$ are compact and have the point a in common. Denote by d the last point of ab in $\bar{M} - M$. It is immediate that the subarc db of ab lies in $M \cup d$, and that $d \in V \subset U$. This completes the proof.

COROLLARY 6.32. Let M be a region in a Peano space T, and a, b distinct points of $\bar{M} - M$. Then, if U, V are neighborhoods of a, b, respectively, there exist points $d \in U \cap (\bar{M} - M)$, $e \in V \cap (\bar{M} - M)$ and an arc de in $M \cup d \cup e$.

PROBLEMS

48. Let T be a metric space with metric ρ, and K a subset of T which has property S with respect to the metric ρ. Prove that the metric ρ, restricted to K, is a B metric.

49. Prove Theorem 6.4.

50. In Example 6.5, show that T is complete in the metric $\rho^\star$.

51. Prove Theorem 6.6.

52. Prove Theorem 6.7.

53. Prove Theorem 6.8.

54. Show that $\rho^\star$, as defined in the proof of Theorem 6.13, is indeed a metric for the space T having the stated properties.

55. Let K be a compact subset of a metric space T, W a metric space, and $f: K \to W$ a continuous mapping. Show that, if K has property S, then $f(K)$ has property S.

56. Prove Theorem 6.19.

57. Consider Example 6.1. Prove that, for $\varepsilon = \frac{1}{8}$, there is an ε growth H of the subset U of the space K, such that H is an open subset of K, and H has property S.

58. Verify that K, as defined in the proof of Theorem 6.20, is an ε growth of H.

59. Prove Theorem 6.22.

60. Let T be a metric space, and p a point of T. Show that

$$S_\varepsilon(p) \subset S_\varepsilon(S_\varepsilon(p)) \subset S_{2\varepsilon}(p)$$

Give examples that illustrate each of the following possibilities.

(*a*) $S_\varepsilon(p) = S_\varepsilon(S_\varepsilon(p)) = S_{2\varepsilon}(p)$.

(*b*) $S_\varepsilon(p) \neq S_\varepsilon(S_\varepsilon(p))$ and $S_\varepsilon(S_\varepsilon(p)) \neq S_{2\varepsilon}(p)$.

(*c*) $S_\varepsilon(p) \neq S_\varepsilon(S_\varepsilon(p))$ and $S_\varepsilon(S_\varepsilon(p)) = S_{2\varepsilon}(p)$.

61. Let T be a space. Prove that, if T is locally arcwise connected, then T is locally connected.

62. In Example 6.30, show that neither M nor K has property S.

7. The Cyclic Connectivity Theorem

We have seen that the unit square in E^2 is a Peano space. One way in which it differs from an arc is that every two points of the unit square lie on a simple closed curve in the square, whereas an arc contains no simple closed curves at all. This is closely related to the fact that every point of the square is a non-cut point of the square. In fact, the main theorem of this section is that every point of a Peano space T is a non-cut point of T if and only if, given *any* two distinct points a and b of T, there exists a simple closed curve in T containing a and b. This result is known as the cyclic connectivity theorem.

Definition 7.1. A space T is said to be cyclic if and only if every point of T is a non-cut point of T.

Definition 7.2. Let a and b be points of a space T. Then a is said to be conjugate to b if and only if

(i) $a \neq b$.

(ii) No point of T separates T between a and b.

The following theorem is immediate.

Theorem 7.3. Let T be a space. Then T is cyclic if and only if, given any pair of distinct points $a, b \in T$, then a is conjugate to b.

In a space T, the relationship between two points which we have called conjugate is not an equivalence relation. However, it is symmetric, and a slightly weakened form of transitivity holds.

THEOREM 7.4. Let T be a space. Then

(i) For any two points $a, b \in T$, if a is conjugate to b, then b is conjugate to a.

(ii) For any three points $a, b, c \in T$, if a is conjugate to b and b is conjugate to c, then a is conjugate to c, provided b is a non-cut point of T, and $a \neq c$.

We give an example of a case in which transitivity does not hold. See Figure 21.

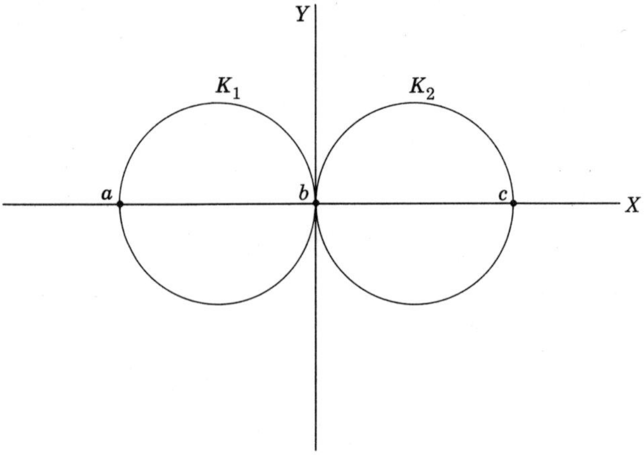

Figure 21

EXAMPLE 7.5. Let K_1, K_2 be the subspaces of E^2 consisting of all points $(x, y) \in E^2$ satisfying $(x + 1)^2 + y^2 = 1$ and $(x - 1)^2 + y^2 = 1$, respectively, and define K as the union of K_1 and K_2. Take a, b, c as the points $(-2, 0)$, $(0, 0)$, $(2, 0)$, respectively. It is clear that a is conjugate to b, b is conjugate to c, but a is not conjugate to c. It should be noted that each pair of distinct points of K_1 is conjugate, and each pair of distinct points of K_2 is conjugate. However, if $x \in K_1 - K_2$ and $y \in K_2 - K_1$, then x is not conjugate to y. Thus K_1 and K_2 have the nature of equivalence classes under the pseudoequivalence relation of conjugacy. Such sets as K_1 and K_2 we call proper cyclic elements.

DEFINITION 7.6. A subset E of a space T is said to be a proper cyclic element of T if and only if

(i) E contains two conjugate points a and b.

(ii) $E - (a \cup b)$ consists of all points x of T, such that x is conjugate to a in T, and x is conjugate to b in T.

The following useful theorem is immediate.

THEOREM 7.7. Let E be a proper cyclic element of a space T. Then
(i) If $p, q \in E$ and $p \neq q$, then p is conjugate to q in T.
(ii) If $p, q \in E$ and $p \neq q$, then $E - (p \cup q)$ consists of all points x of T such that x is conjugate to p in T, and x is conjugate to q in T.

The two subsets K_1 and K_2 of K in Example 7.5 were proper cyclic elements of K.

In our study of equivalence relations, we found that, if two equivalence classes of a set have a point in common, then they are identical. The proof of this result depended on the transitivity of the equivalence relation.

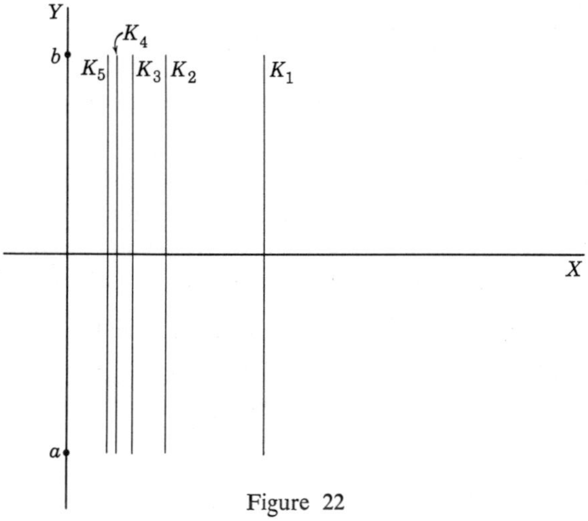

Figure 22

For proper cyclic elements of a space, we find that, if they intersect in more than one point, then they are identical. This result is immediate from Theorem 7.7ii. Note that, in our example, $K_1 \neq K_2$; yet K_1 and K_2 have the point (0, 0) in common.

THEOREM 7.8. Let C_1 and C_2 be proper cyclic elements of a space T. If $C_1 \cap C_2$ is non-degenerate, then $C_1 = C_2$.

We are going to prove that any proper cyclic element C of a Peano space T is connected. Although it is immediate that, if $T - p = A \cup B$ is a separation, and p is not in C, then $C \subset A$ or $C \subset B$, this by no means implies in itself that C is connected. An example will make this clear. See Figure 22.

EXAMPLE 7.9. For each $n \in I$, let K_n be the subset of E^2 consisting of all points (x, y) satisfying $x = 1/n$, $-1 \leq y \leq +1$, and define D to be the subset of E^2 consisting of the two points $a = (0, -1)$, $b = (0, +1)$.

Let $K = D \cup \left(\bigcup_{n \in I} K_n\right)$. Then any point $p \in K - D$ is a cut point of K. Suppose p is such a point, and $K - p = A \cup B$ a separation with $a \in A$. The set A is a neighborhood of a in K; hence there exists a real number $\varepsilon > 0$, such that $K \cap S_\varepsilon(a) \subset A$. We know that there is a $j \in I$ such that $p \in K_j$. Let $n_0 \in I$ be chosen, such that $n_0 > j$, and $1/n_0 < \varepsilon$. It is immediate that, for $n > n_0$, the connected set K_n is a subset of $K - p$ containing the point $(1/n, -1)$ of A. Consequently, A contains K_n for every $n > n_0$. However, b is a limit point of $\bigcup_{n \in I} K_n$, and b is not a limit point of K_n for any $n \in I$. We conclude that $b \in A$; therefore $D \subset A$. But D is not connected. Furthermore, D is a proper cyclic element of K. This example shows the need for some restriction on the space T if we are to conclude that every proper cyclic element of T is connected. We note that the space K of this example is not a Peano space.

We shall show that every proper cyclic element C of a Peano space T is connected. In our proof we make use of a special kind of mapping called a retraction of T onto C. In order to define this mapping we first establish the fact that C is closed, and that each component A of $T - C$ has the property $\bar{A} - A$ is a single point of C.

THEOREM 7.10. Let C be a proper cyclic element of a T_1 space T. If T is locally connected, then C is closed.

Proof. Denote by a, b two distinct points of the non-degenerate set C. We shall show that $T - C$ is open, and conclude that C is closed. Let $p \in T - C$. Then p cannot be conjugate to both a and b, and we choose the notation in such a way that p is not conjugate to b. This means that there exists a cut point q of T and a separation

$$T - q = A \cup B, \qquad p \in A, \qquad b \in B$$

Since T is locally connected and T_1, there exists a region U containing p which does not contain q. Thus $U \subset A$. But $C \subset B \cup q$, which is disjoint with A. Hence $U \cap C = \emptyset$. Thus $T - C$ is open, and C is closed.

THEOREM 7.11. Let C be a proper cyclic element of a Peano space T, and A a component of $T - C$. Then $\bar{A} - A$ consists of exactly one point; this point lies in C and is a cut point of T.

Proof. Since T is connected, it is clear that $\bar{A} - A$ is a non-empty subset of C. Moreover, if $\bar{A} - A$ consists of exactly one point p, then p lies in C and p is a cut point of T, since A is a proper subset of $T - p$ which is both open and closed in $T - p$. To complete the proof, we shall show that $\bar{A} - A$ cannot contain two distinct points. Suppose, otherwise, that a, b are distinct points of $\bar{A} - A$. By Corollary 6.32, there exists an arc cxd in T, having its interior in A, and its end points c and d in C. Let q be an interior point of cxd. Then, since $q \in T - C$, q cannot be conjugate both to the point c and to the point d. We suppose that the notation is chosen in such a way that q is not conjugate to c. Then there exists a point r which separates T between q and c. Clearly r must be an interior point of the subarc cq of cxd. It follows at once that r separates T between c and d; hence c is not conjugate to d. This is a contradiction, since both of these points lie in C. Thus $\bar{A} - A$ consists of exactly one point.

We now define the concept of a retraction of a space onto a subspace, and then prove that every proper cyclic element of a Peano space T is a retract of T.

DEFINITION 7.12. Let S be a space and T a subspace of S. Then T is said to be a retract of S if and only if there exists a continuous mapping $f: S \to T$ of S onto T with the property that $f(x) = x$ for every $x \in T$. The mapping f is called a retraction of S onto T.

THEOREM 7.13. Let C be a proper cyclic element of a Peano space T. Then C is a retract of T.

Proof. Define a mapping $f: T \to C$ in the following way. For each point $x \in C, f(x) = x$. For any point $y \in T - C$, let A be the component of $T - C$ which contains y. Then $\bar{A} - A$ is a unique point z of C, and we define $f(y) = z$. This defines a mapping $f: T \to C$ of T onto C. Since the components of $T - C$ are open in T, it is immediate that f is continuous and, therefore, a retraction of T onto C. Thus C is a retract of T.

By use of the retraction f of T onto C established in the proof just given, the following results are immediate.

THEOREM 7.14. Let T be a Peano space and C a proper cyclic element of T. Then

(i) If H is any connected subset of T, then $H \cap C$ is connected.

(ii) C is a cyclic Peano space.

In our proof of the cyclic connectivity theorem, we shall establish the fact that any two distinct points a and b of a cyclic Peano space T can be joined by two independent arcs in T, and hence lie on a simple closed curve in T. As a first result in this direction, we show that, if A and B are disjoint, closed, non-degenerate subsets of a Peano space T, then there exist two disjoint arcs in T, each having one end point in A and the other in B. The space K of Example 7.5 is a Peano space that does not have this property; it is, of course, not cyclic. Define A as the subset of K consisting of the two points $(-1, -1)$, $(-1, +1)$, and B as the subset of K consisting of the two points $(1, -1)$, $(1, 1)$. Clearly there do not exist such arcs in K joining the disjoint, closed, non-degenerate subsets A and B of K.

THEOREM 7.15. Let A and B be non-degenerate closed and disjoint subsets of a cyclic Peano space T. Then there exist two disjoint arcs C and D in T, such that each one has one end point in A, the other end point in B, and no interior point in $A \cup B$.

Proof. Define the subset E of T to consist of the set A and all points w of $T - A$ such that

(i) There is an arc F in T with one end point in A, the other end point in B, and no interior point in $A \cup B$.

(ii) There is an arc G in $T - F$ having w as one end point, and intersecting A in precisely one point, which is its other end point.

We shall show that E is both open and closed in T. Since T is connected, and $E \neq \emptyset$, this means that $E = T$. Thus, in particular, $B \subset E$; hence the required arcs exist.

We first show that E is open. For $z \in E$ we consider the two cases $z \in A$ and $z \in E - A$. In case $z \in A$, there is an arc F satisfying (i) contained in the region $T - z$, since z is a non-cut point of T, and A is non-degenerate. There is a region U in T containing z and disjoint with F. If $U \subset A$, then $U \subset E$. If U is not contained in A, let w be a point of $U - A$, and wz an arc in U. On wz let q be the first point of the closed set A. Then the subarc $G = wq$ of wz satisfies (ii). Thus $w \in E$, and we have proved that E is open if $z \in A$. In case $z \in E - A$, there exist arcs F and $G = pz$ satisfying (i) and (ii). There exists a region U containing z and disjoint with $F \cup A$. Let $w \in U - z$. There exists an arc wz in U, and we may find, in the union of this arc and the arc pz, an arc wz satisfying (ii). Thus $w \in E$; hence $U \subset E$. This completes the proof that E is an open set.

To complete the proof we shall show that E is closed in T. (See Figure 23.) Assume that this is not the case: i.e., that there exists a point $w \in T - E$ which is a limit point of E. Since T is cyclic and B is non-degenerate, there exists an arc F_1 in $T - w$ satisfying (i). (We need the non-degeneracy of B in case $w \in B$.) Let U be a region in T containing the point w and having the property that $\bar{U}$ is a Peano space satisfying $\bar{U} \cap [F_1 \cup A] = \emptyset$. There exists a point $z \in U \cap E$, since w is a limit point of E. From the fact that z is not a point of A, we see that there exist arcs F_2 and $G = pz$ satisfying (i) and (ii), respectively. Denote by y the first point of pz in $\bar{U}$ and by py the subarc of pz. There is an arc yw in the Peano space $\bar{U}$, and the union of the arcs py and yw is an arc pyw. Both $pyw \cap F_1 \neq \emptyset$ and $pyw \cap F_2 \neq \emptyset$ must hold, since otherwise w would belong to E, contrary to our assumption. Notice that $py \cap F_1 \neq \emptyset$, $yw \cap F_1 = \emptyset$, $py \cap F_2 = \emptyset$, and $yw \cap F_2 \neq \emptyset$. The last statement implies that $F_2 \cap \bar{U} \neq \emptyset$. Now F_2 is an arc ab, where $a \in A$, $b \in B$. Let x be the first point of ab in the compact set $\bar{U}$. There is an arc xw in $\bar{U}$, and the union of this arc with the subarc ax of F_2 is an arc axw. Certainly $axw \cap F_1 \neq \emptyset$, since otherwise $w \in E$, contrary to our assumption. Notice, then, that $ax \cap F_1 \neq \emptyset$, while $xw \cap F_1 = \emptyset$. The set F_1 is an arc cd, where $c \in A$, $d \in B$, and we denote by q the last point of cd which lies in the union of py and ax.

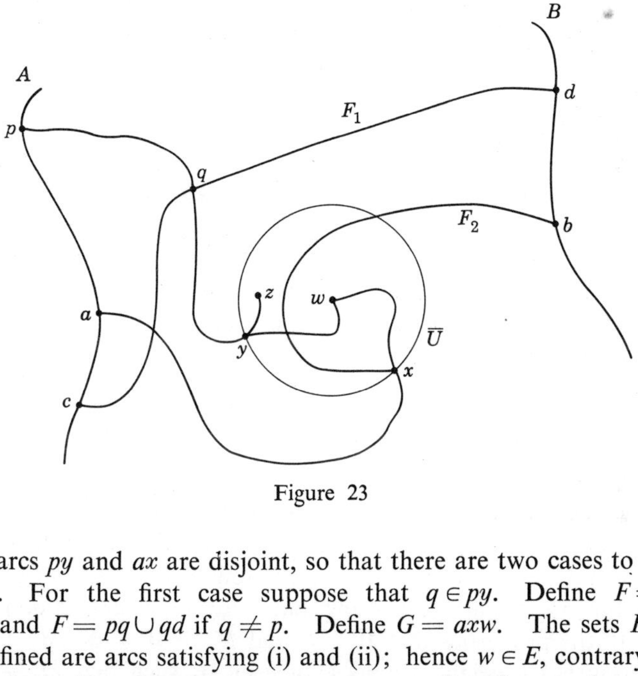

Figure 23

The arcs py and ax are disjoint, so that there are two cases to be considered. For the first case suppose that $q \in py$. Define $F = qd$ if $q = p$, and $F = pq \cup qd$ if $q \neq p$. Define $G = axw$. The sets F and G thus defined are arcs satisfying (i) and (ii); hence $w \in E$, contrary to our assumption. For the remaining case $q \in ax$. In this case, by a similar construction, we can also show that $w \in E$. Thus, in any event, the assumption that w is not a point of E leads to a contradiction. We have thus proved that E is closed in T, and the proof of the theorem is complete.

The result just established is now used to prove that any point of a non-degenerate cyclic Peano space is an interior point of some arc of that space. Our proof makes use of the three theorems that follow. The simple proof of the first of these theorems is left as an exercise for the student.

THEOREM 7.16. Let p be a point of a Peano space T. Then p is an interior point of an arc in T if and only if p is a cut point of some Peano space M contained in T.

THEOREM 7.17. Let p be a non-cut point of a non-degenerate Peano space T, and U a neighborhood of p. Then there exists a neighborhood V of p such that $p \in V \subset \bar{V} \subset U$ and $T - U$ is contained in a single component of $T - V$.

Proof. Suppose that a neighborhood U of p is given. There exists a neighborhood W of p such that $T - \bar{W}$ is non-empty and $\bar{W} \subset U$. The set $T - \bar{W}$ is open, and its components form an open covering of the compact set $T - U$. Let $A_1, A_2, \cdots, A_n$ denote the finite collection of these components which covers $T - U$. Now each of the sets A_i, $i = 1, 2, \cdots, n$, lies in the region $T - p$ of T. Thus there exists, for each $j = 2, 3, \cdots, n$, an arc F_j joining a point of A_1 to a point of A_j in $T - p$. The set F defined by $F = \bigcup_{j=2}^{n} F_j$ is closed and does not contain the point p. Consequently, $V = W - F$ is a neighborhood of p. However, the set $K = F \cup \left(\bigcup_{i=1}^{n} A_i\right)$ is a connected subset of $T - V$ containing the set $T - U$. It follows that $T - U$ lies in a single component of $T - V$, and the proof of the theorem is complete.

THEOREM 7.18. Let T be a non-degenerate cyclic Peano space, p a point of T, and U a neighborhood of p having the property that $M_1 = \bar{U}$ is a Peano space having p as a non-cut point. Then there exists a proper cyclic element of M_1 which contains p.

Proof. The set M_1 is non-degenerate. Thus, in order to establish the existence of a proper cyclic element of M_1 containing p, it suffices to show that M_1 contains a point distinct from p which is conjugate to p. To show this, we begin by observing that there is a neighborhood W of p in M_1 with the properties that $\bar{W} \subset U$ and $M_1 - W \neq \emptyset$. By Theorem 7.17, there exists a neighborhood V of p in M_1 such that $M_1 - W$ lies in a single component A of $M_1 - V$. There is an arc py in M_1 such that $py \cap A = y$, since A is closed in M_1. We now show that the assumption that y is not conjugate to p in M_1 leads to a contradiction. Under this assumption, there exists a point $z \in M_1$ such that z separates M_1 between p and y. Thus z is an interior point of the arc py, and we see at once that z separates M_1 between the point p and the set A. Therefore, every arc in M_1 joining p and a point of A must contain the point z. We arrive at a contradiction by showing the existence of such an arc which does not contain z.

Since z is a non-cut point of T, and $T - W$ is closed in T, there exists an arc pq in $T - z$ such that $pq \cap [T - W] = q$; hence $pq \subset \bar{W} \subset U \subset M_1$. Now $q \in \bar{W} - W \subset M_1 - W \subset A$, and, by construction, z is not in the arc pq. This contradiction proves that y is conjugate to p in M_1; hence there must exist a proper cyclic element of M_1 which contains the point p.

THEOREM 7.19. Every point p of a non-degenerate cyclic Peano space T is an interior point of some arc in T.

Proof. We shall show that the assumption that no arc in T has p as an interior point leads to a contradiction. By Theorem 7.16, our assumption is equivalent to the assumption that p is a non-cut point of every non-degenerate Peano subspace of T which contains it.

The space T is non-degenerate and connected; hence it must contain infinitely many distinct points. Let a, b be distinct points of $T - p$, and define $\varepsilon_n = 1/n$, for $n \in I$. There exists a neighborhood U_1 of p such that $M_1 = \bar{U}_1$ is a non-degenerate Peano space satisfying $M_1 \subset S_{\varepsilon_1}(p) \cap [T - (a \cup b)]$. By Theorem 7.18, there is a proper cyclic element E_1 of M_1 which contains p. By Theorem 7.15, there exist disjoint arcs aa_1 and bb_1 in T, such that $aa_1 \cap E_1 = a_1$, $bb_1 \cap E_1 = b_1$. If $p = a_1$ or $p = b_1$, then the union of the arc aa_1, the arc bb_1, and an arc a_1b_1 in the Peano space E_1 is an arc in T having p as an interior point. This is impossible by our assumption, hence $p \neq a_1$, $p \neq b_1$.

Let U_2 be a neighborhood of p in the Peano space E_1 whose closure M_2 in E_1 is a non-degenerate Peano space satisfying $M_2 \subset S_{\varepsilon_2}(p) \cap [E_1 - (a_1 \cup b_1)]$, and denote by E_2 the proper cyclic element of M_2 which contains p. By Theorem 7.15, there exist disjoint arcs a_1a_2, b_1b_2 in E_1, such that $a_1a_2 \cap E_2 = a_2$, $b_1b_2 \cap E_2 = b_2$. Clearly $p \neq a_2$, $p \neq b_2$. Also $aa_1 \cup a_1a_2$ is an arc aa_1a_2, and $bb_1 \cup b_1b_2$ is an arc bb_1b_2.

Continuing the same process inductively, it is apparent that there exists a sequence $\{E_n\}$ of Peano spaces, each containing p, such that $E_{n+1} \subset E_n$, and $E_n \subset S_{\varepsilon_n}(p)$ for $n \in I$. Also, for each $n \in I$, there exist disjoint arcs aa_n and bb_n such that

$$aa_n \cap E_n = a_n, \qquad bb_n \cap E_n = b_n$$

$$aa_n \subset aa_{n+1}, \qquad bb_n \subset bb_{n+1}$$

for $n \in I$. Let $A = \left(\bigcup_{n \in I} aa_n\right) \cup p \cup \left(\bigcup_{n \in I} bb_n\right)$. It is clear that p is a limit point of each of the connected sets $\bigcup_{n \in I} aa_n$ and $\bigcup_{n \in I} bb_n$. Thus A is connected. It is easily seen that every point of $A - (a \cup b)$ is a cut point of A, whereas each of the points a, b is a non-cut point of A. Since A is compact, this means that A is an arc. Clearly p is an interior point of the arc A. Thus we have obtained a contradiction to our original assumption, and the proof of the theorem is complete.

We now prove the cyclic connectivity theorem.

THEOREM 7.20. Let T be a Peano space. Then T is cyclic if and only if, given any two distinct points $a, b \in T$, there exists a simple closed curve in T containing a and b.

Proof. Suppose first that, given any pair of distinct points a, b of T, there exists a simple closed curve in T containing a and b. From this condition, it is clear that a is conjugate to b, for every pair of distinct points a, b of T. Thus T is a proper cyclic element of itself; hence T is cyclic by Theorem 7.14.

To prove the converse, suppose that T is a cyclic Peano space and that two distinct points a, b of T are given. We shall show that there exists a simple closed curve in T containing a and b. By Theorem 7.19, there is an arc xay in T. Since T is cyclic, $T - a$ is a region; hence there exists an arc xwy in $T - a$. Let s denote the last point on xwy in the arc xa, and denote by sy the subarc of xwy with end points s and y. Let t be the first point of the arc sy in the arc ay. Then the union of the subarc st of xwy and the subarc tas of yax is a simple closed curve in T containing the point a. We have thus shown that every point of T lies on a simple closed curve in T.

Let C_a, C_b be simple closed curves in T containing the points a, b, respectively. The proof of the theorem is complete if either $a \in C_b$ or $b \in C_a$. If neither of these situations occur, three cases remain for consideration in order to complete the proof:

(i) $C_a \cap C_b = \emptyset$.

(ii) $C_a \cap C_b$ is a single point z.

(iii) $C_a \cap C_b$ contains more than one point.

In case (i) there exist, by Theorem 7.15, two disjoint arcs A and B, each having exactly one end point in C_a and one end point in C_b. The union of the four sets C_a, C_b, A, B then clearly contains a simple closed curve containing a and b. In case (ii) there exists an arc A in $T - z$, joining a point of $C_a - z$ to a point of $C_b - z$, and having its interior disjoint with $C_a \cup C_b$. In this case it is clear that the union of the three sets C_a, C_b, A contains a simple closed curve containing a and b. In case (iii) the union of C_a and C_b contains a simple closed curve containing a and b. The proof of the cyclic connectivity theorem is thus complete.

An alternative characterization of cyclic Peano spaces is the three-point theorem which we next establish.

THEOREM 7.21. Let T be a Peano space. Then T is cyclic if and only if, given any three distinct points a, b, c of T, there exist arcs abc, acb, bac in T.

Proof. Suppose first that, given any three distinct points a, b, c of T, there exist arcs abc, acb, bac in T. We shall show that T is cyclic. If this is not the case, then there exists a cut point a of T which separates T between two points b and c. But then the existence of the arc abc in T gives us an immediate contradiction. We conclude that T is cyclic.

Suppose, on the other hand, that T is cyclic and that three distinct points a, b, c of T are given. We shall show that there exists an arc abc in T. The proof of the existence of arcs acb and bac in T is similar. By the cyclic connectivity theorem, there exist simple closed curves C_{ab} and C_{bc} in T containing $a \cup b$ and $b \cup c$, respectively. It is easily seen that $C_{ab} \cup C_{bc}$ contains an arc abc. This completes the proof.

We conclude this section with the proof of another useful result concerning Peano spaces.

THEOREM 7.22. Let T be a cyclic Peano space, K a cyclic Peano space contained in T, and A a component of $T - K$. Then $A \cup K$ is a cyclic Peano space.

Proof. We first prove that $A \cup K$ is a Peano space, and complete the proof by showing that $A \cup K$ is cyclic. Since each of the sets A and K is connected, and since $\bar{A} - A$ is a non-empty subset of K, it follows that the set $A \cup K = \bar{A} \cup K$ is both compact and connected. Since $A \cup K$ is clearly locally connected at each point of the subsets A and $K - \bar{A}$ of $A \cup K$, it suffices to establish that $A \cup K$ is locally connected at each point $p \in \bar{A} \cap K$ in order to conclude that $A \cup K$ is locally connected. Let p be any point of $\bar{A} \cap K$, and U any neighborhood of p in $\bar{A} \cup K$. Then $U \cap K$ is a neighborhood of p in K, and thus contains a connected neighborhood V of p in K, since K is locally connected. Denote by W the union of those components of $U \cap A$ which, in the space $A \cup K$, have limit points in V. Clearly, $V \cup W$ is connected, contains p, and is contained in U. Now $K - V$ is closed in K and hence in $A \cup K$. Also, $A - W$ has no limit points in the space $A \cup K$ which are in $V \cup W$. It follows that

$$(A \cup K) - (V \cup W) = (K - V) \cup (A - W)$$

is closed in $A \cup K$. Hence $V \cup W$ is open, and thus a connected neighborhood of p (in the space $A \cup K$) contained in U. We have thus proved that $A \cup K$ is a non-empty, compact, connected, locally connected, metric space, and thus a Peano space.

It remains to be shown that $A \cup K$ is cyclic. Suppose, assuming $A \cup K$ not cyclic, that p is a cut point of $A \cup K$. The proof divides conveniently into two cases.

Case 1. $p \in K$.

Since K is cyclic, $K - p$ is connected. Since p is a cut point of $A \cup K$, it follows at once that $\bar{A} - A = p$. But this implies that the proper subset A of $T - p$ is both open and closed in the connected set $T - p$, which is impossible.

Case 2. $p \in A$.

Let $[A \cup K] - p = D \cup F$ be a separation. The set K is connected and does not contain p; hence we may choose our notation in such a way that $K \subset F$; thus K contains no point or limit point of D. From this fact it is easily deduced that D is a non-empty proper subset of the connected set $T - p$ which is both open and closed in $T - p$. This contradiction completes the discussion of case 2. Since we have arrived at a contradiction in either case, the proof of the theorem is complete.

PROBLEMS

63. Prove Theorem 7.3.

64. Prove Theorem 7.4.

65. Show that the subsets K_1 and K_2 of the space K of Example 7.5 are proper cyclic elements of K.

66. Prove Theorem 7.7.

67. Prove Theorem 7.8.

68. Show that the subspace K defined in Example 7.9 is neither compact nor locally connected.

69. Prove that the mapping f defined in the proof of Theorem 7.13 is continuous.

70. Prove Theorem 7.14.

71. Let L be the subset of E^2 consisting of all points (x,y) such that $0 \leq x \leq 1$ and $y = x$. Let M be the subset of E^2 consisting of all points (x,y) such that $x = 0$ and $0 \leq y \leq 1$. For each $n \in I$, denote by K_n the subset of E^2 consisting of all points (x,y) such that $y = 1/n$ and $0 \leq x \leq 1/n$. Define $K = L \cup M \cup (\bigcup_{n \in I} K_n)$. Prove that K is a cyclic Peano subspace of E^2. Also prove that there is an arc ap in K such that no arc in K contains the arc ap and has p as an interior point.

72. Draw sketches showing, in the part of the proof of Theorem 7.15 in which the set E is shown to be closed, possible configurations for the arcs and sets involved.

73. Prove Theorem 7.16.

74. Show in detail why the set A defined in the proof of Theorem 7.19 is an arc having the point p as an interior point.

75. Let a and b be distinct points which lie on a simple closed curve in a space T. Show that a is conjugate to b in the space T.

76. Supply the details to support the statements made in the last paragraph of the proof of Theorem 7.20.

77. In the proof of Theorem 7.21, show that $C_{ab} \cup C_{bc}$ contains an arc abc.

78. Let T be a Peano space, and K a Peano space contained in T. Denote by W the union of K and any collection of components of $T - K$. Show that W is a Peano space. Also show that if T and K are each cyclic, then W is cyclic.

79. Let n be a positive integer, $n > 1$. Suppose that a and b are points of a Peano space T, and that no set containing fewer than n points separates T between a and b. Prove that there exists in T a collection of n independent arcs joining a and b. Note that for $n = 2$, this result reduces to the cyclic connectivity theorem. [Hint. See G. T. WHYBURN, On n-arc connectedness, *Transactions of the American Mathematical Society*, vol. 63, pp. 452–456, 1948.]

80. Write a short paper on Peano spaces, including proofs of the results about to be stated. In the event that you are unable to prove any one of the statements listed, assume it to be true, and prove the remaining ones. Let a and b be distinct points of a Peano space T and $\{ax_\alpha b\}_{\alpha \in \Delta}$ the collection of all arcs in T having a and b as end points. Define $H = \bigcup_{\alpha \in \Delta} ax_\alpha b$.

(*a*) Let y be any point of T such that there exists an arc rys such that $r \cup s \subset \bar{H}$. Prove that there exists an arc r_1ys_1 in T such that $r_1 \cup s_1 \subset H$. Then prove that $y \in H$.

(*b*) If pxq is any arc in T such that $p \cup q \subset \bar{H}$, prove that $pxq \subset H$.

(*c*) Let $\{H_\alpha\}_{\alpha \in \Lambda}$ be the collection of all closed subsets of T satisfying the following conditions:

(i) $a \cup b \subset H_\alpha$, for every $\alpha \in \Lambda$.

(ii) If $p, q \in H_\alpha$, and pxq is any arc in T, then $pxq \subset H_\alpha$, and this result holds for every $\alpha \in \Lambda$.

Define $H_0 = \bigcap_{\alpha \in \Lambda} H_\alpha$. Prove that H_0 is a closed set satisfying (i) and (ii), and contained in $\bar{H}$. In your proof show that the collection $\{H_\alpha\}_{\alpha \in \Lambda}$ is non-empty.

(*d*) Show that, for any component A of $T - H_0$, $\bar{A} - A$ is a single point.

(*e*) Prove that H_0 is a Peano space.

(*f*) Suppose there exists a proper cyclic element E of H_0. Show that

(i) If $a \cup b \subset E$, then $H_0 = E$.

(ii) If $H_0 - E$ is non-empty and connected, then E contains exactly one of the points a and b.

(iii) If $H_0 - E$ is non-empty and not connected, then it is the union of exactly two components, one containing a, the other containing b.

(*g*) Each of the points a and b is a non-cut point of H_0.

(*h*) If p is a cut point of H_0, then $H_0 - p$ consists of exactly two components, one containing a, the other containing b.

(*i*) $H_0 = H$.

8. Local Compactness and the Relationship between E^2 and the Sphere

The basic theorem in this section is to the effect that every locally compact space T can be considered as a subspace of a compact space W such that $W - T$ is a single point.

THEOREM 8.1. Let T be a locally compact space. Then there is a compact space W containing T, such that $W - T$ is a single point, and the topology of T as a subspace of W is the same as the given topology of T.

Proof. Let p denote an object not in T, and consider p as a point. Define $W = T \cup p$. We now define a basis for a topology for W and shall show that, in this topology, W is compact, and that the topology for T as a subspace of W is the same as the original topology of T. Let the collection $K = \{V_\alpha\}_{\alpha \in \Lambda}$ consist of all those open subsets of T having compact sets as their closures in T. The collection $M = \{W - \bar{V}_\alpha\}_{\alpha \in \Lambda}$ then has the property that, for each element $(W - \bar{V}_\alpha)$ of this collection, $(W - \bar{V}_\alpha) \cap T = T - \bar{V}_\alpha$ is an open subset of T. (Note that V_α is a subset of T and that by $\bar{V}_\alpha$ we denote the closure of V_α in the space T.) The collection $N = K \cup M$ of subsets of W is defined to be a basis for the open sets of W. It is easily verified, using Theorem III, 1.6, that N is a basis for a topology for W. It is also clear that the subspace T of W and the given space T are the same, and that $W - p = T$.

It remains to be shown that W is compact. Let $\{U_\alpha\}$ be any open covering of W. There exists an index α_0 such that $p \in U_{\alpha_0}$, and thus there exists a set V of the basis N such that $p \in V \subset U_{\alpha_0}$. From the definition of N, it follows that $W - V$ is compact. Thus a finite subcollection $U_{\alpha_1}, U_{\alpha_2}, \cdots, U_{\alpha_n}$ of $\{U_\alpha\}$ covers $W - V$. Consequently, the finite subcollection $U_{\alpha_0}, U_{\alpha_1}, \cdots, U_{\alpha_n}$ of $\{U_\alpha\}$ covers W. Therefore, W is compact, and the theorem is proved.

We will refer to the space W, defined in the proof just given, as a one-point compactification of T.

DEFINITION 8.2. Let T be a locally compact space. Then a space W is said to be a one-point compactification of T if and only if W is compact, $T \subset W$, $W - T$ is a single point p, and the topology for W has as a basis the collection N described in the proof of Theorem 8.1.

The one-point compactification W of the space R of all real numbers is frequently considered in analysis, and the point $W - R$ is usually denoted by the symbol ∞. This has the virtue that in W every sequence has a convergent subsequence. It has the disadvantage that ∞ does not tie in with the usual algebraic operations in R very nicely.

Another compactification of R used in analysis is a space Z containing R as a subspace such that $Z - R$ consists of exactly two points. In this case the points of $Z - R$ are usually denoted as $+\infty$ and $-\infty$. The points $+\infty$ and $-\infty$ of Z are assigned bases of the form $\{(+\infty) \cup K_n\}$ and $\{(-\infty) \cup L_n\}$, respectively. For each $n \in I$, K_n is an open-right-ray with end point n, and L_n is an open-left-ray with end point $-n$. Again the points $+\infty$ and $-\infty$ do not tie in conveniently with the usual algebraic operations in R.

Our interest in one-point compactification is directed in a different direction. We shall show that a one-point compactification of R is a simple closed curve, and that a one-point compactification of E^2 is homeomorphic to the unit sphere Σ of E^3. (The unit sphere Σ will be defined presently.) Our proofs are based on the two theorems that follow.

THEOREM 8.3. Let T be a subspace of a compact Hausdorff space W. If $W - T$ is a single point p, then W is a one-point compactification of T.

Proof. Denote by p the single point $W - T$, and by $K = \{V_\alpha\}_{\alpha \in \Delta}$ the collection of all open subsets of T whose closures in T are compact. Throughout this proof the symbol $\bar{V}_\alpha$ will denote the closure of the set V_α in the space T. Denote by M the collection $\{W - \bar{V}_\alpha\}_{\alpha \in \Delta}$, and by N the collection $K \cup M$. We shall show that N is a basis for W, and thus that W is a one-point compactification of T.

We first observe that, since W is Hausdorff, p is closed in W, and, consequently, V_α open in $T = W - p$ implies V_α open in W for every $\alpha \in \Delta$. Also, $\bar{V}_\alpha$ is a compact subset of T and thus a compact subset of W, for every $\alpha \in \Delta$. Since W is Hausdorff, $\bar{V}_\alpha$ is closed in W, so that $W - \bar{V}_\alpha$ is open in W for every $\alpha \in \Delta$. Therefore, every element of N is open in W. It remains to be shown that, given any $q \in W$ and any neighborhood U of q in W, then there exists an element V of N such that $q \in V \subset U$. Suppose, therefore, that a point q of W and a neighborhood U of q in W have been given. If $q \in T$, then there exist neighborhoods U_1, U_2 of q, p, respectively, such that $q \in U_1 \subset U$, while $U_1 \cap U_2 = \emptyset$, since W is Hausdorff. The closure of U_1 in W is the same as its closure in T, and is compact, since W is compact. Thus there exists $\alpha_0 \in \Delta$ such that $U_1 = U_{\alpha_0}$. If, on the other hand, $q = p$, then $W - U$ is a compact subset of $T = W - p$. Consequently, a finite subcollection V_{α_1}, V_{α_2}, $\cdots$, V_{α_n} of elements of the collection K covers $W - U$. It is immediate that the set Y, defined as the union of this finite collection of elements of K, is itself an element of K. Thus $W - \bar{Y}$ is an element of M, and $W - \bar{Y}$ is a neighborhood of $q = p$ contained in U. Consequently, N is a basis for W.

THEOREM 8.4. Let S and T be homeomorphic locally compact spaces. If W and Z are one-point compactifications of S and T, respectively, then W and Z are homeomorphic.

Proof. Suppose that we are given a homeomorphism $f: S \to T$. Define a mapping $g: W \to Z$ by

$$g(x) = f(x) \quad \text{if} \quad x \in S$$

$$g(W - S) = Z - T$$

It is clear that g is a homeomorphism.

THEOREM 8.5. Let W be a one-point compactification of the space R of all real numbers. Then W is a simple closed curve.

Proof. Let C be a simple closed curve, and p a point of C. Then C is a one-point compactification of $C - p$, by Theorem 8.3. Now W is a one-point compactification of R. Thus, in order to prove that W is homeomorphic with C, it is sufficient, by Theorem 8.4, to prove that $C - p$ is homeomorphic with R. Let q be a point of $C - p$; then C is the union of two independent arcs qxp and qyp. There is a homeomorphism g of qxp onto the closed interval $[0, \frac{1}{2}]$ such that $g(p) = 0$, $g(q) = \frac{1}{2}$, and there is a homeomorphism h of qyp onto the closed interval $[\frac{1}{2}, 1]$ such that $h(q) = \frac{1}{2}$, $h(p) = 1$. Clearly, $C - p = [(qxp) - p] \cup [(qyp) - p]$, and the mapping $f: C - p \to (0, 1)$, defined by $f(z) = g(z)$ for $z \in (qxp) - p$, $f(z) = h(z)$ for $z \in (qyp) - p$, is a homeomorphism. Thus $C - p$ is homeomorphic to the open interval $(0, 1)$ of R, which we already know to be homeomorphic to the space R itself. This completes the proof.

DEFINITION 8.6. The unit sphere is the subspace of E^3 consisting of all points (x, y, z) that satisfy the equation $x^2 + y^2 + z^2 = 1$. We shall usually use the Greek letter Σ to denote the unit sphere.

THEOREM 8.7. Let W be a one-point compactification of E^2. Then W is homeomorphic to the unit sphere Σ.

Proof. We first prove that Σ is compact. Thus, by Theorem 8.3, Σ is a one-point compactification of $\Sigma - (0, 0, 1)$. Then we show that $\Sigma - (0, 0, 1)$ is homeomorphic to the subset M of E^3 consisting of all points (x, y, z) of E^3 satisfying $z = 0$. Since M is homeomorphic to E^2, it then follows, by Theorem 8.4, that W is homeomorphic to Σ.

The space Σ is metric. Thus, to show that Σ is compact, it suffices to show that it is countably compact. Let T be any infinite subset of Σ. Then there is a sequence $\{(x_n, y_n, z_n)\}$ of distinct points of T. Since each of the sequences $\{x_n\}$, $\{y_n\}$, $\{z_n\}$ has a convergent subsequence, there is a subsequence $\{(x_{n_i}, y_{n_i}, z_{n_i})\}$ of $\{(x_n, y_n, z_n)\}$ which converges. Let

(x_0, y_0, z_0) be the point to which this subsequence converges. Since the mapping $f: E^3 \to R$ defined by $f[(x, y, z)] = x^2 + y^2 + z^2$ is continuous, and $f[(x_{n_i}, y_{n_i}, z_{n_i})] = 1$, for every $n \in I$, it follows that $f[(x_0, y_0, z_0)] = 1$. Thus $(x_0, y_0, z_0) \in \Sigma$; hence Σ is compact.

We next observe that the mapping

$$f: [\Sigma - (0, 0, 1)] \to M$$

defined by

$$f[(x, y, z)] = (x, y, 0) \quad \text{if} \quad z \leq 0,$$

$$f[(x, y, z)] = (x/(1 - z), \quad y/(1 - z), 0) \quad \text{if} \quad z \geq 0$$

for every point $(x, y, z) \in [\Sigma - (0, 0, 1)]$ is one-to-one and onto. The mappings f and f^{-1} are seen to be continuous by consideration of convergent sequences. The algebraic details of the one-to-one onto property of f, and the continuity of f and f^{-1} are left to the student. It then follows, as indicated, that W is homeomorphic to Σ.

The fact that a one-point compactification of E^2 is homeomorphic to Σ enables us to deduce many important properties of Σ.

THEOREM 8.8. The unit sphere Σ is a cyclic Peano space.

Proof. Let W be a one-point compactification of E^2, and denote by p the single point $W - E^2$. The space W is compact and metrizable since it is homeomorphic to Σ. Thus W is Hausdorff; hence $W - p$ is an open subset of W. This means that any neighborhood U of a point $q \in E^2$ is also a neighborhood of q in the space W. Consequently, W is locally connected at every point q of $W - p$. To show that W is locally connected at p, let V be any neighborhood of p in W. The set $W - V$ is a compact subset of E^2, and thus a bounded set in the standard metric for E^2. In this metric there thus exists $\varepsilon > 0$ such that $W - V \subset S_\varepsilon[(0, 0)]$. Now the set $\overline{S_\varepsilon[(0, 0)]}$ is compact, hence $W - \overline{S_\varepsilon[(0, 0)]}$ is open in W. We shall show that this set is connected, and thus that it is a connected neighborhood of p contained in V. This will complete the proof that W is locally connected. We note first that $\overline{S_\varepsilon[(0, 0)]} - S_\varepsilon[(0, 0)]$ is a simple closed curve having the set $E^2 - \overline{S_\varepsilon[(0, 0)]}$ as its unbounded component. Consequently, $E^2 - \overline{S_\varepsilon[(0, 0)]}$ is a connected set. Clearly, p is a limit point of this connected set, since otherwise we have the contradiction that $W - p = E^2$ is compact. It follows at once that the set $W - \overline{S_\varepsilon[(0, 0)]} = p \cup (E^2 - \overline{S_\varepsilon[(0, 0)]})$ is connected. This completes the proof that W is locally connected. Clearly, W is connected. We have thus proved that W is a Peano space. To prove that W is cyclic, note that, if $q \in W - p$, then $W - q = p \cup [E^2 - q]$; hence $W - q$ is connected. But $W - p$ is connected, since $W - p = E^2$. Thus W is a cyclic Peano space; hence its homeomorph Σ is likewise a cyclic Peano space. This completes the proof.

In order to establish the fact that every simple closed curve H in Σ has the property that $\Sigma - H$ is the union of two components A and B such that $\bar{A} - A = \bar{B} - B = H$, we first show that, for any $p \in \Sigma$, the set $\Sigma - p$ is homeomorphic to E^2.

THEOREM 8.9. If p is any point of Σ, then $\Sigma - p$ is homeomorphic to E^2.

Proof. Denote by W a one-point compactification of E^2, and by p the single point $W - E^2$. Clearly the theorem is established if we show that, for any point $r = (x_0, y_0) \in W - p$, the set $W - r$ is homeomorphic to E^2.

We first show that $W - r$ is homeomorphic to E^2 if and only if $W - (0, 0)$ is homeomorphic to E^2. To show this, it suffices to show that $W - (0, 0)$ is homeomorphic to $W - r$. The mapping $f: W \to W$ defined by $f(p) = p$, $f[(x, y)] = (x - x_0, y - y_0)$ is a homeomorphism with the property that $f(r) = (0, 0)$. Consequently, $W - r$ is homeomorphic to $W - (0, 0)$.

We next show that $W - p$ is homeomorphic to $W - (0, 0)$. It will then follow that $W - r$ is homeomorphic to $W - p = E^2$, and the proof will be complete. The mapping $g: W \to W$ defined by $g(p) = (0, 0)$, $g[(0, 0)] = p$, $g[(x, y)] = [x/(x^2 + y^2),\ y/(x^2 + y^2)]$ for $(x, y) \neq p$ and $(x, y) \neq (0, 0)$, is easily seen to be a homeomorphism. Thus $W - p$ is homeomorphic to $W - (0, 0)$, and the proof of the theorem is complete.

It now follows easily that no two points of Σ cut Σ.

THEOREM 8.10. Let q, r be distinct points of Σ. Then $\Sigma - (q \cup r)$ is connected.

Proof. By the previous theorem, $\Sigma - q$ is homeomorphic to E^2. Since no point of E^2 cuts E^2, it is immediate that $(\Sigma - q) - r = \Sigma - (q \cup r)$ is connected.

THEOREM 8.11. If T is homeomorphic to the unit sphere Σ, then

(i) T is a Peano space.

(ii) T contains at least one simple closed curve.

(iii) If H is any simple closed curve in T, then $T - H$ consists of exactly two components.

(iv) If A and B are the components of $T - H$, where H is a simple closed curve in T, then $\bar{A} - A = \bar{B} - B = H$.

Proof. The space T is homeomorphic to the unit sphere Σ; hence T is a non-degenerate cyclic Peano space. Thus T contains at least one simple closed curve, by the cyclic connectivity theorem. Consequently, (i) and (ii) are satisfied.

Let H be an arbitrary simple closed curve in Σ. Then $\Sigma - H \neq \emptyset$, since no two distinct points of Σ cut Σ, and every two distinct points of H cut H. Denote by p a point of $\Sigma - H$; then $H \subset \Sigma - p$. Since $\Sigma - p$ is homeomorphic to E^2, it follows from the Jordan curve theorem that $(\Sigma - p) - H$ consists of exactly two components D and E, and that p is a limit point of exactly one of the sets D, E. We suppose that the notation has been chosen in such a manner that p is a limit point of D. Then $\Sigma - H = A \cup B$ is a separation, where each of the sets $A = D \cup p$ and $B = E$ is connected. Since T is homeomorphic to Σ, it now follows that T satisfies (iii). To see that (iv) holds, observe that, since $A = D \cup p$ and $B = E$, $\bar{A} - A = \bar{B} - B = H$ inasmuch as a similar property holds for D and E in the space $\Sigma - p$. This completes the proof.

In the next section we shall prove that the converse of this theorem holds, and thus have the conclusion that a space T is homeomorphic to Σ if and only if it satisfies conditions (i)–(iv) of Theorem 8.11. Indeed, we shall prove the stronger result that a space T is homeomorphic to Σ if and only if it has property Δ (see Definition 9.2).

PROBLEMS

81. Let W be a one-point compactification of a locally compact space T. Prove that W is Hausdorff if and only if T is Hausdorff.

82. Prove that the mapping g of the proof of Theorem 8.4 is a homeomorphism.

83. Let W be a one-point compactification of a metric space T. Show that W is metrizable if and only if T is second countable and there exists a countable basis at the point $p = W - T$ in W.

84. Show that the one-point compactification W of a metric space T in whose metric T is unbounded can never be a completion of T. Conclude, therefore, that the one-point compactifications of the metric space R and the metric space E^2 are not completions of either R or E^2.

85. Show that the mapping $f\colon [\Sigma - (0, 0, 1)] \to M$ defined in the proof of Theorem 8.7 is a homeomorphism.

86. Show that each of the mappings f and g defined in the proof of Theorem 8.9 is a homeomorphism.

87. Show that the one-point compactification of E^n for each $n \in I$ is a cyclic Peano space. Show also that, for every $n \in I$, the one-point compactification of E^n is homeomorphic to the subset Σ^n of E^{n+1} consisting of all points $(x_1, x_2, \cdots, x_{n+1})$ satisfying $x \; + x_2^2 + \cdots + x_{n+1}^2 = 1$.

88. Let T be a locally compact, connected Hausdorff space, which is not compact. Show that, if W is any one-point compactification of T, then W is connected.

89. In the proof of Theorem 8.11 give the details of the argument that $\bar{A} - A = \bar{B} - B = H$.

9. Simple Closed Surfaces

In previous sections, characterizations of the arc and simple closed curve have been given. It is the purpose of this section to give the classical characterization, due to Leo Zippin, of the simple closed surface, as well as the Jordan-Schoenflies theorem.

DEFINITION 9.1. A space is said to be a simple closed surface if and only if it is homeomorphic with the unit sphere Σ.

We shall prove that a space is a simple closed surface if and only if it has property Δ. Thus property Δ will characterize simple closed surfaces.

DEFINITION 9.2. A space T is said to have property Δ if and only if the following conditions hold:

(i) T is a Peano space.

(ii) T contains at least one simple closed curve.

(iii) If H is any simple closed curve in T, then $T - H$ is not connected.

(iv) If K is any arc in T, and K is contained in some simple closed curve in T, then $T - K$ is connected.

As a first step, we shall prove that a space has property Δ if and only if it has the Jordan curve property. It then follows, from Theorem 8.11, that every simple closed surface has property Δ.

DEFINITION 9.3. A space T is said to have the Jordan curve property if and only if the following conditions hold:

(i) T is a Peano space.

(ii) T contains at least one simple closed curve.

(iii) If H is any simple closed curve in T, then $T - H$ consists of exactly two components.

(iv) If A and B are the components of $T - H$, where H is a simple closed curve in T, then $\bar{A} - A = \bar{B} - B = H$.

We divide the proof that property Δ implies the Jordan curve property into several steps.

THEOREM 9.4. Every space having property Δ is cyclic.

Proof. Let T be a space having property Δ, and H a simple closed curve in T. Clearly, every pair of points of H are conjugate; thus H

lies in a proper cyclic element C of T. If $C = T$, then T is cyclic, and the theorem is proved. Otherwise, there exists a point $a \in T - C$. Let A be the component of $T - C$ containing a. By Theorem 7.11, the set $\bar{A} - A$ consists of exactly one point p. This point p lies in C and is a cut point of T. By the cyclic connectivity theorem, there exists a simple closed curve H_1 in C containing the point p. This is a contradiction to property Δ since, clearly, every arc of H_1 containing the point p separates T.

THEOREM 9.5. Let T be a space having property Δ, H a simple closed curve in T, and A a component of $T - H$. Then $\bar{A} - A = H$.

Recalling the proof of the corresponding result which occurred in the proof of the Jordan curve theorem, the student may easily prove Theorem 9.5.

The idea of an arc spanning a set will occur several times in this section.

DEFINITION 9.6. Let T be a space, K a subset of T, A an arc in T with end points a and b. Then A is said to span K (in the space T) if and only if $A \cap K = a \cup b$. The arc A is said to be a spanning arc of K.

DEFINITION 9.7. Let H be a simple closed curve in a space T, and A, B disjoint arcs, each spanning H in T. Then we say that the end points of A separate the end points of B on H if and only if the end points of B lie in different components of $H - A$.

THEOREM 9.8. Let T be a space having property Δ, and H a simple closed curve in T. Then T cannot contain three disjoint arcs A_1, A_2, B satisfying the following conditions:

(i) Each of the arcs A_1, A_2, B spans H.

(ii) A_1' and A_2', the interiors of A_1 and A_2, respectively, lie in different components of $T - H$.

(iii) The end points of B separate the end points of A_1, and also separate the end points of A_2, on H.

Proof. We show that the assumption of the existence of three such arcs A_1, A_2, B leads to a contradiction. Denote by b_1, b_2 the end points of B, and by H_1, H_2 the components of $H - (A_1 \cup A_2)$ containing b_1 and b_2, respectively. These sets are interiors of arcs of H that are joined by the arc B. (See Figure 24.) Denote by K the connected set which is the union of H_1, H_2, and B, and define $C = A_1 \cup A_2 \cup (H - K)$. Clearly C is a simple closed curve in T containing both the arc A_1 and the arc A_2. Since T has property Δ, $T - C$ is not connected, and hence must contain a component D not intersecting K. Thus D is disjoint with H, and hence contained in a single component Q of $T - H$. Since

Q contains D, it must contain the connected subset $D \cup A_1' \cup A_2'$ of $T - H$, as we see by making use of Theorem 9.5. This contradiction to condition (ii) of the theorem completes the proof.

THEOREM 9.9. A space T has property Δ if and only if it has the Jordan curve property.

Proof. It is clear that, if T has the Jordan curve property, then T has property Δ. Thus we assume that T has property Δ, and shall show that T has the Jordan curve property; i.e., that conditions (i), (ii), (iii), (iv) of Definition 9.3 hold. Conditions (i), (ii), and (iv) are immediate from

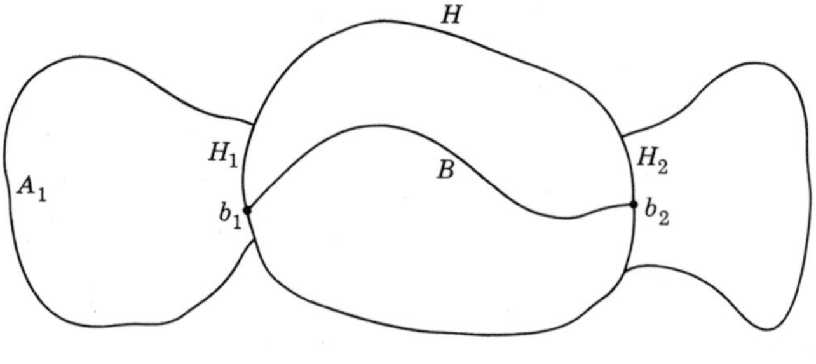

Figure 24

Definition 9.2 and Theorem 9.5. Thus the proof that (iii) holds will be complete if we show that there cannot exist a simple closed curve H in T such that $T - H$ has at least three distinct components. Suppose that such a simple closed curve H exists, and that A, B, C are distinct components of $T - H$. Then, using Theorems 9.4, 9.5, and Corollary 6.32, we see that there exist arcs $A_1 = axb$, $A_2 = cyd$ spanning H in T, such that $A_1' \subset A$, $A_2' \subset C$, where the four points a, b, c, d are distinct, and the end points of A_1 do not separate the end points of A_2 on H. Denote by C_1 the arc ab of H which contains neither c nor d, and C_2 the arc cd of H which contains neither a nor b. Using Theorems 9.4, 9.5, and Corollary 6.32 again, we see that there is an arc B_1 spanning H, such that $B_1' \subset B$, and the end points of B_1 lie in C_1' and C_2'. The three arcs A_1, A_2, B_1 satisfy the conditions of Theorem 9.8. Thus T cannot have property Δ. From this contradiction, we see at once that $T - H$ must have exactly two components; hence T satisfies (iii). This completes the proof.

Combining the results of Theorems 8.11 and 9.9, we see that every simple closed surface has property Δ. We also see that, in order to prove that every space having property Δ is a simple closed surface, it suffices to prove that every space having the Jordan curve property is a simple closed surface. The next theorem shows that we can prove that every space having the Jordan curve property is a simple closed surface, provided that we can establish the following statement.

Statement I. Let T be a space having the Jordan curve property, and H a simple closed curve in T. Suppose that A is a component of $T - H$, Q is the special rectangle in E^2 with vertices $(-1, -1)$, $(-1, 1)$, $(1, 1)$, $(1, -1)$, M is the interior of Q, and $h: H \to Q$ is a homeomorphism of H onto Q. Then there is a homeomorphism $f: (H \cup A) \to (Q \cup M)$ of $H \cup A$ onto $Q \cup M$ with the property that $f(x) = h(x)$ for every x in H.

THEOREM 9.10. Suppose that statement I is true and that T is a space having the Jordan curve property. Then T is a simple closed surface.

Proof. Since T has the Jordan curve property, it contains a simple closed curve H, and the set $T - H$ consists of exactly two components A and B. Let Q be the special rectangle in E^2 described in statement I, and denote by M the interior of Q. Since Q is a simple closed curve, there exists a homeomorphism $h: H \to Q$ of H onto Q. By statement I, there exist homeomorphisms

$$f_1: (H \cup A) \to (Q \cup M), \qquad f_2: (H \cup B) \to (Q \cup M)$$

satisfying $f_1(x) = h(x) = f_2(x)$ for every $x \in H$.

Denote by Z the subset of E^3 consisting of all points (x, y, z) satisfying the two conditions $x^2 + y^2 + z^2 \leq 1$, $z = 0$, and define a mapping

$$g: (Q \cup M) \to Z$$

in the following way:

$$g[(0, 0)] = (0, 0, 0)$$

$$g[(x, y)] = (x/\alpha, y/\alpha, 0) \quad \text{if} \quad x \neq 0 \quad \text{and} \quad |y| \leq |x|$$

where $\alpha = (1 + y^2/x^2)^{1/2}$,

$$g[(x, y)] = (x/\beta, y/\beta, 0) \quad \text{if} \quad y \neq 0 \quad \text{and} \quad |x| \leq |y|$$

where $\beta = (1 + x^2/y^2)^{1/2}$.

It is easily verified that the mapping g is a homeomorphism of $Q \cup M$ onto Z.

Denote by Σ_N, Σ_S the subsets of Σ consisting of all points (x, y, z) of Σ satisfying $z \geq 0$ and $z \leq 0$, respectively. It is easily shown, as in the proof of Theorem 8.7, that the mappings

$$g_N: \Sigma_N \to Z$$

$$g_S: \Sigma_S \to Z$$

defined by

$$g_N[(x, y, z)] = (x, y, 0) \quad \text{for} \quad (x, y, z) \in \Sigma_N$$

$$g_S[(x. y, z)] = (x, y, 0) \quad \text{for} \quad (x, y, z) \in \Sigma_S$$

are homeomorphisms. Furthermore, $g_N[(x, y, z)] = g_S[(x, y, z)]$ for every $(x, y, z) \in \Sigma_N \cap \Sigma_S$.

It now follows at once that the mapping

$$k: T \to \Sigma$$

is a homeomorphism of T onto Σ, where

$$k(x) = g_N^{-1}[g(f_1[x])] \quad \text{for} \quad x \in H \cup A$$

and

$$k(x) = g_S^{-1}[g(f_2[x])] \quad \text{for} \quad x \in H \cup B$$

This completes the proof of Theorem 9.10.

In order to establish the truth of statement I, there are several things that must be proved concerning the set $H \cup A$. Inasmuch as our proofs depend on the fact that T has the Jordan curve property, they hold equally well for $Q \cup M$, since the one-point compactification of E^2 likewise has the Jordan curve property.

THEOREM 9.11. Let a, b be distinct points of a space T which has the Jordan curve property, and A, B, C three independent arcs joining a and b in T. Then, if the notation is properly chosen, we may write

$$T - (A \cup B \cup C) = T_A \cup T_B \cup T_C$$

where T_A, T_B, T_C are components of $T - (A \cup B \cup C)$, and $\bar{T}_A - T_A = B \cup C$, $\bar{T}_B - T_B = C \cup A$, $\bar{T}_C - T_C = A \cup B$.

Proof. Define three subsets C_A, C_B, C_C of T as follows: $C_A = B \cup C$, $C_B = C \cup A$, $C_C = A \cup B$. Clearly each of these sets is a simple closed curve and hence has a complement in T which is not connected. We may thus denote by T_A, T_B, T_C the components of $T - C_A$, $T - C_B$, $T - C_C$ not containing A', B', C', respectively. From the definition of T_A, T_B, and T_C, it is clear that each of these sets is a component of $T - (A \cup B \cup C)$. Moreover, by the Jordan curve property, $\bar{T}_A - T_A = B \cup C$, $\bar{T}_B - T_B = C \cup A$, $\bar{T}_C - T_C = A \cup B$. From these results, it follows at once that T_A, T_B, T_C are distinct components of $T - (A \cup B \cup C)$. The proof of the theorem will thus be complete if we show that the set $T - (A \cup B \cup C)$ has no component T_D distinct from T_A, T_B, T_C. Suppose, on the other hand, that T_D is such a component. It is easily seen that T_D must lie in the component of $T - C_A$ containing A and in the component of $T - C_B$ containing B. Thus there exists an arc β spanning C_C such that β' lies in T_D and the end points e, f of β are interior points of A and B respectively. Since $\bar{T}_B - T_B = C \cup A$, there exists an arc α_1 spanning $C \cup A$, having both its end points in A', such that the point e lies between these end points on the arc A, and, moreover, $\alpha_1' \subset T_B$. Similarly, since $\bar{T}_A - T_A = B \cup C$, there exists an arc α_2, with $\alpha_2' \subset T_A$, spanning $B \cup C$, having both its end points in B', and such that the point f lies between these end points on the arc B. Since the arcs α_1, α_2, β satisfy the conditions of Theorem 9.8, we have a contradiction, and the proof is complete.

Our next result is a special case of Theorem 7.22.

THEOREM 9.12. Let W be a simple closed curve in a space T having the Jordan curve property. Then the union of W with any component of $T - W$ is a cyclic Peano space.

In our proof of the characterization of the arc, we found a countable dense subset of the space to be shown an arc and also one of the space J. This was done in such a way that the dense subsets were naturally ordered, and there was a one-to-one order-preserving mapping of the sets one onto the other. This mapping satisfied the further condition that both it and its inverse were uniformly continuous. The desired homeomorphism was then obtained as the continuous extension of this mapping on the dense subset.

In our present problem, we wish to use a similar device. We shall find countable dense subsets Y and Z of $H \cup A$ and $Q \cup M$, respectively, such that $Y \cap H$ and $Z \cap Q$ are countable dense subsets of H and Q, respectively. Then we shall define a one-to-one mapping $g: Y \to Z$ of Y onto Z such that both g and g^{-1} are uniformly continuous. It will be such that $g(x) = h(x)$, for $x \in Y \cap H$, and $g^{-1}(y) = h^{-1}(y)$, for $y \in Z \cap Q$, where h is the given homeomorphism of H onto Q. Our desired result, and the proof of statement I, will then follow by the continuous extension of g. It should be noted that, in the present case, we do not have the concept of natural order to help us.

In our construction of the countable dense sets Y and Z, we use Theorem 6.31 and its corollary to establish the existence of certain spanning arcs of H and Q. We then employ Theorem 9.11 to identify the components of the complements (in $H \cup A$ and $Q \cup M$) of the union of H with a certain collection of its spanning arcs, and of Q with a certain collection of its spanning arcs, respectively. Thus we use spanning arcs of H to break $H \cup A$ into "small pieces" in much the same way that the cut points of an arc were used to break the arc into "small pieces." The proof that the pieces can be made "small" is harder in the present case than it was in the case of the arc.

We do not know that, if a and b are distinct points of H, then there is a spanning arc of H in $H \cup A$ having end points a and b. Our only information in this direction is that there exist distinct points a', b' of H as close to a and b as we desire, and a spanning arc of H in $H \cup A$ having

a' and b' as its end points. It is for this reason that some care must be used in the construction of spanning arcs. The theorems we use concerning spanning arcs are now given. A lemma will be convenient for the proof of the first of these theorems.

LEMMA 9.13. Let W be a simple closed curve in a space T having the Jordan curve property, and F a component of $T - W$. Then, for any two points x, $y \in F$, there exists a spanning arc D of W in $\bar{F} = F \cup W$ such that $x \cup y \subset D$.

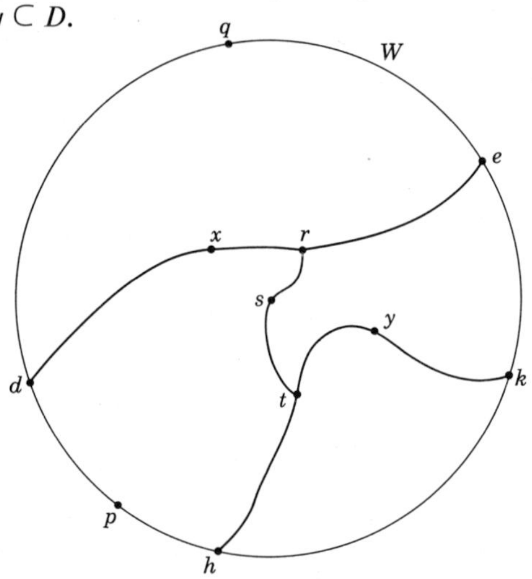

Figure 25

Proof. By Theorem 9.12, $\bar{F}$ is a cyclic Peano space. Let a, b be distinct points of W. By the three-point theorem, there exists an arc axb in $\bar{F}$, and we see at once that there is a subarc dxe of axb such that dxe spans W. Suppose that y is not a point of dxe. The simple closed curve W is the union of two independent arcs dpe and dqe. Using Theorem 9.11, and choosing the notation properly, there must exist a component G of $T - (W \cup dxe)$ such that $\bar{G} - G = dxe \cup dpe$, and $y \in G$. In $\bar{G}$ there must exist a simple arc hyk spanning $dxe \cup dpe$. Several cases may now arise. First of all, suppose that neither h nor k is an interior point of dxe. (See Figure 25.) In this case, both h and k lie on the arc dpe, and we may choose the notation in such a way that h precedes k in dpe. We see, from Theorem 9.11, that there exists a unique component G_1 of $\bar{G} - (dxe \cup dpe \cup hyk)$ such that $\bar{G}_1 - G_1 \supset dxe \cup hyk$. By Corollary 6.32, there exists an arc rst in $\bar{G}_1$ spanning $\bar{G}_1 - G_1$, such that r and t are interior points of dxe and hyk, respectively. Clearly the union of the three arcs dxe, hyk, rst contains an arc in $\bar{F}$ spanning W and containing both x and y.

Second, suppose one end point of hyk is interior to dxe while the other is interior to dpe. In this case the union of the three arcs dxe, dpe, hyk contains an arc in $\bar{F}$ spanning W and containing both x and y.

Finally, suppose both h and k lie in dxe. The existence of the required arc in $\bar{F}$ spanning W and containing both x and y is obvious, except in the single case in which, by a suitable choice of notation, h lies in the subarc dx of dxe, k lies in the subarc xe of dxe, $h \neq x$, $k \neq x$. In this case, we let G_2 be the component of $T - (hxk \cup hyk)$ which does not contain the point p, where hxk denotes a subarc of dxe. An argument involving G_2 quite similar to the one we used involving G_1 now yields the required arc. Thus the proof of the Lemma is complete.

THEOREM 9.14. Let F be a component of the complement of a simple closed curve W in a space T having the Jordan curve property, and let x, y be distinct points of $\bar{F} = F \cup W$. Then there exists in $\bar{F}$ a spanning arc D of W which separates $\bar{F}$ between x and y.

Proof. Let r, s be an arbitrary pair of distinct points of W, and p, q an arbitrary pair of distinct points of F. Using Lemma 9.13, there exists, with the proper choice of notation, an arc $apqb$ in $\bar{F}$ which spans W. Denote by afb and agb the two independent arcs of W joining a and b. Using Theorem 9.11 in the manner to which we have become accustomed, we may find, by modifying $apqb$, a spanning arc $a'pqb'$ of W in $\bar{F}$, with the property that the two independent arcs of W joining a' and b' are $a'rb'$ and $a'sb'$. Define D_1 as the arc $a'pqb'$.

Let G_1 and G_2 be the components of $T - (D_1 \cup a'rb' \cup a'sb')$ satisfying $\bar{G}_1 - G_1 = a'rb' \cup D_1$, $\bar{G}_2 - G_2 = a'sb' \cup D_1$. There exists an arc eh in $\bar{G}_1$ spanning $\bar{G}_1 - G_1$, such that e is an interior point of the subarc $a'r$ of $a'rb'$, and h is an interior point of the subarc pq of D_1. There exists an arc kt in $\bar{G}_2$ spanning $\bar{G}_2 - G_2$, such that k is an interior point of the subarc hq of D_1, and t is an interior point of the subarc sb' of $a'sb'$. Define D_2 as the union of the arc eh, the arc kt, and the subarc hk of D_1.

If x and y both lie on W, take $r = x$, $s = y$, and note that D_2 separates $\bar{F}$ between x and y. If exactly one of the points x, y lies on W, denote that point by r and the other point by p, and note that D_2 separates $\bar{F}$ between x and y. Finally, if neither x nor y lies on W, let $p = x$, $q = y$, and note that again D_2 separates $\bar{F}$ between x and y. This completes the proof of Theorem 9.14.

Our next theorem is a version of the theorem just proved as it applies to simple polygons* in E^2. In essence, the proof is the same as the proof

*We shall frequently use the term "simple polygon" for the term "polygon" to emphasize the important fact that the polygons with which we deal are simple closed curves.

just given. It uses the important additional fact that, if P is a simple polygon in E^2 with interior N, then, given any two distinct points p and q of P, there is a polygonal arc in $P \cup N$ spanning P and having the points p and q as its end points. This fact can be easily established, using the methods and previous results obtained for polygons and polygonal arcs in our sections on the Jordan curve theorem.

THEOREM 9.15. Let P be a simple polygon in E^2, and denote by N the interior of P. Suppose that B is a dense subset of P and that x, y are distinct points of $\bar{N} = P \cup N$. Then there exists in $\bar{N}$ a polygonal arc D spanning P, having both its end points in the subset B of P, and separating $\bar{N}$ between x and y.

The next theorem concerns breaking a simple closed curve and a component of its complement, in a space T having the Jordan curve property, into "small pieces."

THEOREM 9.16. Let T be a space having the Jordan curve property, W a simple closed curve in T, and F a component of $T - W$. Let α denote the diameter of F. Then, for any real number $\delta > 0$, there exists a finite collection $F_1, F_2, \cdots, F_n$ of closed subsets of $\bar{F} = F \cup W$, each of diameter less than the smaller of the two numbers $\delta/2$ and $\alpha/4$, and a finite collection of arcs $T_1, T_2, \cdots, T_m$, such that the following conditions hold:

(i) At least two of the sets F_i, $i = 1, 2, \cdots, n$, are disjoint.

(ii) $\bar{F} = \bigcup_{i=1}^{n} F_i$.

(iii) Each T_i spans W in $\bar{F}$, $i = 1, 2, \cdots, m$.

(iv) If $D = \bigcup_{i=1}^{m} T_i$, then every component of $\bar{F} - D$ has diameter less than δ.

(v) Given r, s such that $F_r \cap F_s = \emptyset$, and given points $x \in F_r$, $y \in F_s$, then there exists an integer j such that $1 \leq j \leq m$, and T_j separates $\bar{F}$ between x and y.

Proof. The set F is non-degenerate, since it is a region in the Peano space T. Denote by β the smaller of the two numbers $\alpha/4$ and $\delta/2$. Thus $\beta > 0$. The set $\bar{F}$, being compact, is totally bounded; hence there exist closed subsets $F_1, F_2, \cdots, F_n$ of $\bar{F}$, such that condition (ii) of the theorem is satisfied, and each F_i, $i = 1, 2, \cdots, n$, has diameter less than β. Since $\beta \leq \alpha/4$, condition (i) is immediate.

Consider a fixed integer r satisfying $1 < r < n$ and a fixed point $p \in \bar{F} - F_r$. For each point $x \in F_r$, there exists, by Theorem 9.14, an arc spanning W which separates $\bar{F}$ between the point x and the point p. Since this arc is a compact set not containing the point x, there must exist a neighborhood V_x of x such that this arc separates $\bar{F}$ between the point p and the neighborhood V_x. The collection $\{V_x\}_{x \in F_r}$ is an open covering of the compact set F_r, and hence contains a finite subcollection which covers F_r. The finite collection of arcs spanning W associated with this finite collection has the following property:

(*a*) Given $y \in F_r$, at least one arc of this finite collection separates F between the point y and the point p.

Consider now two disjoint sets F_r and F_s. For each $p \in F_s$, we can find a finite collection of arcs spanning W and satisfying condition (*a*). Actually, since the union of these arcs is compact and disjoint with p, there is a neighborhood V_p of p which lies on the same side of any separation of $\bar{F}$ by one of these arcs as the point p. A finite subcollection of $\{V_p\}_{p \in F_s}$ covers the compact set F_s. Thus there exists a finite collection of arcs spanning W in $\bar{F}$, such that, given $x \in F_r$, $y \in F_s$, some element of this finite collection separates $\bar{F}$ between x and y.

There are only a finite number of pairs of disjoint sets F_r, F_s. Thus there exists a finite collection of arcs $T_1, T_2, \cdots, T_m$ satisfying conditions (iii) and (v).

To complete the proof we must establish condition (iv). To do this, it suffices to show that, given any point $a \in \bar{F} - D$, the component Z of $\bar{F} - D$ containing a has diameter less than δ. Let $b \in Z$. Then, since a and b cannot be separated in $\bar{F}$ by any T_i, $i = 1, 2, \cdots, m$, it follows that a and b either lie in the same F_i, $i = 1, 2, \cdots, n$, or in two sets F_i and F_j which intersect. Since the diameter of every F_i is less than $\delta/2$, it is immediate that $\rho(a, b) < \delta$. Consequently, Z has diameter less than δ. This completes the proof of (iv) and of the theorem.

A similar theorem holds for simple polygons in E^2.

THEOREM 9.17. Let F be the interior of a simple polygon W in E^2, and B a dense subset of W. Let α denote the diameter of F. Then, for any real number $\delta > 0$, there exists a finite collection $F_1, F_2, \cdots, F_n$ of closed subsets of $\bar{F} = F \cup W$, each of diameter less than the smaller of the two numbers $\delta/2$ and $\alpha/4$, and a finite collection of polygonal arcs $T_1, T_2, \cdots, T_m$, such that conditions (i)–(v) of Theorem 9.16 hold. In addition, the end points of each arc T_i, $i = 1, 2, \cdots, m$, lie in B.

An important property of collections of arcs such as those whose existence is asserted in Theorems 9.16 and 9.17 will now be given.

THEOREM 9.18. Let T be a space having the Jordan curve property, W a simple closed curve in T, and F a component of $T - W$. Let $T_1, T_2, \cdots, T_m$ be a finite collection of arcs, each of which spans W in $\bar{F}$, and denote by ε a positive real number. If G is a Peano subspace of $\bar{F}$ then for each j satisfying $1 \leq j \leq m$, there exist at most a finite number of subarcs of T_j spanning G, such that the end points of each such subarc cannot be joined in G by an arc of diameter less than ε.

Proof. Suppose that $\varepsilon > 0$ and the Peano subspace G of $\bar{F}$ are given. Our proof is by contradiction. Assume that, for some k, $1 \leq k \leq m$, there are infinitely many subarcs of T_k spanning G, such that the end points of each subarc of this infinite collection cannot be joined in G by an arc of diameter less than ε. Let $\{a_ib_i\}$ be a sequence of distinct elements of this infinite collection. For each $i \in I$, a_ib_i is a subarc of T_k spanning G. Thus no two arcs a_ib_i, a_jb_j, $i \neq j$, can have an interior point of one in common with the other.

The set G is compact and $a_i \in G$, $b_i \in G$, for $i \in I$. Thus there exists a subsequence $\{a_{i_j}b_{i_j}\}$ of $\{a_ib_i\}$ and points p, q of G, such that $\{a_{i_j}\}$ converges to p, and $\{b_{i_j}\}$ converges to q. Since the arcs a_ib_i and a_jb_j, $i \neq j$, have disjoint interiors and are subarcs of T_k, at most a finite number of them have diameters greater than or equal to any preassigned real number $\mu > 0$. From this, it follows that $p = q$. Since G is a Peano space, there exists a region V of G containing the point p and having diameter less than ε. Since $\{a_{i_j}\}$ and $\{b_{i_j}\}$ both converge to p, this region V contains the end points of an arc a_ib_i for some i. Since V is a region in G, there must exist an arc a_ixb_i in V. This arc is of diameter less than ε, which is a contradiction to our assumption. The proof of the theorem is thus complete.

It is clear that the result of Theorem 9.18 holds for simple polygons and their interiors in E^2, where T_i, $i = 1, 2, \cdots, m$ are polygonal arcs. In this special case, the T_i, $i = 1, 2, \cdots, m$, may be chosen with their end points in a preassigned dense subset of the given simple polygon.

We now define an inductive process which we shall call a Z process. The justification of this process is contained in the theorems that have just been proved.

DEFINITION 9.19. Let T be a space having the Jordan curve property, W a simple closed curve in T, F a component of $T - W$, and $\delta > 0$ a real number. For this $\delta > 0$, let $F_1, F_2, \cdots, F_n$; $T_1, T_2, \cdots, T_m$ be the sets defined in the proof of Theorem 9.16 or 9.17 and satisfying conditions (i)–(v) of that theorem. Then the Z process for the given $\delta > 0$, the given collection $F_1, F_2, \cdots, F_n$, and the given collection $T_1, T_2, \cdots, T_m$ is the inductive definition of Peano subspaces G_i of $\bar{F}$, for $i \in I$, as follows. Define $G_1 = G_2^1 = W \cup T_1$. If G_i^j has been defined, and $j < m$, define G_i^{j+1} to be the union of G_i^j and the finite collection of subarcs of T_{j+1} spanning G_i^j and having the property that their end points cannot be joined in G_i^j by an arc of diameter less than $1/i$. If G_i^j has been defined, and $j = m$, we define $G_{i+1} = G_i^m = G_{i+1}^1$.

Before proving our principal theorem, we establish certain properties of the sequence $\{G_i\}$ defined above. To save space in the statement and proof of these properties, we assume in full the preliminary statements and notation of Definition 9.19, as well as the notation of the statement of the Z process.

THEOREM 9.20. For the sequence $\{G_i\}$ the following properties hold:

1. Each G_i is the union of a finite collection K_i of arcs such that

(*a*) No point of any arc of K_i is an interior point of any other arc of this collection.

(*b*) Every end point of an arc of K_i is an end point of some other arc of K_i.

2. Each component N of $\bar{F} - G_i$ has the property that $\bar{N} - N$ is a simple closed curve in G_i, and this simple closed curve is the union of arcs of K_i.

3. There exists $k \in I$ such that

(*a*) Every component of $\bar{F} - G_k$ has diameter less than δ.

(*b*) The end points of the arcs in K_k form a δ net in $\bar{F}$.

(*c*) The end points of the arcs in K_k which lie in W form a δ net in W.

Proof. It is immediate, following the steps of the induction defining $\{G_i\}$, that (1) and (2) hold. To prove (3), we first observe that, for $i \in I$, G_i contains at most a finite number of spanning arcs of W. We shall show that there exists $k \in I$, such that the finite collection of spanning arcs of W lying in G_k has the property that, for any pair of points $x, y \in \bar{F}$ such that $x \in F_r$, $y \in F_s$, $F_r \cap F_s = \emptyset$, there exists an arc of this finite collection which separates $\bar{F}$ between x and y. It will turn out that the integer k is the one required for the proof of (3).

Suppose that no such integer k exists. Then, for each $j \in I$, there exists a pair of points x_j, y_j of $\bar{F}$, such that $x_j \in F_{r_j}$, $y_j \in F_{s_j}$, $F_{r_j} \cap F_{s_j} = \emptyset$, and no arc of G_j spanning W separates $\bar{F}$ between x_j and y_j. Since the collection of sets $F_1, F_2, \cdots, F_n$ is finite, there must exist integers r_0, s_0 and subsequences $\{x_{j_i}\}$ and $\{y_{j_i}\}$ of $\{x_j\}$ and $\{y_j\}$, respectively, such that $x_{j_i} \in F_{r_0}$, $y_{j_i} \in F_{s_0}$ for every $i \in I$. Moreover, $F_{r_0} \cap F_{s_0} = \emptyset$. Since the sets F_{r_0} and F_{s_0} are compact, the sequences $\{x_{j_i}\}$ and $\{y_{j_i}\}$ may be chosen so as to converge to points $x \in F_{r_0}$ and $y \in F_{s_0}$, respectively. Since $G_{i+1} \supset G_i$, for every $i \in I$, it follows that no arc of any G_i spanning W separates $\bar{F}$ between x and y. We know, however, that there exists an integer q, $1 < q \leq m$, such that T_q separates $\bar{F}$ between x and y. This T_q will be kept fixed for the remainder of this portion of the proof. There exist arcs X, Y in $\bar{F}$ such that $x \in X$, $y \in Y$, $[X \cup Y] \cap T_q = \emptyset$, $X \cap W \neq \emptyset$, $Y \cap W \neq \emptyset$. Denote by k the least integer satisfying $1/k < \rho[T_q, X \cup Y]$, and by U the set $\overline{S_{1/k}(T_q)} \cap \bar{F}$. Define $V = U \cap G_{k+1}$, and let a, b be the end points of T_q.

The set V may have the property that a and b lie in a connected subset of V. Then G_{k+1} contains an arc spanning W and lying in V. It is easily seen that this arc separates $\bar{F}$ between X and Y and, a fortiori, between x and y. This is a contradiction to our assumption.

In the remaining case, no connected subset of V contains both a and b. Thus there is a separation $V = V_a \cup V_b$, $a \in V_a$, $b \in V_b$, where V_a and V_b are disjoint closed subsets of $\bar{F}$. On the arc $T_q = ab$, let p be the last point of V_a. On the subarc pb of ab, let z be the first point of V_b. Clearly the subarc pz of T_q spans $G_{k+1} = G_k^m$. This means that there must exist a subarc uv of T_q which contains pz and spans G_k^{q-1}. If u and v can be joined in G_k^{q-1} by an arc J of diameter less than $1/k$, then J lies in G_{k+1} and hence in V, which is impossible. On the other hand, if u and v cannot be joined in G_k^{q-1} by an arc of diameter less than $1/k$, then uv belongs to G_{k+1}, which is again a contradiction.

Since we have obtained a contradiction, in any event, to our assumption that no integer k with the required properties exists, it follows that there is a $k \in I$ such that the finite collection of spanning arcs of W lying in G_k has the property that, for any pair of points $x, y \in \bar{F}$ such that $x \in F_r$, $y \in F_s$, $F_r \cap F_s = \emptyset$, there exists an arc of this finite collection which separates $\bar{F}$ between x and y. It now follows, exactly as in the proof of Theorem 9.16iv, that (3*a*) holds.

To see that (3*b*) holds, note that every point of $\bar{F}$ lies in a component of $T - G_k$ or in the closure of such a component. Each component of $T - G_k$ has diameter less than δ by (3*a*), so the closure of each component thus has diameter less than δ. The closure of the component contains an arc of K_k and hence both end points of this arc. Thus the end points of the arcs of K_k form a δ net in $\bar{F}$.

To see that (3*c*) holds, note that every point of W lies in the closure of some component of $\bar{F} - G_k$ and thus lies on an arc of K_k which lies in W and in this closure. This arc has diameter less than δ. Thus the end points of arcs of K_k which lie in W form a δ net in W. The proof of Theorem 9.20 is thus complete.

We now prove the principal theorem of this section, Zippin's characterization of simple closed surfaces.

THEOREM 9.21. A space T is a simple closed surface if and only if it satisfies the following conditions:

(*a*) T is a Peano space.

(*b*) T contains at least one simple closed curve.

(*c*) If H is any simple closed curve in T, then $T - H$ is not connected.

(*d*) If K is any arc in T, and K is contained in some simple closed curve in T, then $T - K$ is connected.

Proof. An equivalent statement of the theorem is that a space T is a simple closed surface if and only if it has property Δ. But a space T has property Δ if and only if it has the Jordan curve property. Thus, since every simple closed surface has the Jordan curve property (see Theorem 8.11), we need only show that every space T having the Jordan curve property is a simple closed surface. Using Theorem 9.10, this result will be established if we show that statement I is true.

Suppose that our notation is chosen to agree with the formulation of statement I. Define $\delta_n = 1/n$ for every $n \in I$. For $\delta = \delta_1$ there exist sets $F_1, F_2, \cdots, F_n$ and arcs $T_1, T_2, \cdots, T_m$ in $H \cup A$ as described in Theorem 9.16, and thus a Z process for $F_1, F_2, \cdots, F_n$; $T_1, T_2, \cdots, T_m$; and $\delta = \delta_1$ in $\bar{A}$ yielding a sequence $\{G_i\}$. Let k be the integer given by Theorem 9.20.

We shall construct a copy M_k in $\bar{M} = Q \cup M$ of the set G_k in $\bar{A} = H \cup A$. To do this, we first construct a copy M_1 in $\bar{M}$ of the set G_1 in $\bar{A}$. For each $x \in H$, the point $h(x)$ will be called the corresponding point of Q. In the same way, for each $q \in Q$, the point $h^{-1}(q)$ will be called the corresponding point of H. Now $G_1 = H \cup T_1$. Define $M_1 = Q \cup S_1$, where S_1 is defined to be a polygonal arc spanning Q in $\bar{M}$ such that the end points of S_1 correspond to the end points of T_1. We shall say that the arc S_1 corresponds to the arc T_1. The end points of T_1 divide H into two independent arcs. We shall say that each of these arcs corresponds to the arc of Q which is its image under h.

The sets $\bar{A} - G_1$ and $\bar{M} - M_1$ have the same number of components. Every such component Y has the property that $\bar{Y} - Y$ is a simple closed curve. A component Y of $\bar{A} - G_1$ is said to correspond to a component Z of $\bar{M} - M_1$ if and only if the simple closed curve $\bar{Y} - Y$ is the union of a collection of arcs, each of which corresponds to an arc in $\bar{Z} - Z$.

In this way, we have defined a one-to-one correspondence between the components of $\bar{A} - G_1$, arcs of G_1, and end points of arcs of G_1, with the components of $\bar{M} - M_1$, arcs of M_1, and end points of arcs of M_1, respectively. [The arcs of G_1 and M_1 referred to satisfy (1) of Theorem 9.20.] This correspondence is such that

(i) If a component X of $\bar{A} - G_1$ and a component Y of $\bar{M} - M_1$ correspond, then the arcs of G_1 in $\bar{X} - X$ correspond to the arcs of M_1 in $\bar{Y} - Y$.

(ii) If an arc of G_1 corresponds to an arc in M_1, then the end points of these arcs correspond.

(iii) If the end point of an arc of G_1 lies on H, then it corresponds to an end point of an arc of M_1 lying on Q, and conversely. Moreover, this correspondence is the same as that given by h.

Paralleling the inductive definition of G_k in terms of G_1, we next show how to construct a copy M_2^2 in $\bar{M}$ of the set G_2^2 in $\bar{A}$, in such a manner that we extend our one-to-one correspondence so that it satisfies conditions (i), (ii), (iii) above. The set G_2^2 is constructed from $G_1 = G_2^1$ by augmenting G_2^1 by a finite number of subarcs of T_2. Let B be such an arc which is to be adjoined to G_2^1. Then B spans G_2^1 in $\bar{A}$. Consequently, the interior of B lies in a component X of $\bar{A} - G_2^1 = \bar{A} - G_1$. We augment $M_2^1 = M_1$ by a polygonal arc D whose interior lies in the component Y of $\bar{M} - M_2^1 = \bar{M} - M_1$ corresponding to X. The arc D is chosen so that it spans $\bar{Y} - Y$, and its end points lie on the arcs of $\bar{Y} - Y$ corresponding to the arcs of $\bar{X} - X$ which contain the end points of B. If the end points of B lie on the same arc of $\bar{X} - X$, then we define the correspondence between the end points of B and those of D so as to preserve the order on the corresponding arc of $\bar{Y} - Y$. Otherwise, we assign the correspondence of the end points to agree with the correspondence of the respective arcs on which they lie. And in any event, if an end point of B lies on H, then the corresponding end point of D is the image of this end point of B under the homeomorphism h. The arcs B and D are said to correspond. The correspondence between the components of $\bar{A} - [G_2^1 \cup B]$ and those of $\bar{M} - [M_2^1 \cup D]$ is defined to agree with the correspondence of the components of $\bar{A} - G_2^1$ and those of $\bar{M} - M_2^1$ where the components are the same. Otherwise, the correspondence is defined to satisfy (i), above, as applied to the new situation. This, it is clear, can be done in one and only one way. Continuing, in this manner by induction, we construct successively copies M_2^2 of G_2^2, $\cdots, M_2^m = M_3^1$ of $G_2^m = G_3^1$ until, after a finite number of steps, we have the desired copy M_k of G_k together with a one-to-one correspondence of regions, arcs, and end points of arcs satisfying (i), (ii), (iii) as applied to M_k and G_k.

Owing to the choice of the integer k, we know that no component of $\bar{A} - G_k$ has diameter greater than $\delta = \delta_1$, as well as the fact that the end points of the arcs of G_k form a $\delta = \delta_1$ net in $\bar{A}$. We also know that the end points of the arcs of G_k which lie in H form a $\delta = \delta_1$ net in H. The same is not necessarily true of the components of $\bar{M} - M_k$ and the end points of the arcs of M_k. However, if it is true for the components, then the conditions on the end points of the arcs of M_k as nets are also satisfied. If a component Y of $\bar{M} - M_k$ has diameter greater than or equal to $\delta = \delta_1$, then, for $\delta = \delta_1$, there exist finite collections of sets $\{F_r\}$ and arcs $\{T_i\}$ in $\bar{Y}$ as described in Theorem 9.17, and thus a Z process for $\{F_r\}$, $\{T_i\}$, and $\delta = \delta_1$ in $\bar{Y}$ yielding a sequence $\{H_i\}$. Let k_1 be the integer given by Theorem 9.20 for this sequence. We construct a copy Z_k of H_k in $\bar{X}$, where X is the component of $\bar{A} - G_k$ corresponding to Y, as we did above. In order that this may be done, it is important that we specify that the collection of sets $\{T_i\}$ in $\bar{Y}$ be polygonal arcs whose end points on $\bar{Y} - Y$ lie in the dense subset B of $\bar{Y} - Y$, where B is defined to consist of the union of the interiors of all arcs of $\bar{Y} - Y$ that do not lie in $(\bar{Y} - Y) \cap Q$, together with the union of those dense subsets of $(\bar{Y} - Y) \cap Q$ that have the property that each point of any one of these sets is the image under h of a point of H which can be joined to a point of X by an arc of $\bar{X}$ having its interior in X. This can be done by Theorem 9.17. It is now clear that, with this specification, the desired copying can be accomplished. Since there are only a finite number of components of $\bar{M} - M_k$, we obtain, in a finite number of steps, a Peano subspace $M_1^\star$ of $\bar{M}$, consisting of a finite collection of arcs, and a Peano subspace $G_1^\star$ of $\bar{A}$, which also consists of a finite collection of arcs. These subspaces are copies, the one of the other, and we obtain a one-to-one correspondence between components of $\bar{A} - G_1^\star$, arcs of $G_1^\star$, end points of arcs of $G_1^\star$, and components of $\bar{M} - M_1^\star$, arcs of $M_1^\star$, end points of arcs of $M_1^\star$, respectively. This correspondence satisfies (i), (ii), (iii) as applied to $G_1^\star$ and $M_1^\star$. Furthermore, the end points of arcs of $G_1^\star$ and $M_1^\star$ form $\delta = \delta_1$ nets in $\bar{A}$ and $\bar{M}$, respectively. Those end points that lie on H and Q also form $\delta = \delta_1$ nets in H and Q, respectively.

By our methods of construction and copying, we divide each component of $\bar{M} - M_1^\star$ and $\bar{A} - G_1^\star$ to obtain $M_2^\star$ and $G_2^\star$ and a one-to-one correspondence. Continuing this process, we obtain sequences $\{M_i^\star\}$ and $\{G_i^\star\}$ such that, for each $i \in I$, $M_i^\star$ is a copy of $G_i^\star$. These sequences have the following properties:

(iv) $M_i^\star \subset M_{i+1}^\star$, for $i \in I$.

(v) $G_i^\star \subset G_{i+1}^\star$, for $i \in I$.

(vi) Every component of $\bar{M} - M_i^\star$ has diameter less than δ_i, for $i \in I$.

(vii) Every component of $\bar{A} - G_i^\star$ has diameter less than δ_i, for $i \in I$.

(viii) The end points of the arcs of $M_i^\star$, $G_i^\star$ form δ_i nets in $\bar{M}$, $\bar{A}$, respectively.

(ix) The end points of arcs of $M_i^\star$, $G_i^\star$ which lie on Q, H form δ_i nets on Q, H, respectively.

(x) There is a one-to-one correspondence between components of $\bar{M} - M_i^\star$, arcs of $M_i^\star$, end points of arcs of $M_i^\star$ and components of $\bar{A} - G_i^\star$, arcs of $G_i^\star$, end points of arcs of $G_i^\star$, respectively. This correspondence satisfies (i), (ii), (iii) when applied to $G_i^\star$, $M_i^\star$. Moreover, the correspondence for $M_{i+1}^\star$ and $G_{i+1}^\star$ is an extension of the correspondence for $M_i^\star$ and $G_i^\star$.

Denote by K_i, L_i the sets of end points of arcs of $G_i^\star$, $M_i^\star$, respectively, and define $K = \bigcup_{i \in I} K_i$, $L = \bigcup_{i \in I} L_i$. Our one-to-one correspondence between K_{i+1} and L_{i+1}, which is an extension of the correspondence between K_i and L_i, enables us to define a one-to-one mapping of K onto L, and we denote this mapping by

$$g: K \to L$$

The sets K, L, $K \cap H$, $L \cap Q$ are dense in $\bar{A}$, $\bar{M}$, H, Q, respectively. Moreover, $g(K \cap H) \subset Q$, $g^{-1}(L \cap Q) \subset H$, $g(x) = h(x)$ for $x \in K \cap H$, and $g^{-1}(y) = h^{-1}(y)$ for $y \in L \cap Q$.

We shall now show that each of the mappings

$$g: K \to L$$

$$g^{-1}: L \to K$$

is uniformly continuous. Since the arguments are essentially the same, we give only that for g. Suppose $\varepsilon > 0$ is given. Choose $n \in I$ such that $1/n < \varepsilon$, and consider the set K_{8n}. For each point p of K_{8n}, define $V_1(p)$ to be the union of the closures of all components of $\bar{A} - G_{8n}^\star$ that contain the point p. Clearly the diameter of $V_1(p)$ is less than $1/4n$. Define $V_2(p) = \bigcup_{q \in V_1(p) \cap K_{8n}} V_1(q)$. Clearly the diameter of $V_2(p)$ is less than $1/n$. Now $V_1(p)$ is a compact subset of $V_2(p)$, and there exists an open subset $U(p)$ of $\bar{A}$ such that $V_1(p) \subset U(p) \subset V_2(p)$. This means that there exists a real number $\alpha_p > 0$ such that $q \in K \cap V_1(p)$ implies $S_{\alpha_p}(q) \subset V_2(p)$. Since K_{8n} is a finite set, there exists a real number α which is the least α_p for $p \in K_{8n}$. Moreover, $\alpha > 0$. Now every point of K lies in $V_1(p) \subset V_2(p)$ for some $p \in K_{8n}$. Furthermore, the diameter of $g(V_2(p))$ for any $p \in K_{8n}$ is less than $1/n < \varepsilon$. Thus, given r, $s \in K$, then $s \in S_\alpha(r)$ implies that the distance from $g(s)$ to $g(r)$ is less than ε. Consequently, g is uniformly continuous. In the same manner it can be shown that g^{-1} is uniformly continuous. By Problem IV,79, there is then a unique mapping $f: (H \cup A) \to (Q \cup M)$ which is a homeomorphism, and satisfies the conditions

$$f(x) = g(x) \quad \text{for} \quad x \in K$$

$$f^{-1}(y) = g^{-1}(y) \quad \text{for} \quad y \in L$$

Since the mapping f is uniquely determined and $g(x) = h(x) = f(x)$ for $x \in K \cap H$, it follows that $h(x) = f(x)$ for all $x \in H$. Thus statement I is established, and the proof of the theorem is complete.

There are several important corollaries to this theorem.

COROLLARY 9.22. Let T be a space having the Jordan curve property, H a simple closed curve in T, A a component of $T - H$, Q a special rectangle in E^2 with vertices $(-1, -1)$, $(-1, 1)$, $(1, 1)$, $(1, -1)$, M the interior of Q, and $h: H \to Q$ a homeomorphism of H onto Q. Then there is a homeomorphism $f: (H \cup A) \to (Q \cup M)$ of $H \cup A$ onto $Q \cup M$ with the property that $f(x) = h(x)$ for every $x \in H$.

This is, of course, statement I which was established in the proof of Theorem 9.21.

COROLLARY 9.23. Let T be a simple closed surface, H, J two simple closed curves in T, $h: H \to J$ a homeomorphism of H onto J. Then there exists a homeomorphism $f: T \to T$ of T onto T with the property that $f(x) = h(x)$ for every $x \in H$. Thus $f(H) = J$.

Proof. There is a homeomorphism g of J onto Z, where Z is the subset of Σ consisting of all points (x, y, z) of E^3 such that $x^2 + y^2 + z^2 = 1$, $z = 0$. Using Corollary 9.22 and the methods of Theorem 9.10, there is then a homeomorphism k of T onto Σ such that $k(x) = g(h(x))$ for every $x \in H$. Similarly, there is a homeomorphism m of T onto Σ such that $m(x) = g(x)$ for every $x \in J$. Thus the mapping

$$f: T \to T$$

defined by $f(x) = m^{-1}(k(x))$ for every $x \in T$ is the desired homeomorphism, for $f(x) = m^{-1}(k(x)) = g^{-1}(g(h(x))) = h(x)$, for every $x \in H$.

With only a slight modification of the previous argument Corollary 9.24 follows at once.

COROLLARY 9.24. Let H and J be simple closed curves in simple closed surfaces T and S, respectively, and suppose that $h: H \to J$ is a homeomorphism of H onto J. Then there exists a homeomorphism $f: T \to S$ of T onto S such that $f(x) = h(x)$, for every $x \in H$. Thus $f(H) = J$.

We next obtain a result known as the Jordan-Schoenflies theorem.

COROLLARY 9.25. Let H and J be simple closed curves in E^2, and $h: H \to J$ a homeomorphism of H onto J. Then there exists a homeomorphism $f: E^2 \to E^2$ of E^2 onto E^2 such that $f(x) = h(x)$, for every $x \in H$. Thus $f(H) = J$.

Proof. There exists a homeomorphism k of J onto Q, the special rectangle in E^2 with vertices $(-1, -1)$, $(-1, 1)$, $(1, 1)$, $(1, -1)$ and interior M. Let H_1, H_2 denote the interior and exterior of H in E^2, respectively, and J_1, J_2 the interior and exterior of J in E^2, respectively. Denote by $W = E^2 \cup p$ a one-point compactification of E^2. By Corollary 9.22, there exists a homeomorphism m of $H \cup H_1$ onto $Q \cup M$ such that

$m(x) = k(h(x))$, for every $x \in H$, a homeomorphism n of $H \cup (H_2 \cup p)$ onto $Q \cup M$ such that $n(x) = k(h(x))$, for every $x \in H$, a homeomorphism r of $J \cup J_1$ onto $Q \cup M$ such that $r(x) = k(x)$ for every $x \in J$, and a homeomorphism s of $J \cup (J_2 \cup p)$ onto $Q \cup M$ such that $s(x) = k(x)$ for every $x \in J$. There exist homeomorphisms u and v of $Q \cup M$ onto $Q \cup M$ such that $u(x) = v(x) = x$, for every $x \in Q$, and $u(n(p)) = v(s(p)) = (0, 0)$ (see Problem 99).

We now define a mapping

$$g: W \to W$$

by

$$g(x) = r^{-1}(m(x)) \qquad \text{for} \quad x \in H \cup H_1$$

$$g(x) = s^{-1}v^{-1}(u(n(x))) \quad \text{for} \quad x \in H \cup (H_2 \cup p)$$

This mapping is clearly a homeomorphism of W onto W with the property that $g(p) = p$, and $g(x) = h(x)$ for every $x \in H$. The desired homeomorphism f of E^2 onto E^2 is then defined by $f(x) = g(x)$ for every $x \in E^2$. This completes the proof.

Definition 9.26. Let S be a space, B a subset of S, p a point of S. Then the point p is said to be arcwise accessible from B if and only if, there exists an arc $W = pb$ in $B \cup p$. A subset A of S is said to be arcwise accessible from B if and only if every point of A is arcwise accessible from B.

Since every point of the special rectangle Q is arcwise accessible from its interior M, the homeomorphisms which we have established give us at once the following corollary to Theorem 9.21.

Corollary 9.27. Every simple closed curve W in a simple closed surface T is arcwise accessible from each component of $T - W$. A simple closed curve in E^2 is arcwise accessible both from its interior and from its exterior.

Notice that, in our proof of Zippin's theorem, we were continually faced with the question of arcwise accessibility. For simple polygons in E^2, we had the result. But, for simple closed curves in a space having the Jordan curve property, we depended on Corollary 6.32, which gave us only a dense subset of each simple closed curve that was arcwise accessible from the particular component of its complement which we had under consideration.

A better characterization of simple closed surfaces has been given by R. H. Bing (The Kline Sphere Characterization Problem, *Bulletin American Mathematical Society*, vol. 25, pp. 644–653, 1946. See also

Complementary Domains of Continuous Curves, *Fundamenta Mathematicae*, vol. 36, p. 316, 1949.) Bing's statement is the following:

A space T is a simple closed surface if and only if

(i) T is a Peano space.

(ii) Given any two points a, b of T, the set $T - (a \cup b)$ is connected.

(iii) If H is a simple closed curve in T, then $T - H$ is not connected.

One of the proofs which Bing gives for this result is based on the concept of a partitionable space. We develop some results on partitionable spaces in the next chapter, although we do not go so far as proving the theorem just stated.

PROBLEMS

90. In the last part of the proof of Theorem 9.4, why would every arc of H_1 containing p separate T?

91. Prove Theorem 9.5.

92. Prove that the mappings g, g_N, g_S, and k defined in the proof of Theorem 9.10 are homeomorphisms.

93. Draw a sketch to illustrate the proof of Lemma 9.13.

94. Show how the modification $a'pqb'$ of $apqb$ may be obtained in the proof of Theorem 9.14. Also draw a sketch to illustrate the proof of this theorem.

95. Prove that, if P is a simple polygon in E^2 with interior N and exterior K, and if p and q are distinct points of P, then there is a polygonal arc in E^2 spanning P and having p, q as its end points. Show that this arc may be chosen in such a manner that its interior lies in K.

96. Prove Theorem 9.15.

97. In the proof of Theorem 9.20, must $u \cup v = p \cup z$?

98. In the last part of the proof of Theorem 9.21, why must there exist an open set $U(p)$ such that $V_1(p) \subset U(p) \subset V_2(p)$?

99. Let Q denote the special rectangle in E^2 with vertices $(-1, -1)$, $(-1, 1)$, $(1, 1)$, $(1, -1)$, and M the interior of Q. For any given point $(x_0, y_0) \in M$, show that there exists a homeomorphism g of $Q \cup M$ onto itself such that $g(x) = x$ for $x \in Q$, and $g[(x_0, y_0)] = (0, 0)$.

100. Outline the main steps in the proof of Zippin's theorem.

101. Let S be the subspace of E^2 consisting of four independent arcs of E^2 joining two distinct points a and b. Show that S satisfies conditions (i), (ii), (iii) of property Δ, but does not satisfy condition (iv). What conditions of the Jordan curve property does S satisfy? What conditions of Bing's characterization of simple closed surfaces does it satisfy?

102. Prove that the interior of any simple closed curve in E^2 has property S. Also prove that each component of the complement of a simple closed curve in a simple closed surface has property S.

103. Let S be a Peano space which is not separated by any pair of its points, and which contains no primitive skew curve of type I. Prove that S contains no primitive skew curve of type II. (See Problems 39 and 40 for the definitions of primitive skew curves of types I and II.)

104. Using the steps outlined below, prove the following theorem:

There is no non-degenerate Peano space S such that, for a given integer $k > 2$, it is true that every k points of S cut S but no $k - 1$ points of S cut S.

Assuming that such a Peano space S exists, a contradiction can be reached by establishing the following statements:

(*a*) S is cyclic and contains a simple closed curve J.

(*b*) $S - J \neq \emptyset$, and, if D is a component of $S - J$, then $\bar{D} - D$ contains at least k points of J.

(*c*) There exists a point $p \in D$, and k distinct points $x_1, x_2, \cdots, x_k$ of J such that, for each i, $i = 1, 2, \cdots, k$, there is an arc px_i in $D \cup x_i$.

(*d*) Let $b_1, b_2, \cdots, b_k$ be points of J such that $x_1b_1x_2b_2 \cdots x_kb_k$ is an arc contained in J. Then $S - \bigcup_{i=1}^{k} b_i$ has a component E not containing the point x_1.

(*e*) E is a component of $S - J$ and $\bar{E} - E = \bigcup_{i=1}^{k} b_i$.

(*f*) There are uncountably many points that could play the role of b_1. Thus $S - J$ has uncountably many distinct components. This is the desired contradiction.

Bibliographical Comment

The student who desires to continue his explorations of the new territory introduced in this chapter should consult the following books and papers.

Books

T. Rado, Length and Area, *American Mathematical Society Colloquium Publications*, vol. 30, New York, 1948.

G. T. Whyburn, Analytic Topology, *American Mathematical Society Colloquium Publications*, vol. 28, New York, 1948.

R. L. Wilder, Topology of Manifolds, *American Mathematical Society Colloquium Publications*, vol. 32, New York, 1949.

Papers

W. L. Ayres, Concerning the Arc Curves and Basic Sets of a Continuous Curve, *First and Second Papers*, *Transactions American Mathematical Society*, vol. 30, pp. 567–578, 1928, and vol. 31, pp. 595–612, 1929.

R. H. Bing, Complementary Domains of Continuous Curves, *Fundamenta Mathematicae*, vol. 36, p. 316, 1949.

R. H. Bing, The Kline Sphere Characterization Problem, *Bulletin American Mathematical Society*, vol. 52, pp. 644–653, 1946.

E. R. van Kampen, On some Characterizations of 2-Dimensional Manifolds, *Duke Mathematical Journal*, vol. 1, pp. 74–93, 1935.

J. R. Kline, What is the Jordan Curve Theorem?, *American Mathematical Monthly*, vol. 49, pp. 281–286, 1942.

G. T. Whyburn, On *n*-arc Connectedness, *Transactions American Mathematical Society*, vol. 63, pp. 452–456, 1948.

G. T. Whyburn, What is a Curve?, *American Mathematical Monthly*, vol. 49, pp. 493–497, 1942.

R. L. Wilder, The Origin and Growth of Mathematical Concepts, *Bulletin American Mathematical Society*, vol. 59, pp. 423–448, 1953.

Leo Zippin, On Continuous Curves and the Jordan Curve Theorem, *American Journal of Mathematics*, vol. 52, pp. 331–350, 1930.

C H A P T E R S I X

PARTITIONABLE SPACES

1. Introduction to Partitionability

In Chapter V, we introduced the concept known as property S for metric spaces, and noted that a subset K of a metric space T has property S if and only if, for any $\varepsilon > 0$, K can be expressed as the union of a finite number of connected sets, each having diameter less than ε. We have seen that property S is useful in studying Peano spaces, perhaps its greatest utility arising from the fact that every Peano space has a basis, each element of which has property S.

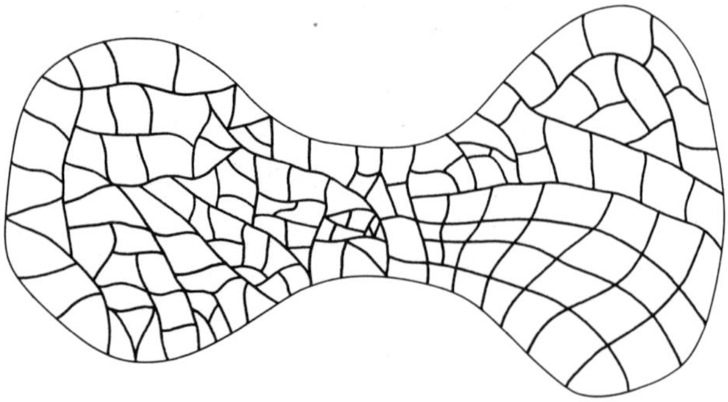

Figure 26

In our characterization of the simple closed surface, we proved that, if T is a space having the Jordan curve property, if W is a simple closed curve in T, and if F is a component of $T - W$, then, given $\varepsilon > 0$, we can find, by using the Z process, a Peano subspace G of $\bar{F}$ such that $\bar{F} - G$ has a finite number of components $R_1, R_2, \cdots, R_n$. Each of these components is open in $\bar{F}$ and has diameter less than ε. We proved further that $\bar{F} = \bigcup_{i=1}^{n} \bar{R}_i$. Such a finite collection of disjoint regions in $\bar{F}$ is called a simple ε partitioning of $\bar{F}$. In Figure 26, a simple ε partitioning of a dumbbell-shaped region in the plane is shown.

DEFINITION 1.1. A subset K of a metric space T is said to be simply partitionable if and only if, for each real number $\varepsilon > 0$, there exists a finite collection $R_1, R_2, \cdots, R_n$ of disjoint regions in K, such that $K = K \cap \bigcup_{i=1}^{n} \bar{R}_i$, and each of the regions R_i, $i = 1, 2, \cdots, n$, has diameter less than ε. The particular collection of regions chosen for a given $\varepsilon > 0$ is called a simple ε partitioning of K, and each R_i, $i = 1, 2, \cdots, n$, is referred to as an element of this simple ε partitioning.

Returning, for a moment, to the discussion of $\bar{F}$ and the sets $R_1, R_2, \cdots, R_n$, we notice that we also proved that each of the R_i, $i = 1, 2, \cdots, n$, constructed by the Z process had property S.

DEFINITION 1.2. A subset K of a metric space T is said to be partitionable if and only if, for each real number $\varepsilon > 0$, there exists a simple ε partitioning of K, every element of which has property S. Such a simple ε partitioning of K is called an ε partitioning of K.

The concept of partitioning is due to R. H. Bing. The important result, due to Bing, is that a metric space T is partitionable if and only if T has property S. This result is remarkable in the generality of its hypotheses, in contrast with our very restrictive hypothesis that T have the Jordan curve property. It is the purpose of this chapter to prove Bing's result on partitioning.

Our next theorem is an immediate consequence of the definitions involved. It establishes the easy half of Bing's theorem.

THEOREM 1.3. (*a*) If a subset K of a metric space T is partitionable, then K is simply partitionable.

(*b*) If a subset K of a metric space is simply partitionable, then K has property S.

The argument proving that partitionability is equivalent to property S has two main steps, corresponding to parts (*a*) and (*b*) of Theorem 1.3. These steps are, in their logical order, the proofs that the converses of (*b*) and of (*a*) hold. The two arguments are exactly the same, except in their first steps, and the difference between these first steps is that one has the hypothesis of property S whereas the other has the hypothesis of simple partitionability. Even in their first steps, which we give in this section, the two arguments are similar. In the next section, the proof that simple partitionability is equivalent to property S is completed. Using this result, and Theorem 1.5 of this section, the same argument yields the equivalence of simple partitionability and partitionability, thus giving us the end result that property S and partitionability are equivalent.

THEOREM 1.4. Let T be a compact connected metric space having property S, H and K two disjoint subsets of T such that, if both H and

K are non-empty, then $\rho(H, K) > 0$. Then there exist disjoint open subsets U and V of T satisfying the following conditions:

(*a*) $U \supset H$, $V \supset K$.

(*b*) U has property S.

(*c*) V has at most a finite number of components.

(*d*) $T = \bar{U} \cup \bar{V}$.

Proof. If either of the sets H, K is empty, then either $U = T$, $V = \emptyset$ or $U = \emptyset$, $V = T$ satisfies the conclusion of the theorem. In the remaining case $H \neq \emptyset$, $K \neq \emptyset$, and $\rho(H, K) > 0$ by hypothesis. Define, for this case, $\alpha =$ minimum $[1, \rho(H, K)/3]$.

The space T is a Peano space and hence has a basis, every element of which has property S, and the closure of every non-empty element of which is a Peano space. Inasmuch as T is compact, it can be covered by a finite collection of elements of this basis, each having diameter less than $\alpha/2$. Pick a point from each element of this finite covering whose closure does not intersect $H \cup K$, and define K_1 to be the union of K, the finite collection of points just chosen, and all elements of the finite covering that intersect K. Define $H_1 = H$.

The sets H_1 and K_1 have some properties that should be noted:

(i) $H \subset H_1$, $K \subset K_1$.

(ii) $\rho(H_1, K_1) > 0$.

(iii) K_1 has a finite number of components.

(iv) Given $p \in T$, there exists an arc pq in T intersecting $H_1 \cup K_1$ and having diameter less than $\alpha/2$.

Let $a \in H_1$, $b \in K_1$. These points will remain fixed throughout the remainder of the proof. Using Corollary V,5.8, and a basis for T, the closure of each element of which is a Peano space of diameter less than $\alpha/2^i$, for a fixed $i \in I$, we can find a finite simply ordered subset R_i of T satisfying the following conditions:

(v) The first and last points of R_i are a and b, respectively.

(vi) Given $p \in T$, there exists $q \in R_i$ and an arc pq in T having diameter less than $\alpha/2^{i+1}$.

(vii) Every pair of adjacent points of R_i has distance apart less than $\alpha/2^{i+1}$.

We now define, by induction, two sequences $\{H_i\}$ and $\{K_i\}$ of subsets of T having the following properties for each $i \in I$:

(viii) $H \subset H_i$, $K \subset K_i$.

(ix) $\rho(H_i, K_i) > 0$.

(x) The number of components of K_i does not exceed that of K_1.

(xi) Given $p \in T$, there exists an arc pq in T intersecting $H_i \cup K_i$ and having diameter less than $\alpha/2^i$.

To start the induction, we note that, for $i = 1$, conditions (viii)–(xi) reduce to conditions (i)–(iv) which are known to be true. We suppose, therefore, that we have defined the sets H_i, K_i for all i satisfying $1 \leq i \leq k$, and show how to define H_{k+1} and K_{k+1}.

Consider the set R_k. We shall express this set as the union of two disjoint finite sets X_k and Y_k (either of which may be empty). A point $p \in R_k$ is a point of X_k if and only if there exists an arc pq in $T - \bar{H}_k$ such that $q \in K_k$. The set Y_k is defined by $Y_k = R_k - X_k$. Thus $R_k = X_k \cup Y_k$.

For each point p of X_k, there exists an arc disjoint with $\bar{H}_k$ which joins p to a point of K_k. We denote by K_k' the union of K_k and one such arc for each point of X_k. It is clear from this definition that the set K_k' has the following properties:

(xii) $\rho(K_k', \bar{H}_k) > 0$.

(xiii) The number of components of K_k' does not exceed the number of components of K_1 [recall that (x) holds for $i = k$].

(xiv) $K_k' \supset X_k$.

Next consider a point $s \in Y_k$. By (xi), for $i = k$, there exists an arc st in T having diameter less than $\alpha/2^k$, such that $t \in H_k \cup K_k$. Let sz be the unique subarc of st such that $sz \cap (\bar{H}_k \cup \bar{K}_k)$ is the single point z. Since $s \in Y_k = R_k - X_k$, we must have $z \in \bar{H}_k$. Using (xii), we may find a region V in $T - K_k'$ containing the point z. The union of the arc sz and the region V contains an arc su disjoint with K_k', having diameter less than $\alpha/2^k$, such that $u \in H_k$. We define H_k' to be the union of H_k and one such arc from each point of Y_k. It is clear from the definition that the set H_k' has the following properties:

(xv) $\rho(H_k', K_k') > 0$.

(xvi) $H_k' \supset Y_k$.

Define $3\delta_k = \rho(H_k', K_k')$. From (xv), we have at once that $\delta_k > 0$. From (xiv) and (xvi) we see that $R_k \subset H_k' \cup K_k'$. Thus, since R_k is a simply ordered set from the point a of H_k' to the point b of K_k', there must exist two consecutive points of R_k, one of which lies in H_k' and the other in K_k'. Consequently, by (vii),

(xvii) $\delta_k < \alpha/3(2^{k+1})$.

We define H_{k+1} as the union of H_k' and all arcs of T having diameter less than δ_k that intersect H_k'. We define K_{k+1} as the union of K_k' and all arcs of T having diameter less than δ_k that intersect K_k'. We have at once

(xviii) $H \subset H_k \subset H_{k+1}$, $K \subset K_k \subset K_{k+1}$.

(xix) $\rho(H_{k+1}, K_{k+1}) > 0$.

(xx) $R_k \subset H_{k+1} \cup K_{k+1}$.

From (xx) and the definition of R_k, we have

(xxi) Given $p \in T$, there exists an arc pq in T intersecting $H_{k+1} \cup K_{k+1}$ and having diameter less than $\alpha/2^{k+1}$. From (xiii) and the definition of K_{k+1}, we have

(xxii) The number of components of K_{k+1} does not exceed the number of components of K_1.

We thus see that the sets H_{k+1} and K_{k+1} satisfy conditions (viii)–(xi) for $i = k + 1$. Thus, by induction, the sequences $\{H_i\}$ and $\{K_i\}$ are defined.

We define $U = \bigcup_{i \in I} H_i$, $V = \bigcup_{i \in I} K_i$. The remainder of the proof will be the demonstration that U and V are disjoint open sets satisfying the conclusions (*a*), (*b*), (*c*), and (*d*) of the theorem.

From the local arcwise connectivity of T, it follows by the definition of H_{k+1} and K_{k+1} that these sets are open for every $k \in I$. Consequently, using (xviii), each of the sets U, V is open, and $H \subset U$, $K \subset V$.

To show that U and V are disjoint, note that any point that belongs to both U and V must belong to both H_i and K_i for i sufficiently large. This is impossible, by (ix).

From (x), it is immediate that V has only a finite number of components, as a matter of fact not more than K_1 has. This proves (*c*), while (*d*) is immediate from the fact that (xx) holds for every $k \in I$.

The proof will be complete if we show that U has property S. In view of Theorems V,6.22 and V,6.23 and the fact that T has property S, we can prove this by showing that U is an ε-sequential growth of $H_1 = H$. Define the sequence $\{\varepsilon_i\}$ of real numbers by $\varepsilon_i = 7\alpha/3(2^{i+1})$, and note that $\sum_{i \in I} \varepsilon_i = 7\alpha/6$. Thus the proof will be complete if we show that, for each $k \in I$, H_{k+1} is an ε_k growth of H_k. To do this we must establish three things:

(xxiii) $H_k \subset H_{k+1}$.

This is (xviii).

(xxiv) If $x \in H_{k+1}$, then there exists a connected subset of H_{k+1} containing x, intersecting H_k, and having diameter less than ε_k.

Since $x \in H_{k+1}$, there exists an arc xy in T of diameter less than δ_k, where $y \in H_k'$. There exists an arc yz of diameter less than $\alpha/2^k$, where $z \in H_k$, by the definition of H_k'. The union of these arcs is a connected subset of H_{k+1}, containing x, intersecting H_k, and having diameter less than $\delta_k + \alpha/2^k < \alpha/3(2^{k+1}) + \alpha/2^k = 7\alpha/3(2^{k+1}) = \varepsilon_k$ [see (xvii)].

(xxv) There exists $\sigma_k > 0$ such that any connected subset of T of diameter less than σ_k that intersects H_k is contained in H_{k+1}.

It follows easily from the definition of H_{k+1} that $\bar{H}_k \subset H_{k+1}$. Define

σ_k as the distance between the two non-empty, disjoint, compact sets $\bar{H}_k$ and $T - H_{k+1}$. Clearly, (xxv) is satisfied.

This completes the proof of the theorem.

We will use Theorem 1.4 to obtain the equivalence of simple partitionability and property S. The only reason our argument will *not* give the equivalence of partitionability and property S is that we cannot assert, in the conclusion of Theorem 1.4, that both U and V have property S. Let us examine our proof to see why this is so. We see that we obtained H_k' from H_k by adjoining to H_k a finite collection of arcs, each having diameter less than $\alpha/2^k$. This was vital to our proof that the set $U = \bigcup_{k \in I} H_k$ was an ε-sequential growth and thus had property S. On the other hand, K_k' was obtained from K_k by adjoining to K_k arcs of unknown diameter. Thus we could not conclude that the set $V = \bigcup_{k \in I} K_k$ was an ε-sequential growth and thus had property S. Moreover, it appears impossible, with the information at hand, to control the diameters both of the arcs adjoined to H_k and of the arcs adjoined to K_k in forming H_k', K_k', respectively, and at the same time meet the requirement that $\rho(H_k', K_k') > 0$, which was also vital to our proof.

However, if we knew the additional fact that the space T of Theorem 1.4 was simply partitionable, then we would have means of controlling the diameters both of the arcs adjoined to H_k and of those adjoined to K_k in forming H_k' and K_k'. Thus, by replacing the hypothesis "T has property S" by "T is simply partitionable" in Theorem 1.4, we obtain the following theorem. Since its proof differs from that of Theorem 1.4 only in the construction of the sets H_k', K_k' from H_k, K_k, respectively, we confine our proof to the details of this construction.

THEOREM 1.5. Let T be a compact connected metric space that is simply partitionable, and H, K two disjoint subsets of T such that, if both H and K are non-empty, then $\rho(H, K) > 0$. Then there exist disjoint open subsets U and V of T satisfying the following conditions:

(*a*) $U \supset H$, $V \supset K$.

(*b*) Each of the sets U, V has property S.

(*c*) $T = \bar{U} \cup \bar{V}$.

Proof. As in the proof of Theorem 1.4, we may assume that each of the sets H, K is non-empty. We define $\alpha =$ minimum $[1, \rho(H, K)/3]$. The space T being simply partitionable, we denote by G_i, for each $i \in I$, a simple $\alpha/2^{i+1}$ partitioning of T. For a fixed integer i, consider the simple partitioning G_{i+1}. This simple partitioning consists of a finite number of disjoint non-empty regions $h_1, h_2, \cdots, h_{n_{i+1}}$ of T. For a fixed integer j

satisfying $1 \leq j \leq n_{i+1}$, the set h_j is non-empty and open. Consequently, since the union of the elements of the simple partitioning G_i is dense in T, there must be a point $p_j \in h_j$ such that p_j lies in some element of the simple partitioning G_i. Define R_{i+1} to be the finite collection of points $p_1, p_2, \cdots, p_{n_{i+1}}$ thus determined. We have thus defined, for each integer i, a finite subset R_{i+1} of T having the following properties:

(i) Each element of G_{i+1} contains at least one point of R_{i+1}.

(ii) Each point of R_{i+1} belongs to some element of G_i.

We define R_1 to be a finite subset of T, such that every element of G_1 contains a point of R_1, and every point of R_1 lies in an element of G_1. Define H_1 to be the union of H and all points of R_1 that do not lie in the set K. Define $K_1 = K$. The sets H_1, K_1 satisfy the following conditions:

(iii) $\rho(H_1, K_1) > 0$.

(iv) Every element of G_1 intersects $H_1 \cup K_1$.

Suppose that the sets $H_1, H_2, \cdots, H_k$; $K_1, K_2, \cdots, K_k$ have been defined satisfying

(v) $\rho(H_i, K_i) > 0$, $i = 1, 2, \cdots, k$.

(vi) $H_i \cup K_i$ intersects each element of G_i, for $i = 1, 2, \cdots, k$.

(vii) $R_i \subset H_i \cup K_i$ for $i = 1, 2, \cdots, k$.

We shall show how to define H_{k+1}, K_{k+1} satisfying these same conditions. The remainder of the proof follows exactly the lines of the proof of Theorem 1.4. Here, however, both U and V are ε growths, and hence have property S.

Let p be an arbitrary point of R_{k+1}. Then there exists a unique element h_p of G_k containing p. Now h_p is a region in T which intersects $H_k \cup K_k$. If there exists an arc pq in h_p that is disjoint with $\bar{H}_k$, where $q \varepsilon K_k$, let A_p be such an arc. Find such an A_p for each point p of R_{k+1} that lies in h_p and for which such an arc exists. Denote by A_{h_p} the union of the arcs of the collection thus formed. There are at most a finite number of arcs in this collection. Clearly, every arc of this collection has diameter less than the diameter of h_p, and hence diameter less than $\alpha/2^{k+1}$. Denote by K_k' the union of K_k and all the sets A_{h_p} for $h_p \in G_k$.

Considering h_p again, if there exists a point r of R_{k+1} that lies in h_p, but not in A_{h_p}, then it is impossible to join r to a point of K_k' by an arc in h_p that does not intersect H_k. From this, it is immediate that there exists an arc B_r in h_p joining r to a point of H_k but disjoint with K_k'. We denote by B_{h_p} the union of the arcs of the collection thus formed. There are at most a finite number of these arcs. Denote by H_k' the union of H_k and all the sets B_{h_p} for $h_p \in G_k$. Clearly the diameter of each arc thus adjoined to H_k is less than $\alpha/2^{k+1}$.

The remainder of the definition of H_{k+1} and K_{k+1} is precisely the same as in Theorem 1.4. By the methods used in that theorem, it follows that the sets H_{k+1} and K_{k+1} are ε_k growths of H_k, K_k, respectively, where $\varepsilon_k = 7\alpha/3(2^{k+1})$. It then follows at once that both the sets U and V have property S. This completes the proof.

PROBLEM

1. Let p be a point of a non-degenerate Peano space T, and U a neighborhood of p. Prove that there exists a neighborhood V such that $p \in V \subset \bar{V} \subset U$, and $T - U$ is contained in the union of a finite number of components of $T - V$. Compare this result with Theorem V,7.17.

2. The Equivalence of Partitionability and Property S

The proof that simple partionability and property S are equivalent which we now give is based on Theorem 1.4. The proof being completed, then, following the same steps, but using Theorem 1.5 in place of Theorem 1.4, we obtain the proof that partitionability and property S are equivalent. The details, using Theorem 1.4, are as follows.

THEOREM 2.1. Let W be a region with property S in a metric space T, H and K subsets of W with $\rho(H, K) > 0$. Then there exist open sets U, V such that

1. $H \subset U \subset W$, $K \subset V \subset W$, $U \cap V = \emptyset$.
2. $W \subset (\bar{U} \cup \bar{V})$.
3. U has property S and V has a finite number of components.

Proof. The subspace W of T has an M metric ρ_1 which is equivalent, relative to property S, to the restriction of the given metric for T to W. Thus, in the metric ρ_1, W has property S. Let Z denote the completion of W in the M metric ρ_1. Then Z is compact, by Theorems V,6.4 and 6.7. Also, Z is connected and has property S. Thus, by Theorem 1.4, since $\rho_1(H, K) > 0$, we have at once that there exist open sets U_Z, V_Z of Z which satisfy

(i) $U_Z \supset H$, $V_Z \supset K$, $U_Z \cap V_Z = \emptyset$.

(ii) $Z = \bar{U}_Z \cup \bar{V}_Z$.

(iii) U_Z has property S, and V_Z has a finite number of components.

Define $U = U_Z \cap W$, $V = V_Z \cap W$. It follows from the above, and Theorem IV,9.13, that U and V are open in W and satisfy (1), (2), and (3). This completes the proof.

When Theorem 1.5 is used and the hypothesis that W has property S is replaced by W is simply partitionable, the theorem just proved takes the following form:

THEOREM 2.2. Let W be a region which is simply partitionable in a metric space T, H and K subsets of W such that $\rho(H, K) > 0$. Then there exist open sets U, V such that

1. $H \subset U \subset W$, $K \subset V \subset W$, $U \cap V = \emptyset$.
2. $W \subset (\bar{U} \cup \bar{V})$.
3. Each of the sets U, V has property S.

THEOREM 2.3. Let T be a metric space. Then T has property S if and only if, for every $\varepsilon > 0$, there exists a simple ε partitioning of T.

Proof. If, for every $\varepsilon > 0$, there exists a simple ε partitioning of T, then T has property S by Theorem 1.3. We thus assume that T has property S, and shall show that T is simply partitionable.

Since T has property S, it has a finite number of components. In order to show that T is simply partitionable, it is sufficient to show that each component of T is simply partitionable. Let M be a component of T. We will consider M, in the metric of T restricted to M, as our entire space for the remainder of the proof. We note that this connected metric space M has property S; hence M is totally bounded. Thus there exists a finite set $p_1, p_2, \cdots, p_n$ of points of M which forms an $\varepsilon/4$ net in M. Define $K_i = S_{\varepsilon/4}(p_i)$, $H_i = M - \overline{S_{\varepsilon/3}(p_i)}$, for $i = 1, 2, \cdots, n$. Certainly $\rho(H_i, K_i) > 0$ for each i such that $1 \leq i \leq n$.

Using Theorem 2.1 and the sets H_i, K_i, $i = 1, 2, \cdots, n$, we will define finite collections $G_1, G_2, \cdots, G_n$ of disjoint regions of M such that

(i) Each element of G_i, $i = 1, 2, \cdots, n$ that has diameter at least ε has property S.

(ii) Each G_i, $i = 1, 2, \cdots, n$, contains a subcollection, each element of which has diameter less than ε. The closure of the union of the elements of this subcollection contains $\bigcup_{j=1}^{i} K_j$.

(iii) The union of the elements of each G_i, for $i = 1, 2, \cdots, n$, is dense in M.

By Theorem 2.1, there exist disjoint open sets U_1, V_1 of M containing H_1, K_1, respectively, such that U_1 has property S, V_1 has a finite number of components, and $U_1 \cup V_1$ is dense in M. Define G_1 to be the collection of all components of $U_1 \cup V_1$. By construction, the set V_1 has diameter less than ε. Thus G_1 satisfies (i), (ii), and (iii). Suppose that G_1, $G_2, \cdots, G_j$ have been defined for $1 \leq j < n$. The set G_{j+1} is then defined as follows. If there exists a finite subcollection of G_j, such that each element of this subcollection has diameter less than ε, and the closure of the union of the elements of the subcollection contains $\bigcup_{i=1}^{j+1} K_i$, then we define $G_{j+1} = G_j$. If no such finite subcollection exists, we let M_1,

$M_2, \cdots, M_t$ be those elements of G_j that have diameter at least ε and contain points of K_{j+1}. It follows that $M_i \cap K_{j+1} \neq \emptyset$, $M_i \cap H_{j+1} \neq \emptyset$ for $i = 1, 2, \cdots, t$. Moreover, each M_i, $i = 1, 2, \cdots, t$, has property S. Apply Theorem 2.1 to M_i, for $i = 1, 2, \cdots, t$, to obtain pairs of disjoint open sets $Q_1, R_1;\ Q_2, R_2; \cdots ;\ Q_t, R_t$ such that

(iv) $Q_i \supset (M_i \cap K_{j+1})$, $R_i \supset (M_i \cap H_{j+1})$.

(v) Q_i has a finite number of components; R_i has property S.

(vi) $Q_i \cup R_i$ is dense in M_i.

Define G_{j+1} to consists of all elements of G_j except the elements M_1, $M_2, \cdots, M_t$, together with all the components of each of the sets $Q_1, R_1; Q_2, R_2; \cdots ; Q_t, R_t$. It follows that G_{j+1} satisfies (i), (ii), and (iii). Thus, by induction, the collection G_n is determined. It follows from conditions (i), (ii), and (iii), and the fact that $p \in M$ implies $p \in K_i$ for some i, that G_n is a simple ε partitioning of M. Since each of the finite number of components of T is thus simply partitionable, it follows that T is simply partitionable.

THEOREM 2.4. Let T be a metric space. Then T has property S if and only if, for every $\varepsilon > 0$, there exists an ε partitioning of T.

Proof. If, for every $\varepsilon > 0$, there exists an ε partitioning of T, then T has property S by Theorem 1.3. On the other hand, if T has property S, then T is simply partitionable by the previous theorem. Thus it follows, using Theorem 2.2 in place of Theorem 2.1 in the proof of Theorem 2.3, that T is partitionable. This completes the proof.

One of the great advantages of partitioning over simple partitioning is that each element of a partitioning has property S and may thus be partitioned into smaller pieces. From this remark, the following useful theorem is immediate.

THEOREM 2.5. Let T be a metric space having property S, and $\{\varepsilon_i\}_{i\in I}$ a sequence of positive numbers. Then there exists a sequence $\{G_i\}_{i\in I}$ with the following properties:

(*a*) For each $i \in I$, G_i is an ε_i partitioning of T.

(*b*) For each $i \in I$, every element of G_{i+1} is contained in some element of G_i.

(*c*) For each $i \in I$, the collection of all elements of G_{i+1} that are contained in a given element of G_i is an ε_{i+1} partitioning of this given element.

A great deal of important work has been done using partitioning, in the past several years. It is possible to partition a metric space in considerably nicer ways than we have done: for example, to have the elements of the partitioning arranged somewhat as bricks are arranged in a wall.

We do not discuss these developments in detail here. It should be pointed out, however, that the idea of partitioning has made possible the solution of problems that for many years had defied the abilities of the best mathematicians. We mention two of these problems here.

The student will have noticed that the characterization of the simple closed surface was far more difficult than that of the simple arc. He would expect, and he would be right, that the characterization of a set like the three-dimensional space E^3 would be even more difficult. Bing has obtained such a characterization in terms of sequences of partitionings such as those described in Theorem 2.5. The details are not given here, but the type of theorem to be expected can be illustrated by a characterization (also due to Bing) of the simple closed curve. It is to the effect that a metric space T is a simple closed curve if and only if it has a sequence of ε_i partitionings $\{G_i\}_{i\in I}$ (see Theorem 2.5), with the following properties:

(*a*) The sequence $\{\varepsilon_i\}$ converges to zero.

(*b*) For each i, every element g of G_i has the property that $\bar{g} - g$ is a pair of distinct points and $\overline{(T - \bar{g})} = T - g$.

(*c*) If g, h, k are three elements of G_i, for any fixed i, then the intersection of $\bar{g} - g$, $\bar{h} - h$, $\bar{k} - k$ is the null set.

Peano spaces may be given metrics with better properties, in some ways, than M metrics by the use of partitionings. As a matter of fact, it can be shown that every Peano space T has a metric ρ with the property that, given any two distinct points a, b of T, there exists a point c of T such that c is halfway between a and b; i.e., such that $\rho(a, c) = \rho(c, b)$ and $\rho(a, b) = \rho(a, c) + \rho(c, b)$. This problem may look simple, but it was proposed by K. Menger in 1928, and remained unsolved until 1949 when the existence of such a "convex" metric was demonstrated by Bing and E. E. Moise.

Esthetically, one of the most satisfying things about partitionability is that it ties in so nicely not only with our experience in the space of real numbers but also with the other concepts we have had for metric spaces. The following theorem summarizes some of these interrelations.

THEOREM 2.6. Let T be a compact metric space, and consider the properties listed below. The space T has all of these properties if and only if it has any one of them.

(*a*) T is locally connected.

(*b*) T has property S.

(*c*) T is simply partitionable.

(*d*) T is partitionable.

Furthermore, every Peano space has all four of these properties.

PROBLEMS

2. For each i, $1 \leq i \leq n$, let $J_i = J$, the closed unit interval considered as a subspace of the space R of all real numbers. Prove that the metric space $J_1 \times J_2 \times \cdots \times J_n$ is partitionable.

3. Let S_1 and S_2 be metric spaces with metrics ρ_1 and ρ_2, respectively. Determine if the partitionability of S_1 and S_2 implies that $S_1 \times S_2$ is partitionable and conversely.

Bibliographical Comment

For references in connection with this chapter, the reader is referred to the following papers of R. H. Bing:

Partitioning a Set, *Bulletin American Mathematical Society*, vol. 55, pp. 1101–1110, 1949.

Partitioning Continuous Curves, *Bulletin American Mathematical Society*, vol. 58, pp. 536–556, 1952.

Complementary Domains of Continuous Curves, *Fundamenta Mathematicae*, vol. 36, pp. 306–318, 1949.

A Convex Metric for a Locally Connected Continuum, *Bulletin American Mathematical Society*, vol. 55, pp. 812–819, 1949.

(With E. E. Floyd) Coverings with Connected Intersections, *Transactions American Mathematical Society*, vol. 69, pp. 387–391, 1950.

A Characterization of 3-Space by Partitionings, *Transactions American Mathematical Society*, vol. 70, pp. 15–27, 1951.

The following papers are also of interest in connection with this chapter.

K. Menger, Untersuchungen über allgemeine Metrik, *Mathematische Annalen*, vol. 100, pp. 81 and 98, 1928.

E. E. Moise, Grille Decomposition and Convexification Theorems for Compact Metric Locally Connected Continua, *Bulletin American Mathematical Society*, vol. 55, pp. 1111–1121, 1949.

E. E. Moise, A Note of Correction, *Proceedings American Mathematical Society*, vol. 2, p. 838, 1951.

CHAPTER SEVEN

THE AXIOM OF CHOICE

1. Partial, Simple, and Well-Orderings

The axiom of choice, as used in our work in previous chapters, has been confined to such acts as selecting from a given infinite set a sequence of distinct points. Some of the more adroit uses of the axiom of choice are based on statements of it in terms of various types of order. This section is devoted to the terminology associated with the concepts of partial orderings and well-orderings as well as their relation with simple orderings. We also prove an important theorem concerning partially ordered sets which are said to be inductive.

DEFINITION 1.1. Let S be a set. If $<$ is a simple ordering defined on S, then $<$ is said to be a well-ordering for S (or S is said to be well-ordered by $<$) if and only if, given any non-empty subset K of S, there exists an element $x \in K$ such that $y \in K - \{x\}$ implies $x < y$. The element x is called the least element of K or the first element of K.

It is clear that a well-ordering of a set is a simple ordering of the set with a special property. The usual simple ordering of the set of all positive integers I is a well-ordering. The usual simple ordering of the set of all real numbers is not a well-ordering.

If K is a subset of a simply ordered set S, then the elements of K are simply ordered by the simple order relation on S restricted to K. To be precise about this, it is convenient to think of the simple order on S as a subset of $S \times S$ and define the order relation on K as a subset of $K \times K$.

DEFINITION 1.2. Let K be a subset of S, and F a binary relation defined on S. Then a binary relation L on K is said to be the restriction of F to K if and only if $L = (K \times K) \cap F$. Unless otherwise stated, the relation considered on a subset K of S is always assumed to be the given relation F on S restricted to K.

The following theorems are easily proved. Note that the statement of each uses the convention stated in the definition just given.

THEOREM 1.3. If K is a subset of a simply ordered set S, then K is simply ordered.

THEOREM 1.4. If K is a subset of a well-ordered set S, then K is well-ordered.

THEOREM 1.5. If K is a finite subset of a simply ordered set S, then K is well-ordered.

DEFINITION 1.6. Let S be a set and G a binary relation defined on S. Then G is said to be a partial order for S (and S is said to be partially ordered by G) if and only if

(i) For $x \in S$, $x \, G \, x$.

(ii) For any pair of elements $x, y \in S$, if $x \, G \, y$ and $y \, G \, x$, then $x = y$.

(iii) For any elements $x, y, z \in S$, if $x \, G \, y$ and $y \, G \, z$, then $x \, G \, z$.

We often denote G by $\leq$ and always understand that $y \geq x$ if and only if $x \leq y$.

Notice that, if G is a partial order for a set S and x, y are points of S, then it may be true that neither $x \, G \, y$ nor $y \, G \, x$ is true.

An important example of partial orderings is provided by sets which are ordered by inclusion.

DEFINITION 1.7. Let M be a collection of subsets of a set S. Then the binary relation $\leq$ defined on M by $A \leq B$ if and only if $A \subset B$, for any two elements A, B of M, is said to be the ordering of M by inclusion.

The next theorem is immediate.

THEOREM 1.8. Let M be a collection of subsets of a set S. Then the ordering of M by inclusion is a partial ordering of M.

Our next definition introduces terminology that is useful in the discussion of subsets of partially ordered sets.

DEFINITION 1.9. Let S be a set that is partially ordered by a binary relation $\leq$, and let T be a subset of S. Then

1. An element $x \in T$ is said to be a minimal element for T if and only if there exists no element $z \in T$ such that $z \neq x$ and $z \leq x$.

2. An element $x \in T$ is said to be a maximal element for T if and only if there exists no element $z \in T$ such that $z \neq x$ and $x \leq z$.

3. An element $x \in S$ is said to be an upper bound for T if and only if $z \leq x$ for every $z \in T$.

4. An element $x \in S$ is said to be a lower bound for T if and only if $x \leq z$ for every $z \in T$.

5. An element $x \in S$ is said to be a least upper bound for T if and only if x is an upper bound for T, and, for any upper bound z of T, we have $x \leq z$.

6. An element $x \in S$ is said to be a greatest lower bound for T if and only if x is a lower bound for T, and, for any lower bound z of T, we have $z \leq x$.

The theorem that follows is easily proved.

THEOREM 1.10. Let S be a partially ordered set, and T a subset of S. Then

(*a*) If T has a greatest lower bound, then T has exactly one greatest lower bound.

(*b*) If T has a least upper bound, then T has exactly one least upper bound.

We give an example to illustrate the terminology that has just been introduced.

EXAMPLE 1.11. Let M be the collection of all *proper* subsets of the set R of all real numbers, and suppose that M is ordered by inclusion. The partially ordered set M (see Theorem 1.8) has infinitely many minimal elements, infinitely many maximal elements, but no upper bounds and no lower bounds.

The subset T_1 of M having as its two elements the closed intervals $[0, 2]$ and $[1, 3]$ has two minimal elements, two maximal elements, infinitely many lower bounds, infinitely many upper bounds. It has $[0, 3]$ and $[1, 2]$ as its least upper bound and greatest lower bound, respectively.

The subset T_2 of M having as its elements all closed intervals of the forms $[n - 1, n + 1]$, $[-n, 2 - n]$, for $n \in I$ has every element as both a maximal element and a minimal element. It has no upper bounds and no lower bounds.

The subset T_3 of M having as its elements all infinite subsets of M that contain the integer 2 has no minimal element, infinitely many maximal elements, no upper bounds, and the integer 2 as its greatest lower bound.

To get at the relationship between partial orderings and simple orderings, we return to our observation that, if S is partially ordered and $x, y \in S$, then it may be that neither $x \leq y$ nor $y \leq x$ hold in S. On the other hand, it may be true that one or the other of these statements always holds, for any $x, y \in S$.

DEFINITION 1.12. Let S be a set and G a partial order for S. Then G is said to be a total partial order for S (and S is said to be totally partially ordered) if and only if, given any $x, y \in S$, either $x \, G \, y$ or $y \, G \, x$ holds.

DEFINITION 1.13. Let S be a set. The identity relation on S is the subset Δ of $S \times S$ defined by $\Delta = \bigcup_{x \in S} (x, x)$. Thus $x \, \Delta \, y$ means that x and y denote the same element of S. We frequently write $x = y$ for $x \, \Delta \, y$.

The connection between total partial orders and simple orders will now be given.

THEOREM 1.14. (*a*) If G is a total partial order for a set S, then $F = G - \Delta$ is a simple order for S.

(*b*) If F is a simple order for a set S, then $G = F \cup \Delta$ is a total partial order for S.

We make use of Theorem 1.14 in a convention which will now be stated.

DEFINITION 1.15. Let K be a subset of a partially ordered set S. Then K is said to be a simply ordered subset of S if and only if the restriction of the partial order of S to K is a total partial order for K, and the simple order for K is that formed from this total partial order as in Theorem 1.14*a*.

Important to our later work is the next theorem which concerns partially ordered sets that are inductive.

DEFINITION 1.16. Let S be a partially ordered set. Then S is said to be inductive (in the given partial order) if and only if every non-empty simply ordered subset of S has a least upper bound in S.

THEOREM 1.17. Let S be a non-empty partially ordered set which is inductive in its partial order relation $\leq$. If $f: S \to S$ is a mapping with the property that $x \in S$ implies $x \leq f(x)$, then there exists at least one element $x_0 \in S$ such that $f(x_0) = x_0$.

Proof. Suppose that S is inductive and that a mapping $f: S \to S$ is given, with the property that $x \leq f(x)$ for every $x \in S$. For some element a of S, consider the collection M of subsets of S defined in the following way. A subset D of S belongs to M if and only if

(i) $a \in D$.

(ii) $x \in D$ implies $f(x) \in D$.

(iii) If E is any non-empty simply ordered subset of D, then the least upper bound of E in S lies in D. (This least upper bound exists since S is inductive.)

We note first of all that M is non-empty since two of its elements are the set S itself and the set $A = \bigcap_{D \in M} D$. We shall show that the set A is simply ordered. The proof is then complete, since the least upper bound x_0 of A in S must lie in A by (iii), and $f(x_0) \in A$ by (ii). Consequently, $f(x_0) = x_0$.

In order to show that A is simply ordered, it suffices to show that A is totally partially ordered. As our first step, we prove that $x \in A$ implies $x \geq a$. Let D_1 be the subset of A consisting of all elements $x \in A$ such that $x \geq a$. Clearly D_1 satisfies (i). If $x \in D_1$, then $f(x) \geq x \geq a$. Since $f(x) \in A$, this implies $f(x) \in D_1$. Thus D_1 satisfies (ii). If E is a non-empty simply ordered subset of D_1, then E has a least upper bound y, and $y \geq a$. Since $E \subset A$, $y \in A$. Thus $y \in D_1$, and we have proved that D_1 satisfies (iii). Therefore, $D_1 \in M$, so that $A \subset D_1$. Since D_1 is a subset of A, this means $D_1 = A$. The proof is thus complete that $x \in A$ implies $x \geq a$.

We shall say that an element $x \in A$ has property Q if and only if, for any $y \in A$, such that $y \neq x$ and $y \leq x$, the relation $f(y) \leq x$ holds. We shall show that, if $x \in A$ and if x has property Q, then $y \in A$ implies $y \leq x$ or $y \geq x$. Our proof will then be completed by showing that every element of A has property Q.

Suppose $x \in A$ and x has property Q. Define D_2 to be the set of all $y \in A$, such that $y \leq x$ or $y \geq f(x)$. We shall show that $D_2 = A$. To do this, it suffices to show that D_2 satisfies (i), (ii), and (iii). Since $a \leq x$, it is clear that D_2 satisfies (i). If $y \in D_2$, then

(*a*) If $y \leq x$, $y \neq x$, then $f(y) \leq x$, since x has property Q. Thus $f(y) \in D_2$.

(*b*) If $y = x$, then $f(y) \geq f(x)$. Hence $f(y) \in D_2$.

(*c*) If $y \geq f(x)$, then $f(y) \geq y \geq f(x)$. Hence $f(y) \in D_2$.

Thus, in any case, if $y \in D_2$, then $f(y) \in D_2$; i.e., D_2 satisfies (ii). Finally, if $E \subset D_2$, $E \neq \emptyset$, and E is simply ordered; then either every $y \in E$ has the property that $y \leq x$, in which case the least upper bound of E is $\leq x$, or there exists an element w of E such that $w \geq f(x)$. In this case the least upper bound b of E must satisfy $b \geq f(x)$, and thus, since $b \in A$, $b \in D_2$. It follows that D_2 satisfies (iii); hence $D_2 = A$. Thus, if $x \in A$, and x has property Q, then, for every $z \in A$, either $z \leq x$ or $z \geq f(x) \geq x$; i.e. $z \leq x$, or $z \geq x$.

It remains to be shown that every element of A has property Q. Define D_3 to be the set of all elements of A that have property Q. We shall show that $D_3 = A$, by showing that D_3 satisfies (i), (ii), and (iii). Since there do not exist any elements $y \in A$, such that $y \leq a$ and $y \neq a$, we see that a has property Q; hence D_3 satisfies (i). If $x \in D_3$, then, to show $f(x) \in D_3$, we must show $z \in A$, $z \leq f(x)$, and $z \neq f(x)$ implies $f(z) \leq f(x)$. Since x has property Q, $z \leq f(x)$ and $z \neq f(x)$ implies $z \leq x$. If $z \leq x$ and $z \neq x$, then $f(z) \leq x \leq f(x)$. If $z = x$, then $f(z) \leq f(x)$. Therefore, D_3 satisfies (ii). Suppose E is a simply ordered non-empty set contained in D_3 and that t is the least upper bound of E. To show that $t \in D_3$, and hence that D_3 satisfies (iii), we must show that t has property Q in A. Suppose $z \in A$, $z \leq t$, and $z \neq t$. Then there exists an element $t_1 \in E$, such that $z \leq t_1$ and $z \neq t_1$, for, otherwise, z would be the least upper bound of E. Since $t_1 \in E$, t_1 has property Q; hence $f(z) \leq t_1 \leq t$. Consequently, t has property Q, and hence satisfies (iii). This proves that $D_3 = A$. Consequently, A is totally partially ordered, and thus simply ordered. The conclusion then follows, as indicated previously, that there exists $x_0 \in S$ such that $f(x_0) = x_0$.

PROBLEMS

1. Let S be a set simply ordered by $<$, and, for each $x \in S$, denote by L_x the set of all $y \in S$ such that $y < x$. Then S is said to have the transfinite induction property if and only if
   (*a*) S has a minimal element x_0.
   (*b*) If S_1 is a subset of S such that $x_0 \in S_1$, and $L_x \subset S_1$ implies $x \in S_1$, then $S_1 = S$.

Prove that a non-empty simply ordered set T is well-ordered if and only if it has the transfinite induction property in its given simple order. Compare this with the principle of induction in the set of all positive integers.

2. Let S be a set simply ordered by $<$. For each $x \in S$, denote by L_x the set of all $y \in S$ such that $y < x$, and by R_x the set of all $y \in S$ such that $x < y$. Consider the following properties for S. Prove that S has both the properties (i) and (ii) if and only if it has all the properties A, B, and C.
   (i) If $a \in S$, then L_a is a finite set.
   (ii) The set S has no least upper bound in S.
   A. The given simple ordering for S is a well-ordering for S.
   B. The set S has no least upper bound in S.
   C. If S_1 is a subset of S which contains the first element of S and has the property that, if $x \in S_1$, then the first element of the set R_x lies in S_1, then $S_1 = S$.

3. Let S be a non-empty set well-ordered by $<$ and having x_0 as its first element. Suppose f and g are each mappings of S into a set T and that these mappings satisfy the following conditions:
   (i) $f(x_0) = g(x_0)$.
   (ii) If $y \in S$ and $f(x) = g(x)$ for every $x \in L_y$ (see Problem 1), then $f(y) = g(y)$.

Prove that f and g are the same mapping.

4. Prove Theorem 1.3.

5. Prove Theorem 1.4.

6. Prove Theorem 1.5.

7. Prove Theorem 1.8.

8. Give all missing details in Example 1.11.

9. Prove Theorem 1.10.

10. Show that the collection of all subsets of a set S ordered by inclusion is inductive.

11. Show that the set of all positive integers in their usual order is not inductive.

12. Show that, if S is a partially ordered set that is inductive, and $a \in S$, then the subset T of S consisting of all $x \in S$ such that $x \geq a$ is inductive. Show also, by example, that a subset of an inductive set need not be inductive.

13. Let S be a partially ordered set. Show that the collection of all simply ordered subsets of S, ordered by inclusion, is inductive.

14. Let S and T be spaces, and M the collection of all continuous mappings of subsets of S into T. If $f, g \in M$ and $f: A \to T$, $g: B \to T$, define $f \leq g$ in M if and only if g is a continuous extension of f from A to B. Prove that this is a partial ordering of M.

15. In the proof of Theorem 1.17, show

(*a*) If $x \in A$, then there is no $y \in A$ such that $y \neq x$, $y \neq f(x)$, and $x \leq y \leq f(x)$.

(*b*) $A \in M$.

16. In the proof of Theorem 1.17, give in detail the argument concerning property (iii) for D_3 that, if $z \in A$, $z \leq t$, and $z \neq t$, then there exists an element $t_1 \in E$, such that $z \leq t_1$ and $z \neq t_1$.

2. Alternative Forms of the Axiom of Choice

The first alternative form of the axiom of choice that we consider is commonly known as Zorn's lemma. It involves the notion of a maximal simply ordered subset.

DEFINITION 2.1. Let S be a set partially ordered by $\leq$, and K a subset of S. Then K is said to be a maximal simply ordered subset of S if and only if K is simply ordered and there exists no element $x \in S - K$ such that $y \leq x$ holds for every $y \in K$.

The sense in which Zorn's lemma is equivalent to the axiom of choice is left to the discussion following our proof of this lemma.

THEOREM 2.2 (*Zorn's lemma*). Let S be a partially ordered set. Then there exists at least one maximal simply ordered subset K of S.

Proof. Let M be the collection of all simply ordered subsets of S. The collection M is non-empty since the null set of S is an element of M. We partially order the collection M of subsets of S by inclusion, and observe that every simply ordered subset F of M has a least upper bound in M, namely the element of M that is the union of all subsets of S that are elements of M and belong to F. Thus M is inductive. Suppose there are no maximal simply ordered subsets of S. Then, given a simply ordered subset B of S, there is a simply ordered subset C of S, such that $B \neq C$ and $B \subset C$. Thus, under our assumption, given any element X of M, there is an element Y of M, such that $X \leq Y$ and $X \neq Y$. For each $X \in M$, denote by $K(X)$ the subset of M consisting of all elements Y of M, such that $X \leq Y$ and $X \neq Y$. For each $X \in M$, the set $K(X)$ is non-empty, by our assumption. Thus, by the axiom of choice, there is a mapping g of the collection $\{K(X)\}_{X \in M}$ into M such that $g[K(X)] = Y \in K(X)$, for every $X \in M$. Define $f\colon M \to M$ by

$$f(X) = g[K(X)] \quad \text{for} \quad \text{every} \quad X \in M$$

Clearly $X \leq f(X)$ and $X \neq f(X)$ for every $X \in M$. This contradicts Theorem 1.17, since we have seen that the set M is inductive in its given partial order. Thus our assumption is false. This means that there must exist a maximal simply ordered subset of S.

To see how close Zorn's lemma comes to implying the axiom of choice, consider the following argument. Let S be a set, and M a collection of non-empty subsets of S. Consider the collection F of all mappings of subsets of M into S, such that a mapping f of the subset A of M into S belongs to F if and only if $f(X) = x \in X$, for every subset X of S that lies in A. (Notice that X is regarded both as a point of M and as a subset of S.) Define a partial order in F by $f \leq g$ if and only if g is an extension of f. Thus, in particular, $f \leq g$ implies that the subset of M on which g is defined contains the subset of M on which f is defined. By Zorn's lemma, there exists a maximal simply ordered subset K of F. We now show that K has a least upper bound in F. Let L be the subset of M which is the union of all subsets of M on which mappings belonging to K are defined. Define a mapping $f\colon L \to S$ by the following rule. If $x \in L$, then $x \in Q$, where Q is a subset of M on which a mapping g of K is defined; define $f(x) = g(x)$. Since K is simply ordered, this does define a mapping, and the mapping f, thus defined, lies in F and is the least upper bound of K in F. Since K is maximal, $f \in K$. We now show that $L = M$. For, if not, then there exists $Y \in M - L$, and we define

$$h\colon L \cup \{Y\} \to S$$

by

$$h(X) = f(X) \quad \text{for} \quad X \in L$$

$$h(Y) = \text{some element of the non-empty subset } Y \text{ of } S$$

The mapping $h \in F$, and $h \geq f$, $h \neq f$. Thus $h \notin K$, and $K \cup \{h\}$ is simply ordered. This contradicts the maximality of K. Hence $L = M$. The existence of the mapping f we have obtained is exactly the content of the axiom of choice.

There is a difficulty with this argument, however. Since we are attempting to prove the axiom of choice, it is not fair to use it in our proof. But we made two choices in the course of our proof. From the non-empty (by Zorn's lemma) collection of maximal simply ordered subsets of F, we chose K. From the element Y of $M - L$, where we regard Y as a subset of S, we chose an element of S. Thus, if we could prove, without using the axiom of choice, the so-called "finite axiom of choice," namely:

"Given a non-empty finite collection N of non-empty subsets of a set T, then an element may be chosen from each of these sets; i.e. there exists a mapping

$$h: N \to T$$

such that $h(X) = x \in X$ for every subset X of T which is an element of N," then our proof would be valid. Since we do not settle this subtle question, we refer to Zorn's lemma as an alternative form of the axiom of choice, rather than as an equivalent form. (For a detailed discussion of the finite axiom of choice see A. Tarksi, *Fundamenta Mathematicae*, vol. 6, pp. 49–95, 1924.)

Our next alternative form of the axiom of choice is known as the well-ordering theorem. We prove it using Zorn's lemma.

THEOREM 2.3. Let S be a set. Then there exists a well-ordering for S.

Proof. Let W be the collection of all well-order relations defined for subsets of S. If $G \in W$, then G may be considered as a subset of $H \times H$, where H is the subset of S on which G is defined. Thus the elements of W are subsets of $S \times S$. Partially order these elements by inclusion. It is then clear that any simply ordered subset of W has a least upper bound in W. By Zorn's lemma, there is a maximal simply ordered subset K of W. By the previous remark, there is an element G of W which is a least upper bound of K in W. Certainly $G \in K$, since K is maximal. The well-ordering G must be defined on all of S, for otherwise there is a point $x_0 \in S$ not in the subset A of S on which G is defined. We can then define a simple order $G^\star$ on $A \cup x_0$ by defining $y\, G^\star\, z$ for $y, z \in A$ if and only if $y\, G\, z$, and $y\, G^\star\, x_0$ for every $y \in A$. Clearly $G^\star$ is a well-ordering. But $G^\star \geq G$, $G^\star \neq G$ in W, thus contradicting the maximality of K. Therefore, G is a well-ordering of S, and the proof is complete.

To see how the well-ordering theorem implies the axiom of choice, consider the following argument. Let S be a set, and M a collection of non-empty subsets of S. There is a well-ordering of S, and the restriction of this well-ordering to any element of M, considered as a subset of S, is a well-ordering of that subset. Define a mapping f of M into S by the following rule. For each element X of M, $f(X)$ is the least element of the well-ordered subset X of S. The existence of such a mapping f is exactly the content of the axiom of choice.

Notice that we use the "finite axiom of choice" to choose a well-ordering of S from the non-empty (by the well-ordering theorem) collection of all well-orderings of S. Thus, as with Zorn's lemma, we refer to the well-ordering theorem as an alternative statement of the axiom of choice.

The next alternative statement of the axiom of choice which we present involves properties of sets that are finite in character.

DEFINITION 2.4. A property P of subsets of a set S is said to be of finite character if and only if, given a subset X of S, then X has property P if and only if every non-empty finite subset of X has property P.

As an example, we show that the finite intersection property (see Definition III, 3.24) is a property of finite character.

EXAMPLE 2.5. Let T be the collection of all subsets of a set S. Denote by P the finite intersection property, and notice that P is a property of subsets of T. If X is a subset of T having property P, then every non-empty finite subset of X clearly has property P. On the other hand, if a subset X of T is such that every non-empty finite subset of X has property P, then, by definition, X has the finite intersection property; i.e., X has property P. Thus P is a property of finite character.

Our alternative form of the axiom of choice in terms of properties of finite character will now be given.

THEOREM 2.6. Let S be a set, and P a property of subsets of S that is of finite character. If there is a subset of S that has property P, then there is a subset A of S such that A has property P, and, if there is a subset B of S such that $B \supset A$ and B has property P, then $B = A$.

Proof. Suppose there is a subset of S that has property P, and denote by M the collection of all subsets of S that have property P. Order the elements of M by inclusion. Any non-empty simply ordered subset F of M has a least upper bound X_F in M; namely that element of M which, when considered as a subset of S, is the union of all elements of F, considered as subsets of S. To see that the subset X_F of S is actually an element of M, notice that any finite subset Q of X_F must lie in a finite number of the subsets of S which are in F. Since F is simply ordered, one of this finite number of subsets of S, say Y, must contain Q. Now Y has property P, and P is a property of finite character; hence Q has property P. We have thus shown that every finite subset of X_F has property P. Thus X_F is an element of M. The additional fact that X_F is the least upper bound of F in M is then immediate. The proof we have given shows that the set M is inductive.

By Zorn's lemma there is a maximal simply ordered subset K of M. The set K has a least upper bound, since M is inductive, and this least upper bound $A \in K$, since K is maximal. Moreover, there is no element B of M such that $B \supset A$ and $B \neq A$. That is, if there is a subset B of S such that $B \supset A$ and B has property P, then $B = A$. This completes the proof.

We shall now see how this theorem on properties of finite character implies the axiom of choice, by considering the following argument showing that the theorem implies Zorn's lemma. Let S be a partially ordered set. If S is empty, it is simply ordered, and thus maximal in itself. Suppose that S is non-empty. Let P be the property of simple ordering for subsets of S, this simple ordering being the restriction of the given partial ordering of S to subsets of S. Property P is of finite character. To see this, note that, if a subset X of S is simply ordered, then every non-empty finite subset of X is simply ordered. Conversely, if a subset X of S has the property that each of its finite subsets is simply ordered, then, in particular, every subset of X consisting of two points is simply ordered. Thus X is totally partially ordered; hence X is simply ordered. This proves that P is a property of finite character. By Theorem 2.6, there exists a subset A of S such that A has property P, and, if there is a subset B of S such that B has property P and $B \supset A$, then $B = A$. It follows that A is a maximal simply ordered subset of S. Thus Zorn's lemma is established.

To get from Zorn's lemma to the axiom of choice, we encountered the finite choice question. It is for this reason that we refer to Theorem 2.6, merely as an alternative form of the axiom of choice.

The final alternative form of the axiom of choice which we consider is the Tychonoff theorem, to the effect that the Cartesian product of any given collection of compact spaces is compact. For this to make sense, we must first define what we mean by the Cartesian product of a given collection of sets, and then show how to introduce a topology into the set that is the Cartesian product in the case that each of the given sets is a space.

DEFINITION 2.7. Let $\{X_\alpha\}_{\alpha\in\Delta}$ be a collection of sets indexed by Δ. Then the Cartesian product $\prod_{\alpha\in\Delta} X_\alpha$ of the sets of this collection is defined to be the set of all mappings $f: \Delta \to \bigcup_{\alpha\in\Delta} X_\alpha$ such that $f(\alpha)$ is an element of the set X_α for every $\alpha \in \Delta$.

The question immediately arises as whether this definition agrees or conflicts, in the case that Δ consists of exactly two elements, with our earlier definition of the Cartesian product of two spaces. It is easily seen that the two definitions are in complete agreement. For suppose $\Delta = \{1, 2\}$. Then $X_1 \times X_2$ consists, by our earlier definition, of all pairs (x, y) where $x \in X_1$, $y \in X_2$. Thus the mapping $f\colon \Delta \to \bigcup_{i=1}^{2} X_i$ defined by $f(1) = x, f(2) = y$ is exactly the element of $\prod_{\alpha \in \Delta} X_\alpha$ that was previously denoted by (x, y), and conversely. The set of *all* such mappings is clearly the set of all points of $X_1 \times X_2$, and conversely. Thus the two definitions of Cartesian product are in complete agreement. This is in accord with the following commonly used convention.

DEFINITION 2.8. Let $\{X_\alpha\}_{\alpha \in \Delta}$ be a collection of sets indexed by Δ. Then the mapping $f\colon \Delta \to \bigcup_{\alpha \in \Delta} X_\alpha$, which is an element of the Cartesian product $\prod_{\alpha \in \Delta} X_\alpha$, is frequently denoted by $\{x_\alpha\}_{\alpha \in \Delta}$, where $f(\alpha) = x_\alpha$ for every $\alpha \in \Delta$.

Indeed, this is, in essence, the device used in our original definition of the Cartesian product of two sets.

Before defining a topology for $\prod_{\alpha \in \Delta} X_\alpha$ in the case that each X_α is a space, we state the following easily proved theorem which justifies our definition.

THEOREM 2.9. Let $\{X_\alpha\}_{\alpha \in \Delta}$ be a collection of spaces indexed by Δ. Denote by σ the collection of all subsets of $\prod_{\alpha \in \Delta} X_\alpha$ that are of the form $\prod_{\alpha \in \Delta} Y_\alpha$ where, for some *finite* subset Σ of Δ, Y_β is an open subset of X_β for every $\beta \in \Sigma$ and $Y_\alpha = X_\alpha$ for every $\alpha \in \Delta - \Sigma$. Then σ is a basis for a topology for the product set $\prod_{\alpha \in \Delta} X_\alpha$.

DEFINITION 2.10. Let $\{X_\alpha\}_{\alpha \in \Delta}$ be a collection of spaces indexed by Δ. The product space $\prod_{\alpha \in \Delta} X_\alpha$ is then the set $\prod_{\alpha \in \Delta} X_\alpha$ with the topology that has as a basis the collection σ of subsets defined in Theorem 2.9.

It is easily shown that the product space defined earlier for the product of two spaces is the same as that given above in the case Δ consists of two elements. The proof is left to the student as an exercise.

We now state and prove the Tychonoff theorem.

THEOREM 2.11. Let $\{X_\alpha\}_{\alpha \in \Delta}$ be a collection of non-empty spaces indexed by Δ. Then the product space $\prod_{\alpha \in \Delta} X_\alpha$ is compact if and only if the space X_α is compact for every $\alpha \in \Delta$.

Proof. Suppose the product space $\prod_{\alpha \in \Delta} X_\alpha$ is compact. We will then show that the space X_α is compact, for every $\alpha \in \Delta$. For each $\beta \in \Delta$, we define a mapping p_β of $\prod_{\alpha \in \Delta} X_\alpha$ into X_β as follows:

$$p_\beta[\{x_\alpha\}_{\alpha \in \Delta}] = x_\beta$$

for each point $\{x_\alpha\}_{\alpha \in \Delta}$ of $\prod_{\alpha \in \Delta} X_\alpha$. The mapping p_β, for $\beta \in \Delta$, is easily seen to be continuous. Moreover, p_β is a mapping onto X_β since (using the axiom of choice), given $r \in X_\beta$, there is a point $\{x_\alpha\}_{\alpha \in \Delta}$ such that $x_\beta = r$. Thus, X_β, for each $\beta \in \Delta$, is the continuous image of the compact space $\prod_{\alpha \in \Delta} X_\alpha$. Consequently, X_β is compact.

To complete the proof, suppose that X_α is compact for every $\alpha \in \Delta$. We will then show that the product space $\prod_{\alpha \in \Delta} X_\alpha$ is compact. To show this, we show that, if $\{F_\nu\}_{\nu \in \Sigma}$ (in general Σ is different from Δ) is any collection of closed subsets of $\prod_{\alpha \in \Delta} X_\alpha$ which has the finite intersection property, then $\bigcap_{\nu \in \Sigma} F_\nu \neq \emptyset$. The fact that the product space is compact then follows at once from Theorem III, 3.25. Let $\{F_\nu\}_{\nu \in \Sigma}$ be such a collection of closed sets, and let property P be the finite intersection property. As seen in Example 2.5, P is a property of finite character. Thus, by Theorem 2.6, there is a collection $G = \{G_\delta\}_{\delta \in \Lambda}$ of subsets (not necessarily all closed) of $\prod_{\alpha \in \Delta} X_\alpha$ which has the finite intersection property, and no collection of subsets of the product space having G as a proper subcollection has the finite intersection property. The collection G must, therefore, satisfy the following conditions:

(i) The intersection of any non-empty finite collection of sets which are elements of G is an element of G.

(ii) If L is a subset of the product space such that L intersects every element of G, then L is an element of G.

Furthermore, G can be taken so that the collection $\{F_\nu\}_{\nu \in \Sigma}$ is a subcollection of the collection G, and we assume that this has been done.

For each $\beta \in \Delta$, denote by Y_β the collection of closed sets $\{\overline{p_\beta(G_\delta)}\}_{\delta \in \Lambda}$. It follows that Y_β is a family of closed subsets of X_β which has the finite intersection property. Thus, for each $\beta \in \Delta$, there exists, since X_β is compact, a point $x_\beta \in \bigcap_{\delta \in \Lambda} \overline{p_\beta(G_\delta)}$. Let $z = \{x_\beta\}_{\beta \in \Delta}$ denote the point of $\prod_{\alpha \in \Delta} X_\alpha$ thus chosen, and let U be any open subset of the product space that contains the point z. We shall show that $U \cap F_\nu \neq \emptyset$ for every $\nu \in \Sigma$. Since U is any open subset of the product space containing z, it follows that z is a limit point of the closed set F_ν for every $\nu \in \Sigma$. Thus $z \in \bigcap_{\nu \in \Sigma} F_\nu$, and our proof is complete. To see that $U \cap F_\nu \neq \emptyset$ for every $\nu \in \Sigma$, notice first that $p_\beta(U) \cap \overline{p_\beta(G_\delta)} \neq \emptyset$, for every $\delta \in \Lambda$ and $\beta \in \Delta$. Since p_β, for every $\beta \in \Delta$, is an open mapping, it then follows that $p_\beta(U) \cap p_\beta(G_\delta) \neq \emptyset$, for every $\delta \in \Lambda$ and $\beta \in \Delta$. Consequently, $[p_\beta^{-1}(p_\beta(U))] \cap G_\delta \neq \emptyset$, for every $\beta \in \Delta$, $\delta \in \Lambda$. It is a consequence of this and property (ii) that $p_\beta^{-1}(p_\beta(U)) \in G$ for every $\beta \in \Delta$. But $U = \bigcap_{\beta \in \Gamma} p_\beta^{-1}(p_\beta(U))$, where Γ is some *finite* subset of Δ. Consequently, by property (i), $U \in G$. Therefore, $U \cap F_\nu \neq \emptyset$ for every $\nu \in \Sigma$, since each F_ν is an element of G. The proof of the theorem is thus complete.

To see how the Tychonoff theorem implies the axiom of choice, consider the following argument. Let $M = \{X_\sigma\}_{\sigma \in \Sigma}$ be a collection of non-empty subsets of set S, and denote by p an object not in any X_σ. Define $Y_\sigma = X_\sigma \cup \{p\}$, for each $\sigma \in \Sigma$. Define a topology for each Y_σ by saying that a subset of Y_σ is open if and only if it is empty or its complement in Y_σ is a finite set. The collection of open subsets of Y_σ thus defined is a topology for Y_σ, and Y_σ is a compact space in this topology. By the Tychonoff theorem, the product space $\prod_{\sigma \in \Sigma} Y_\sigma$ is therefore compact. For each $\beta \in \Sigma$, define Z_β to be the subset of this product space consisting of all points $\{y_\sigma\}_{\sigma \in \Sigma}$ with the property that $y_\beta \in X_\beta$. Clearly each Z_β, for $\beta \in \Sigma$, is a closed set. Furthermore, each Z_β is non-empty. To see this, note that the point $\{y_\sigma\}_{\sigma \in \Sigma}$ satisfying $y_\sigma = p$ for $\sigma \in \Sigma - \{\beta\}$, and $y_\beta \in X_\beta$ lies in Z_β. Here we have used the finite axiom of choice to pick the point y_β of X_β. Moreover, for any finite subset Λ of Σ, $\bigcup_{\beta \in \Lambda} Z_\beta \neq \emptyset$, since, by the finite axiom of choice, there is a point $\{y_\sigma\}_{\sigma \in \Sigma}$ such that $y_\sigma = p$ for $\sigma \in \Sigma - \Lambda$, $y_\beta \in X_\beta$ for $\beta \in \Lambda$, and this point is in the set $\bigcap_{\beta \in \Lambda} Z_\beta$. The set $\prod_{\sigma \in \Sigma} Y_\sigma$ being compact, we must then have $\bigcap_{\beta \in \Sigma} Z_\beta \neq \emptyset$. Since this set is not empty we can find, using the finite axiom of choice, a point $\{x_\sigma\}_{\sigma \in \Sigma}$ which lies in $\bigcap_{\beta \in \Sigma} Z_\beta$ and hence in $\prod_{\sigma \in \Sigma} X_\sigma$. Define $f: M \to S$ by $f(X_\sigma) = x_\sigma$, for every $\sigma \in \Sigma$. The existence of the mapping f is exactly the content of the axiom of choice.

Inasmuch as this argument depends on the assumption of the finite axiom of choice, we refer to the Tychonoff theorem as an alternative statement of the axiom of choice rather than as an equivalent form of this axiom.

PROBLEMS

17. Let S be a partially ordered set, and p an element of S. Prove that there exists at least one maximal simply ordered subset K of S such that $p \in K$.

18. Show that the partially ordered set W in the proof of Theorem 2.3 is inductive.

19. Let S be a set, and P a property of subsets of S that is of finite character. Prove that, if there is a subset D of S that has property P, then there is a subset A of S such that $D \subset A$, A has property P, and, if there is a subset B of S such that $B \supset A$ and B has property P, then $B = A$.

20. Prove Theorem 2.9.

21. Let X_1 and X_2 be two spaces. Show that the product space $X_1 \times X_2$ as defined in Chapter III and the product space $\prod_{\alpha \in \Delta} X_\alpha$, where $\Delta = \{1, 2\}$, defined in this section are the same.

22. Let S and T be two non-empty sets. Show that there exists a one-to-one mapping of one of the sets S and T into the other.

23. Let $\{X_\alpha\}_{\alpha \in \Delta}$ be a collection of Hausdorff spaces. Show that the space $\prod_{\alpha \in \Delta} X_\alpha$ is Hausdorff.

24. Show that the mappings p_β defined in the proof of Theorem 2.11 are both open and continuous.

25. Show that the collection G of the proof of Theorem 2.11 can be chosen so as to contain the collection $\{F_\nu\}_{\nu \in \Sigma}$. Also show that G satisfies conditions (i) and (ii).

26. Show that the collection Y_β of the proof of Theorem 2.11 has the finite intersection property.

27. In the proof of Theorem 2.11, why does

$$U = \bigcap_{\beta \in \Gamma} p_\beta^{-1}[p_\beta(U)]$$

for some finite subset Γ of Δ?

3. Examples of Applications of the Axiom of Choice

As applications of the alternative statements of the axiom of choice, we give a proof of the existence of non-cut points in non-degenerate compact connected Hausdorff spaces, a characterization of completely regular Hausdorff spaces, and an example showing some of the distinctions between completely regular T_1, normal T_1, and completely normal T_1 spaces.

THEOREM 3.1. Let S be a non-degenerate compact connected Hausdorff space. Then S has at least two non-cut points.

Proof. Denote by N the set of non-cut points of S. We shall show that the supposition that N contains at most one point leads to a contradiction. Under this assumption, $S - N \neq \emptyset$. Let $x_0 \in S - N$. Then there is a separation $S - x_0 = A \cup B$, where the notation is so chosen that $N \subset B$. For each $x \in A$, choose a fixed separation $S - x = A_x \cup B_x$, where $x_0 \in B_x$. Clearly, $A_x \cup x \subset A$. Let the collection $\{A_x\}_{x \in A}$ be partially ordered as follows: $A_{x_1} \leq A_{x_2}$ if and only if $A_{x_2} \subset A_{x_1}$. This is easily seen to be a partial order. By Zorn's lemma there exists a maximal simply ordered subcollection $\{A_{x_\sigma}\}$ of the collection $\{A_x\}_{x \in A}$. This simply ordered subcollection has the finite intersection property. Moreover, $\bigcap_\sigma A_{x_\sigma} = \bigcap_\sigma [A_{x_\sigma} \cup x_\sigma]$. Now each of the sets $A_{x_\sigma} \cup x_\sigma$ is closed, and S is compact. Consequently, $\bigcap_\sigma A_{x_\sigma} = \bigcap_\sigma [A_{x_\sigma} \cup x_\sigma] \neq \emptyset$; hence there exists a point $y \in \bigcap_\sigma A_{x_\sigma}$. This contradicts the maximality of the collection $\{A_{x_\sigma}\}$ since $A_y \subset A_{x_\sigma}$ for every σ, and $A_y \neq A_{x_\sigma}$ for any σ, because $y \notin A_y$. Thus N must contain at least two points, and the proof of the theorem is complete.

The student should compare this proof with that of Theorem V,1.3.

Our next theorem shows that every completely regular Hausdorff space is homeomorphic to a subset of a compact Hausdorff space. The construction used in the proof is typical of many constructions used in mathematics.

THEOREM 3.2. Let S be a completely regular Hausdorff space. Then there exists a compact Hausdorff space T and a subset K of T such that S is homeomorphic to K.

Proof. If S consists of a single point, then $K = T = S$ satisfies the conclusion of the theorem. If S is non-degenerate, we proceed as follows. For any point $p \in S$ and any closed subset C of S such that $p \notin C$, there exists a continuous mapping $f: S \to J$, where J is the closed unit interval, such that $f(p) = 0$, and $f(C) = 1$. Let W be the set of all such mappings for all such pairs of points and closed sets. Consider the product space $\prod_{f \in W} J_f$, where J_f, for every $f \in W$, is the closed unit interval of the space R of real numbers. Let T denote this product space. It is immediate from the Tychonoff theorem and Problem 23 that T is a compact Hausdorff space. Thus the proof will be complete if we show that S is homeomorphic to a subset K of T. We define a mapping ψ of S into T by $\psi(x) = \{f(x)\}_{f \in W}$, for every $x \in S$. Notice that we can also write $\{f(x)\}_{f \in W}$ as $\{x_f\}_{f \in W}$. Define $K = \psi(S)$. It is now easily verified that the mapping $h: S \to K$ defined by $h(x) = \psi(x)$, for every $x \in S$, is a homeomorphism. This completes the proof.

Our example showing some of the distinctions between completely regular T_1 spaces and normal T_1 spaces is based on a use of the well-ordering theorem.

DEFINITION 3.3. Let S denote the set of all real numbers. There is a well-ordering α for $S - \{1\}$. Define the well-ordering β of S as follows: If $x, y \in S - \{1\}$ then $x \beta y$ if and only if $x \alpha y$. Also $y \beta 1$ for every $y \in S - \{1\}$. Let ω denote the least element of S in the well-ordering β, such that the set of all y of S that precede ω is countable and infinite. Let ω_1 denote the least element of S in the well-ordering β, such that the set of all y of S preceding ω_1 is uncountable. Denote by S_0 the subset of S consisting of ω and all y of S such that $y \beta \omega$. Let S_1 denote the subset of S consisting of ω_1 and all y of S such that $y \beta \omega_1$. The spaces S_0 and S_1 are defined to be the sets S_0 and S_1 in their interval topologies in the well-ordering which is the restriction of β.

The following theorem is of interest concerning S_0 and S_1.

THEOREM 3.4. Let S_0 and S_1 be the spaces just defined. Then
(i) S_0 is compact and Hausdorff.
(ii) S_1 is compact and Hausdorff.
(iii) $S_1 - \omega_1$ is countably compact, but not compact.

Proof. Consider (i). Let $\{U_\alpha\}$ be any open covering of S_0. There is an index α_0 such that $\omega \in U_{\alpha_0}$. Since $S_0 - U_{\alpha_0}$ is finite, it follows at once that S_0 is compact. It is also clear that S_0 is Hausdorff.

Consider (ii). It is clear that S_1 is Hausdorff. Let $\{U_\alpha\}$ be any open covering of S_1. Then there is an index α_0 such that $\omega_1 \in U_{\alpha_0}$. Furthermore, $U_{\alpha_0} - \omega_1$ is non-empty. Moreover, there exists $p \in U_{\alpha_0} - \omega_1$ such that the set of all $y \in S_1$ such that $p \beta y$ is in U_{α_0}. Since the subset M of S_1 consisting of p and all elements of S_1 preceding p is countable, it follows that, if M is countably compact, then a finite number of elements $U_{\alpha_1}, U_{\alpha_2}, \cdots, U_{\alpha_n}$ of $\{U_\alpha\}$ cover M. Thus $U_{\alpha_0}, U_{\alpha_1}, \cdots, U_{\alpha_n}$ covers S_1. This shows that, in order to prove S_1 compact, it suffices to show that M is countably compact. We shall show that M is countably compact in the following way. Let K be any infinite subset of M, and q the first point of M that is preceded by infinitely many elements of K. It follows that q is a limit point of K which lies in M. Thus M is countably compact, and (ii) is proved.

Consider (iii). The set $S_1 - \omega_1$ is not compact, since the open covering $\{U_\alpha\}$ of $S_1 - \omega_1$, where each U_α is an interval of $S_1 - \omega_1$, open in $S_1 - \omega_1$, cannot contain a finite subcovering of $S_1 - \omega_1$. For suppose it did. Then, since each U_α is countable, it follows that $S_1 - \omega_1$ is countable, which is impossible by the definition of S_1.

To see that $S_1 - \omega_1$ is countably compact, let K be any infinite subset of $S_1 - \omega_1$, and denote by q the first element of S_1 that is preceded by infinitely many elements of K. Clearly, q is a limit point of K in S_1.

To see that $q \in S_1 - \omega_1$, notice that q is preceded by only a countable subset of K. Thus $q = \omega_1$ would imply $S_1 - \omega_1 = \bigcup_{\substack{k \in K \\ k \beta q}} A_k$, where A_k, for each $k \in K$, is the set of all elements of S_1 that precede k in S_1. Since A_k is therefore countable, for every $k \in K$, and there are only countably many elements of K that precede q in S_1, it follows that the equality $S_1 - \omega_1 = \bigcup_{\substack{k \in K \\ k \beta q}} A_k$ is impossible. Thus $q \in S_1 - \omega_1$, and we have completed the proof that $S_1 - \omega_1$ is countably compact.

EXAMPLE 3.5. Let T be the product space $S_1 \times S_0$, where S_1 and S_0 are given by Definition 3.3. The space T is a compact Hausdorff space, and thus a normal space. But every normal Hausdorff space is completely regular; hence T is completely regular. Therefore, every subspace of T is completely regular. In particular, the subspace $W = T - (\omega_1, \omega)$ is completely regular. We shall show that W is not normal. Consider the subsets W_0 and W_1 of W defined by $W_0 = (S_1 - \omega_1) \times \{\omega\}$ and $W_1 = \{\omega_1\} \times (S_0 - \omega)$. Each of these sets is closed in W. Let G denote an open subset of W containing W_1. Then G contains a neighborhood of each of its points. Hence there exists $\alpha \in S_1 - \omega_1$, such that all points (q, r) satisfying q follows α in S_1 and r precedes ω in S_0 lie in G. The closure of G must thus have points in common with W_0; hence W is not normal. As a consequence of this, it follows that T cannot be completely normal. It is also apparent that T is a compact normal Hausdorff space which is not metrizable.

Further examples, such as that of a regular Hausdorff space which is not completely regular may be found in *Treatise on Set Topology* by R. Vaidyanathaswamy, Indian Mathematical Society, 1947, or in other well-known treatises on topology. The Polish journal, *Fundamenta Mathematicae*, is also a rich source of examples.

PROBLEMS

28. Find an example of a compact T_1 space that is not Hausdorff.

29. Let $\{X_\alpha\}_{\alpha \in \Delta}$ be a collection of metric spaces. Show that, in general, the product space $\prod_{\alpha \in \Delta} X_\alpha$ is not metrizable.

30. Justify the existence of ω_0 and ω_1 as asserted in Definition 3.3.

31. Show that the spaces S_0 and S_1 as defined in Definition 3.3 are Hausdorff.

32. In the proof of (iii) of Theorem 3.4, why is q preceded by only a countable subset of K?

33. In Example 3.5, show in detail why the closure of G must have points in common with W_0.

[*Hint.* For each $p \in (S_0 - \omega)$, there exists α_p, β_p, γ_p, such that $(\alpha_p, \omega_1] \times (\beta_p, \gamma_p) \subset G$, where α_p precedes ω_1 and $p \in (\beta_p, \gamma_p) \subset (S_0 - \omega)$. Since $\{\alpha_p\}_{p \in (S_0 - \omega)}$ is a *countable* collection, there exists α preceding ω_1 such that α_p precedes α for all $p \in S_0 - \omega$. Hence $(\alpha, \omega_1] \times (S_0 - \omega) \subset G$. The desired result now follows easily.]

Bibliographical Comment

The proof of Theorem 1.17, as well as the proofs of the alternative forms of the axiom of choice, are based on unpublished lecture notes of H. F. Bohnenblust at Princeton University in 1940. Bohnenblust attributes these proofs to C. Chevalley.

The proof given for the Tychonoff theorem follows that presented by G. S. Young Jr. in one of his classes at the University of Michigan. The discussion of the implication of the axiom of choice by the Tychonoff theorem follows that given by J. L. Kelley. See

J. L. Kelley, The Tychonoff Product Theorem Implies the Axiom of Choice, *Fundamenta Mathematicae*, vol. 37, pp. 75–76, 1950.

Most of the examples were adapted from those given by R. Vaidyanathaswamy. See his textbook:

R. Vaidyanathaswamy, *Treatise on Set Topology*, Part I, Indian Mathematical Society, Madras, 1947.

Other material on the axiom of choice may be found in the following:

G. Birkhoff, Lattice Theory, *American Mathematical Society Colloquium Publications*, vol. 25, New York, 1949.

A. Tarski, Sur les Ensembles finis, *Fundamenta Mathematicae*, vol. 6, pp. 45–95, 1924.

R. L. Wilder, *Introduction to the Foundations of Mathematics*, John Wiley & Sons, New York, 1952.

LIST OF SPECIAL SYMBOLS

$\in$	belongs to, 1
$\notin$	does not belong to, 1
$\supset$	contains, 1
$\subset$	is contained in, 1
$=$	equals, 1, 277
$\emptyset$	Null set, 2
Λ	Usually used to denote an index set, 3
$\{A_\alpha\}$	Collection of sets, (index set understood), 3
$\{A_\alpha\}_{\alpha\in\Lambda}$	Collection of sets indexed by Λ, 3
$\bigcup$	Union, 3
$\cup$	Union, 4
$\bigcap$	Intersection, 3
$\cap$	Intersection, 4
$A - B$	Points of A not in B, 5
$S \times T$	Cartesian product of S and T, 7
(s, t)	Ordered pair of points s and t, 7
R	Usually the set or space of real numbers, 8, 24
$f: S \to T$	Mapping f of S into T, 8
$f(A)$	Image of A under the mapping f, 9
$f^{-1}(B)$	Source (or inverse) of B under the mapping f, 9
f/A	Restriction of the mapping f to A, 10
aFb	a is in the relation F to b, 12
$>, <$	Symbols denoting an order relation, 14, 153
$\{a_n\}$	Sequence whose nth element is a_n, 16
I	Usually denotes the set of all positive integers, 16
$\{p\}$	Set consisting of the single point p, 18
$\lim a_n$	Limit of the sequence $\{a_n\}$, 25
$\lvert A - a_n \rvert$	Absolute value of $A - a_n$, 25
(s, t)	Ordered pair of points, 7
(a, b)	Open-interval with end points a and b, 34, 157
$[a, b]$	Closed-interval, 43, 157
$[a, b)$	Half-open-interval, 157
$(a, b]$	Half-open-interval, 157
(S, σ)	Set S with σ as a basis for the collection of all open subsets of S, 55
$\sigma_1 \times \sigma_2$	Cartesian product of the collections of sets σ_1 and σ_2, 55
$(S_1 \times S_2, \sigma_1 \times \sigma_2)$	Topological product of (S_1, σ_1) and (S_2, σ_2), 56
E^2	Cartesian plane, 58, 145
$\bar{A}$	Closure of the set A, 63
H'	Derived set of the set H, 63
H'	Interior of the arc H, 178
T_1	A T_1 space is one in which every point is closed, 68
$S_r(K)$	All points with distance less than r from at least one point of K, 70, 112
$\sigma \cap K$	Collection of sets each of whose elements is of the form $H \cap K$ where $H \in \sigma$, 71

$\infty, -\infty$	Symbols often used to denote the points of the two-point compactification of R, 90
$\delta(A)$	Diameter of the set A, 90
$\rho(A, B)$	Distance between the sets A and B, 91
$\overline{\text{Lim}}\, A_n$	Limit superior of the sequence of sets $\{A_n\}$, 103
$\underline{\text{Lim}}\, A_n$	Limit inferior of the sequence of sets $\{A_n\}$, 103
$\rho(x, A)$	Distance between the point x and the set A, 91, 112
J	Usually denotes the closed unit interval $[0, 1]$ considered as a subspace of the space R of all real numbers, 120
$B(J)$	Space of all continuous mappings of J into R, 131
$B(S)$	Space of all continuous mappings of S into R, 136
C	Set of all Cauchy sequences of a metric space S, 137
$C^*(S)$	Metric space of equivalent Cauchy sequences of a metric space S, 137, 138
E^n	Euclidean n-space, 145
J^ω	Hilbert cube, 147, 148
E^ω	Hilbert space, 147, 148
$M_1(X), M_2(X)$	Sets of a separation of a space by a subset X, 151
L_α	Union of all elements A_β of $\{A_\alpha\}$ such that $A_\beta < A_\alpha$, 153
R_α	Union of all elements A_β of $\{A_\alpha\}$ such that $A_\alpha < A_\beta$, 153
$K(A,B)$	Non-empty, non-separated collection of cuttings of S, each of which separates A and B in S, 154, 156
$E(a, b)$	Set of all points of S that separate a and b in S, 156
K	Often used to denote the unit circle, 170
$\not\subset$	not contained in, 174
$L(p_1, p_2)$	Line segment joining the points p_1 and p_2 of E^2, 173
A'	Interior of the arc A, 178
$H(g)$	Interior of the special rectangle H with center g, 177
$ap_1p_2 \cdots p_nb$	Denotes the arc with end points a and b such that, in the natural order from a to b, $a < p_1 < p_2 < \cdots < p_n < b$, 187
Σ	Usually denotes the unit sphere in Chapter V, sections 8 and 9, 136
$\geq\ \leq$	Symbols denoting a partial order relation, 275
Δ	Usually denotes the identity relation in Chapter VII, section 1, 277
$x\Delta y$	x and y denote the same element of S, 277
$x = y$	x and y denote the same element of S, 277

INDEX